KU-624-460

Methods in Human Geography

A guide for students doing a research project

Second edition

Edited by Robin Flowerdew and David Martin
University of St Andrews
University of Southampton

Harlow, England • London • New York • Boston • San Francisco • Toronto • Sydney • Singapore • Hong Kong
Tokyo • Seoul • Taipei • New Delhi • Cape Town • Madrid • Mexico City • Amsterdam • Munich • Paris • Milan

Pearson Education Limited
Edinburgh Gate
Harlow
Essex CM20 2JE
England

and Associated Companies throughout the world

Visit us on the World Wide Web at:
www.pearsoned.co.uk

First published 1997
Second edition published 2005

© Pearson Education Limited 1997, 2005

All rights reserved. No part of this publication may be reproduced, stored in a retrieval system, or transmitted in any form or by any means, electronic, mechanical, photocopying, recording or otherwise, without either the prior written permission of the publisher or a licence permitting restricted copying in the United Kingdom issued by the Copyright Licensing Agency Ltd, 90 Tottenham Court Road, London W1T 4LP.

ISBN 0 582 47321 7

British Library Cataloguing-in-Publication Data
A catalogue record for this book is available from the British Library

Library of Congress Cataloging-in-Publication Data
Methods in human geography: a guide for students doing a research project / edited by Robin Flowerdew and David Martin. – 2nd ed.
p. cm.
Includes bibliographical references and index.
ISBN 0-582-47321-7 (pbk.)
1. Human geography–Research. 2. Human geography–Methodology. I. Flowerdew, Robin. II. Martin, David, 1965 July 18–

GF26.M47 2005
304.2′072–dc22

2004057275

10 9 8 7 6 5 4 3 2 1
10 09 08 07 06 05

Typeset in 9.5/12.5pt Stone Serif by 35
Printed and bound by Bell & Bain Limited, Glasgow

The publisher's policy is to use paper manufactured from sustainable forests.

Methods in Human Geography

We work with leading authors to develop the strongest educational materials in geography, bringing cutting-edge thinking and best learning practice to a global market.

Under a range of well-known imprints, including Prentice Hall, we craft high quality print and electronic publications which help readers to understand and apply their content, whether studying or at work.

To find out more about the complete range of our publishing, please visit us on the World Wide Web at: www.pearsoned.co.uk

Contents

List of figures

List of tables

List of boxes

List of contributors

Stuart C Aitken, Department of Geography, San Diego State University, San Diego, California 92182, USA

Paul J Boyle, School of Geography and Geosciences, University of St Andrews, St Andrews, Fife KY16 9AL

Gordon Clark, Department of Geography, Lancaster University, Lancaster LA1 4YB

David Conradson, School of Geography, University of Southampton, Southampton SO17 1BJ

Ian Cook, School of Geography, Earth and Environmental Sciences, The University of Birmingham, Birmingham B15 2TT

James Craine, Department of Geography, California State University, Northridge, 18111 Nordhoff Street, Northridge, CA 91330-8249, USA

Mike Crang, Department of Geography, University of Durham, Science Site, South Road, Durham DH1 3LE

Christine E Dunn, Department of Geography, University of Durham, Science Site, South Road, Durham DH1 3LE

Robin Flowerdew, School of Geography and Geosciences, University of St Andrews, St Andrews, Fife KY16 9AL

A Stewart Fotheringham, National Centre for Geocomputation, National Institute for Regional and Spatial Analysis, National University of Ireland, Maynooth, County Kildare, Ireland

Anthony C Gatrell, Institute for Health Research, Lancaster University, Lancaster LA1 4YT

Elspeth Graham, School of Geography and Geosciences, University of St Andrews, St Andrews, Fife KY16 9AL

Mike Kesby, School of Geography and Geosciences, University of St Andrews, St Andrews, Fife KY16 9AL

Sara Kindon, Institute of Geography, Victoria University of Wellington, Wellington, Aotearoa New Zealand

Andrew A Lovett, School of Environmental Sciences, University of East Anglia, Norwich NR4 7TJ

David Martin, School of Geography, University of Southampton, Southampton SO17 1BJ

Rachel Pain, Department of Geography, University of Durham, Science Site, South Road, Durham DH1 3LE

Julian Parfitt, WRAP, The Old Academy, 21 Horse Fair, Banbury, Oxon. OX16 0AH

Gill Valentine, School of Geography, University of Leeds, Leeds LS2 9JT

Preface

This book is intended for use in two different ways. Firstly, we hope that it will prove a valuable guide to students undertaking independent research projects in human geography in the form of undergraduate or master's-level dissertations. Secondly, the book may be used as an accompanying text for taught courses that aim to introduce students to research methods. The text is unusual in its coverage of both quantitative and qualitative approaches to the subject, and does not set out to endorse any particular set of research methodologies at the expense of any other. Discussion of research approaches is coupled with consideration of practical issues such as choosing a topic and preparing the final report. Our aim has not been to produce a complete step-by-step guide to the completion of a research project, but to provide readers with sufficient introductions to key methodological approaches so that they can identify the approach they wish to take and pursue it independently.

Several new chapters have been added for this second edition, reflecting our assessment of the increasing diversity of research approaches available to students of human geography. The need for such a text was originally recognised by the Quantitative Methods Research Group of the Royal Geographical Society with the Institute of British Geographers, and the project was initially taken forward as an activity of that group, although the contributors have very diverse backgrounds and our approach to the second edition has been very much driven by the helpful feedback received from users of the first.

We have tried as far as possible to ensure that the material is presented in a way that has broad international applicability. Though some of the chapters concentrate on UK examples, in most cases there will be equivalent resources available in other developed countries. All our readers are urged to give careful attention to the local guidance and dissertation regulations issued by their own institutions. While the focus of this text is explicitly on human geography, our experience with the first edition suggests that many of the chapters will have relevance to work in other social sciences and even in physical geography!

Readers familiar with the first edition will find that substantial changes have been made, largely through the introduction of several new chapters. The chapters by David Conradson on focus groups, and Mike Kesby, Sara Kindon and Rachel Pain on participatory methods are entirely new. The material in Stuart Aitken's chapter on analysis of texts has been expanded and split into two chapters, one of which is co-authored with James Craine. The material on writing up has been reorganised, resulting in a chapter by Paul Boyle and Robin Flowerdew on the organisation of a dissertation being separated from one by Paul Boyle on writing skills. Other authors were given the opportunity to make whatever changes they felt were needed. In some cases, little has been done beyond updating the references, while other chapters have been fundamentally reorganised.

Our thanks are due to all our contributors for their ideas and cooperation and to Morton Fuglevand and Andrew Taylor at Pearson Education for their encouragement to consider a second edition of the book and support in preparing it.

Acknowledgements

We are grateful to the following for permission to reproduce copyright material:

Box 15.5 photo courtesy of Portland Cement Association; Figures 15.2 and 15.3 from T. Pringle (1988) The privation of history: Landseer, Victoria and the Highland myth. In D. Cosgrove and S. Daniels (eds) *The Iconography of Landscape: Essays on the symbolic representation, design and use of past environments*, pp. 142–60, with permission from Cambridge University Press; Figure 17.1 from B A Kennedy (1992) First catch your hare . . . research designs for individual projects, in A Rogers, H Viles & A Goudie (eds) *The Student's Companion to Geography*, with permission from Blackwell Publishing; Figure 19.2 from D Weiner, T A Warner, T M Harris & R M Levin (1995) Apartheid representations in a digital landscape: GIS, remote sensing and local knowledge in Kiepersol, South Africa, *Cartography and Geographic Information Systems*, 22, 30–44, with permission from the authors; Figure 19.3 reproduced by permission of Ordnance Survey on behalf of The Controller of Her Majesty's Stationery Office, © Crown Copyright 100018613; Figure 19.6 from National Statistics 2001 *Regional Trends* No 36, Crown copyright material is reproduced with the permission of the Controller of HMSO and the Queen's Printer for Scotland; Figure 19.7b reproduced with kind permission of Michael Friendly, York University, Toronto.

1 Introduction

David Martin and Robin Flowerdew

An individual research project leading to the production of a dissertation or written report forms an important element of many undergraduate and Master's-level geography degree courses. In many institutions this research is introduced by means of a programme of introductory teaching about research methodologies. Doing a research project is quite different to most other components of a conventional taught course of study and, for many students, becomes one of the highlights of the course. In an environment in which both university research and teaching are increasingly scrutinised for quality, it has been repeatedly recognised that good research is strongly linked with good teaching, and that the opportunity to conduct individual research should be a central aspect of contemporary geographical education. Undertaking a research project can develop and exercise many of the key skills that are to be gained from the study of geography. Undoubtedly, some of the best teachers and lecturers in geography are motivated and excited about the subject through their own individual research activities.

Students come to research projects with a wide variety of attitudes and expectations. You may currently view your project as a large, and somewhat frightening, chore or you may welcome it as an opportunity to express originality and get out of the classroom. Both the product and the process of the academic research project have similarities with the requirements of many jobs, and this is reflected in the frequency with which geographers being interviewed for employment will be asked to talk about their research projects in more depth. Research skills and the ability to present findings and conclusions clearly are very definitely valuable personal skills. For many, an undergraduate research project provides a first opportunity to look into the world of academic research that actually creates the knowledge around which conventional class-based teaching is constructed. In a few cases, research projects undertaken as part of degree programmes actually go on to make significant contributions to the subject itself.

It is important to recognise that both the subject and methods of research change quite rapidly. Since we prepared the first edition of this text in 1996, the balance of research methods has changed to the extent that we have felt the need to include several new chapters in this second edition. Established techniques develop and new ones come into fashion. It is all too easy to find methods textbooks that give the impression that there is only one way to tackle a question. Here, we are seeking to provide an overview of a broad range of research methods that are not usually found together, in the belief that this clear presentation of alternatives will lead to more appropriate and more rewarding research. The focus of the book is on research that studies substantive issues, rather than research whose objective is the development of new methodologies.

Although the precise constraints differ according to local circumstances, almost all research is constrained by time, expense and formal requirements of various kinds. This book is particularly aimed at those who are undertaking independent research, usually with the occasional aid of a supervisor, which leads to the production of a substantial written report. The chapters should also provide a useful review for those starting on the path of postgraduate research, but this audience will have the comparative luxury of a less constrained timetable, and will need to consider issues relating to research methodology in much greater depth than is possible here. You will find useful hints here concerning the practicalities of completion and submission of a good research project, but the aim has been primarily to review research methods, and to equip you to think out the most appropriate ways of dealing with your own research problems.

When the first edition of this text was published, there were surprisingly few textbooks on research methods in human geography, but there are now several books (and numerous Master's degree programmes) that cover aspects of the subject. For an informal guide to the basics, see Parsons and Knight (1995). Some of the material provided in Rogers, Viles and Goudie (2002) is relevant to the undergraduate research project as well as to many other aspects of the student experience, while Bell (1999) provides an introductory overview for students in the social sciences. Hoggart, Lees and Davies (2002) and Kitchin and Tate (2000) are perhaps the most directly focused on research methodology and the practice of the dissertation in human geography, although neither takes quite the approach we have adopted here. Raghuram *et al.* (1998) have published a discussion of the implications of feminist methodology for student projects. Clifford and Valentine (2003) have put together a volume that is similar to this one in that it contains chapters on many of the same issues and methods. Some themes in the Clifford and Valentine volume are not covered by this book, especially issues concerning physical geography, while other important topics are covered here but not there.

There may also be material of interest in books intended mainly for research students such as Phillips and Pugh (2000). Each of these has more detailed discussion of certain issues than is attempted here, but it is hoped that the chapters in this book will be particularly relevant for your needs and, at worst, a good starting point for finding material that you will find helpful for specific topics.

It is particularly important that anyone undertaking a research project which will be formally examined as part of a degree programme obtains and studies their own institutional guidelines, which will give precise instructions concerning length, number of copies required, format, coverage, use of supervisors, submission dates and so on. There will almost certainly be warnings about the unacceptability of plagiarism, which is the term for using other people's words and ideas without acknowledging the source. (It is our hope in preparing this book that although we positively encourage you to investigate previous work fully, you will be inspired to develop excellent original material of your own!)

Further, there may be intermediate steps formally required of you, for example the submission of risk assessments, project proposals or literature reviews prior to the completion of the main report. In seeking to achieve the very best result of which you are capable, it is essential that you follow all these instructions precisely. In many places in this book, the reader is referred to local regulations, where our experience suggests that the details differ widely. Although you should read all of your institution's guidelines carefully at the very beginning, do not expect them to tell you everything that you need to know. There will probably be details about the size of print and widths of margins to be used, but such documents rarely tell you how to choose a topic, how to go about studying it and what to put in your final report.

Two other crucial considerations at the very start of your project are, firstly, health and safety and, secondly, ethical issues applying to the research you have in mind. These may be thought of broadly as measures designed for your own protection and those designed for the protection of those affected by your research. Your institution will expect you to take seriously the process of risk assessment for your proposed project, and will probably provide a standard template for this, which is likely to require discussion with your supervisor prior to completion and signature (see Higgitt and Bullard, 1999). The purpose of risk assessment is to ensure that potential hazards are fully recognised and that appropriate actions are taken to mitigate them: in completing the risk assessment you commit yourself to taking these appropriate actions! Despite the obvious importance of personal safety while conducting research, this is another area in which there has been considerable change between the first and second editions of this text, with a far greater emphasis now placed on risk assessment and record-keeping. It is particularly important that these issues are addressed before any work commences.

It is tempting to think that physical risks apply only to physical geographers working in obviously hazardous situations such as mountain or coastal environments. However, the potential dangers of working or travelling alone or conducting interviews in unfamiliar residential areas may be just as great (Lee, 1995). Clearly, different research designs present different challenges. Control measures may include (but are not limited to) making careful prior arrangements with interviewees about access to their premises and topics for discussion; ensuring that a responsible person knows where you are going, can contact you by mobile phone at an agreed time, and is notified of your safe return; dressing appropriately for the research situation or perhaps working in combination with another researcher. More subtly, some types of research design involving personal interaction with others may present emotional and psychological hazards rather than physical ones. All the relevant hazards should be considered and taken into account before you begin. Lastly do not be a hero for the sake of your research project. If you think you are getting into a dangerous situation, physically or emotionally, trust your judgement and remove yourself from the problem area. Geographical research is important, but your own safety and well being is far more so.

Health and safety considerations also stress the avoidance of putting others in danger by your actions or giving offence to interviewees, and this brings us to the ethical considerations applying to human geography research projects (discussed in more detail by Hay, 2003). These will variously include issues of personal and commercial confidentiality, legal responsibilities and forms of representation. At the simplest level, they will include the requirement for adequate acknowledgement of ownership of materials and ideas used in your report and compliance with copyright constraints. Other issues include the necessity to obtain informed consent from the people you are studying, promising anonymity or confidentiality to them and keeping the promise, allowing them to withdraw from the research without explanation if they so wish, and giving them the opportunity to find out the eventual results of the research. Some researchers argue that the subjects of a study should have the opportunity to comment on how the researcher represents the issues facing them. Some also feel that the researcher – subject relationship is exploitative, helping the researcher get a degree but doing nothing for the people studied. As a reaction to this, researchers may feel an obligation to help their subjects, but you should be wary of getting personally involved to an extent of promising more than you can deliver.

Both the British Sociological Association and the British Psychological Society have issued guidelines about ethical issues in research, and much of what they say is also applicable to human geographers. Generally speaking, the issues are most complex and the guidance least specific where the researcher interacts closely with subjects and the topic of the research is potentially sensitive. As with risk assessment, the degree to which you will need to tackle these matters varies enormously with different project designs. Certainly you should discuss these with your supervisor and follow any local guidance carefully. Lee-Treweek and Linkogle (2000) present a wide-ranging collection of commentaries on the risk and ethics issues involved in social research. The broader interrelationships between geographical research and ethical issues are addressed in the collection of papers edited by Proctor and Smith (1999).

Given that there seem to be so many things to think about, you might long for the chance to visit the experts in each of your potential research methodologies and find out what is really involved, before starting on your project at all. You will want to avoid looking back halfway through your project and saying 'I wish someone had told me *that* at the beginning!' At a later stage, you might want very much to get some more specialist advice concerning your analysis or the presentation of your report. In most geography departments you will have a supervisor who is assigned to oversee your project, although the role expected of supervisors varies between institutions. You will also have a range of other academic and specialist staff who will do their best to answer your concerns constructively, but these are all busy (and sometimes inaccessible!) people, and you are unlikely to have the luxury of a couple of hours' chat over a cup of coffee with each one of them individually. That is exactly what we hope this book will provide: we have sought to bring together the group of experts that you might have wished to visit, and encouraged them to write down the kind of advice they would have given you. Each of them has experience of supervising student projects, and knows the most frequently asked questions concerning the different approaches presented. Just like the staff in your own department, they each have individual styles of approach to their subject, they have favourite ways of doing things and they don't always completely agree with one another. Nevertheless, in editing this collection we have tried to preserve as much as possible the individuality of their approaches, while ensuring that all the basic concepts are clearly explained, and that the material is presented in a logical order with suitable cross-references between chapters.

Again, just like the staff in your own department, our contributors do not cover all possible subdivisions of human geography. We hope you will be able to relate comments made about other fields to your own work and we want to get away from the idea that certain techniques are only applicable to certain kinds of problem. However, there are two omissions that spring from our methodology-by-methodology approach that may be important to some students. Most of the chapters make the assumption that you are working close to home or at least in a familiar environment, and little explicit consideration is given to the particular problems that may be encountered in doing research overseas, such as specific cultural, legal and logistical obstacles. If this is particularly relevant to you, there are other references which you may find helpful: for example, Nash (2000a, 2000b) and Smith (2003). Robson and Willis (1997), although intended primarily for postgraduates, is full of useful advice and insights, while Devereaux and Hoddinott (1992) deals specifically with fieldwork in developing countries, and Dixon and Leach (1984) provide a handbook for survey research in developing countries.

Likewise, we have not considered in detail the specific problems of historical research. Advice to new researchers in historical geography is provided by Ogborn (2003) and Black (2003). This topic is particularly challenging to write about because different approaches and different types of source are likely to be appropriate for different periods in the past: researchers in historical geography might find themselves as likely to make use of GIS as of textual analysis. Useful texts include Butlin (1993) and Dodgshon and Butlin (1990); students might also find some volumes in the Historical Association's 'Help for Students of History Series' useful (e.g. Perks, 1992).

The chapters of this book can be read sequentially, but it is not necessary for you to do so. Indeed, for many projects you may find that some of the chapters are of indirect relevance only. In order to help you determine whether a chapter contains what you are looking for, each begins with a brief synopsis, outlining its topic and coverage. Equally, it is not possible for chapters of this size to provide comprehensive detailed advice on a specific topic, and contributors have therefore provided a small selection of useful sources for 'further reading' which are given at the end of each chapter. A full consolidated reference list containing all the material referenced in the book is given at the end. Readers will also find that we have created a companion website for the book, which can be accessed via the publisher's pages at *http://www.pearsoned.co.uk/flowerdew*. This serves as an online reference for the many web addresses that are found in the chapters, and will be maintained and updated while the book is in print.

Chapter 2 deals with philosophies underlying human geography research. This chapter has been deliberately set apart because we feel that the issues it raises are so important. All research projects adopt a philosophical position, even if this is done unconsciously, and what is said here is of relevance whatever the specific detail of your project. Readers who start by looking up some particular methodology-based chapter that is of particular interest should return to Chapter 2 once their ideas have begun to crystallise.

The rest of the book is divided into four sections, which broadly reflect the stages involved in a typical research project. Section A covers preparatory issues such as how to choose a topic, how to find previous work on the topic and how to make the most of secondary data sources. This section addresses important basic issues that will be central to any research project, regardless of topic. For most students giving first consideration to their forthcoming project, this will probably be the starting point, although it should be stressed that it is wise to give some consideration to how you will approach data analysis before and during data collection, and to writing up before and during analysis. Not all researchers work according to the same temporal sequence, and we would not wish to be overly prescriptive.

Section B deals with the collection of primary data by various methods, although not all research involves primary data collection or fieldwork. Students are particularly urged to consult their own department's instructions on these issues, as there may be specific requirements concerning data collection. Chapters in this section cover questionnaire design and sampling, interviewing, ethnographic research and new chapters on focus groups and 'participatory' research methods, which have grown in popularity as research tools for human geography projects.

Section C addresses what might broadly be called 'analysis', beginning with the analysis of numerical and categorical data, followed by the analysis of qualitative research findings, textual and graphical material, and ending with spatial analysis and geographical information systems (GIS). These approaches do not usually appear side by side in the same text, due to the rather different philosophical positions held by many of their proponents. Our aim here, however, has been to present a broad range of potential approaches to your research within a single framework, and we would stress our very broad conception of 'data'.

The final section addresses the process of designing and writing the final report and the use of maps and illustrations, which so frequently form an important part of research reporting in geography. This should again be relevant to all projects. Each individual section is prefaced by some further editorial comments that give an overview of the material covered, and seek to provide some continuity for the reader. These particular section and chapter titles are intended as 'navigational aids' when using the book, and do not reflect a view that research should be undertaken in a particular order, or even that the chapters fit neatly into the sections. For some projects, finding and analysing secondary data sources may comprise the main body of the research, while others might find that GIS is more useful as a way of managing information than as a tool for performing data analysis. Nowhere here will you find an author who claims to tell you 'how to do it'. Indeed, most of them explicitly state that there is no single 'right' way of tackling research problems in human geography, while chapters such as that by Kesby *et al.* (Chapter 9) include examples of projects which use at least three of the different methods described elsewhere in the book. It is this variety that can make each individual research project, at whatever level, an exciting and original experience. We have found that there is always something new and exciting to be learnt from our colleagues' writings about their research approaches. We hope that you will enjoy joining us in this experience, and wish you well as you tackle your own project!

BOX 1.1

Further reading

Bell, J 1999 *Doing your research project: a guide for first-time researchers in education and social science*, Buckingham: Open University Press

Clifford N J and G Valentine (eds) 2003 *Key methods in geography*, London: Sage

Hoggart, K, L Lees and A Davies 2002 *Researching human geography*, London: Arnold

Kitchin, R and N J Tate 2000 *Conducting research into human geography: theory, methodology and practice*, Harlow: Prentice Hall

Parsons, T and P Knight 1995 *How to do your dissertation in geography and related disciplines*, London: Chapman and Hall

Rogers, A, H Viles and A Goudie (eds) 2002 *The student's companion to geography*, 2nd edn, London: Blackwell

2 Philosophies underlying human geography research

Elspeth Graham

Synopsis

This chapter is designed to encourage readers to engage with some of the philosophical debates that underlie research in human geography. Starting from the claim that all knowledge must be warranted, it proceeds to review some of the philosophical choices that such a justification of knowledge in human geography necessitates: between *naturalism* and *anti-naturalism*, *realism* and *anti-realism*. The relationship between human geography and social theory is then elaborated in terms of the contrasting ways in which both social structures and human agency might be represented by Marxists, feminists and humanists, as well as the postmodern challenge to totalising discourses. The links between philosophical positions and empirical research strategies are briefly considered and the chapter ends by introducing the idea of *critical* science and asking about the purpose of research itself.

The discussion does not aim to provide a comprehensive coverage of underlying philosophies nor to resolve those philosophical debates it considers. Rather, its purpose is to raise questions in the reader's mind about what might otherwise be taken for granted, to encourage philosophical thinking about the nature of research in human geography and, above all, to persuade the reader that philosophy matters!

Introduction

The aim of this chapter is twofold. The first is to persuade you, in case you need persuading, that philosophy is relevant to human geography because any piece of geographical research is based on philosophical assumptions or choices. The second is to provide some insight into what these choices are and how they might influence the design of your research. The latter, at least, is no easy task. Contemporary human geography is characterised by methodological diversity, a plethora of methods and approaches which link in complex ways to underlying philosophical debates. Further, the approaches evident in the academic literature are not static collections of ideas but have evolved over time, often in reaction to the perceived shortcomings of other approaches. And they continue to be refined as human geographers examine and re-examine the strengths and weaknesses of the various ways of doing research, along with their similarities and differences. Of necessity, I will pick my way through this methodological maze in a highly selective manner. Before we plunge deeper into the contested and shifting ground of geographical methodology, however, I want to begin by reflecting, in very general terms, on what is involved in doing a dissertation.

The purpose of any piece of research work, including an undergraduate dissertation, is to add to our stock of knowledge in however small a way. By choosing a topic to research and posing questions about it which you hope to answer, you make at least three important assumptions. First, that the questions themselves are intelligible. Secondly that they can, in principle, be answered and thus, thirdly, that the answers will add to our understanding of the topic. Of course, it may turn out for all sorts of reasons that you cannot answer the questions you have asked. The data you seek may simply not be available, for example. Or you may come to realise that the questions you posed are less straightforward than first assumed. Where this is not the case, however, the answers you give must in some way be appropriate or legitimate answers to the questions otherwise they cannot be said to constitute knowledge.

Suppose, for example, you pose a question about the potential impact of a new motorway extension, or the operation of the housing market in relation to an ethnic minority group. An answer which refers solely to the movement of the planets would not be considered either adequate or appropriate. This may seem both obvious and trivial but I use it to highlight the point that *not just any old answer will do*. We have to make judgements about adequacy and appropriateness. We cannot avoid them. To put the point in more philosophical language: for claims about the world to constitute knowledge, they need to be justified in some way or other and this justification will always move beyond the empirical evidence, or data, we use to formulate the answer. Most of us would reject a planetary explanation of the location of Asian groups in Bradford or Birmingham as 'off the wall'. Ask yourself why this is so. Surely it is because we start with a set of beliefs about what sort of things influence what sort of other things and we know, or think we know, that the relative locations of Jupiter, Saturn or Uranus could not possibly be the sort of thing that has any impact on traffic flows or the operation of a housing market. This too, then, is a claim to know. If an astrologer came along to challenge the claim, how could it be justified? I am not looking for a serious answer at present but this is a serious question. If you claim to know something, you must be able to justify that claim when it comes under attack. There are many possible layers of justification but this helps to explain why many philosophers now use the term 'warranted knowledge' to indicate the necessity of justifying the answers you give to research questions. The purpose of your dissertation, then, is to produce *warranted* geographical knowledge.

Why bother with philosophy?

You may already be wondering why you are reading this chapter. After all, what you want to know is how to do a dissertation in human geography. Abstract philosophising may strike you as irritating, distracting you from the main task in hand – doing geography. Why on earth must you bother with philosophy? The short answer is that you cannot avoid it, whether you like it or not! To do a dissertation is to provide answers to questions and by providing such answers you will be claiming to know something. Since any claim to know raises philosophical questions about whether and how knowledge claims are warranted, such questions can only be avoided by turning a blind eye. Of course, it is always possible to refuse to entertain such questions but only at a cost. And, as I hope to demonstrate, at rather a large cost given the nature and current state of human geography.

It should not have escaped your notice that contemporary human geography is infused with '-isms'. Numerous articles and books that you read in the course of your studies make reference to empiricism, positivism, behaviouralism, humanism, structuralism or postmodernism, to name but a few. This signals the first good reason for attending to philosophy, namely that you will be unable to understand fully the substantive literature of human geography without doing so. Over the last three decades or so, since the dominance of spatial science in the 1960s, human geography has taken a strongly philosophical turn as human geographers have sought new approaches to their research. Many books have appeared which have become classics or milestones in the search for new ways of doing human geography. Among them, David Harvey's (1973) *Social justice and the city*, Derek Gregory's (1978) *Ideology, science and human geography* and Ley and Samuels's (1978) edited collection on *Humanistic geography: prospects and problems* stand out as early attempts to change our understanding of what human geography is and thus of the ways in which we do research (both in terms of the questions to be asked and the ways in which we might try to answer them). Despite greater insight into the differences and communalities between various approaches (Gregory 1994a) and a number of attempts to find common ground in order to reshape human geography (Gould and Olsson, 1982; Gregory and Walford, 1989; Kobayashi and Mackenzie, 1989; Macmillan, 1989a), the methodological diversity has not diminished. And, as I hope will become apparent, an appreciation of certain philosophical debates is crucial to any justification of the methodological choices you will have to make when undertaking your own research.

The second reason for taking philosophy seriously is related to the first but draws attention to another aspect of unavoidability. Philosophy is to research as grammar is to language, whether we immediately recognise it or not. Just as we cannot speak a language successfully without following certain grammatical rules, so we cannot conduct a successful piece of research without making certain philosophical choices. Philosophy, like grammar, is always there. We may not be able to articulate the rules of English grammar which allow us to speak, write, be understood and understand others but once they are pointed out we can begin to appreciate how much they influence the practicalities of communication. In a similar manner, I am suggesting, even the most philosophically inarticulate researcher makes philosophical choices simply by doing research. I want to pursue this analogy without, I hope, putting too much strain on it.

Doing research is in some ways like speaking a language. The novice researcher, just like the language learner, has often to master new terms. This involves reading, or listening to, those already versed in the relevant practice (whether it be doing research in particular ways or speaking a particular language). Humanistic geographers often talk about 'taken-for-granted worlds' and 'insiders' and 'outsiders' views; many Marxist geographers refer to the 'substructure' and 'superstructure'; positivist geographers frequently couch their discussions in terms of 'hypotheses', 'tests' and the search for 'laws'; and contemporary cultural geographers speak of 'embodiment', 're-presentation', 'the Other' or 'subaltern geographies'. Such linguistic differences are not accidental but rather part and parcel of particular ways of doing human geography research. There is much more, of course, to doing research than simply mastering new terms but, in so far as language directs our thinking, linguistic differences between the terms used by humanistic, Marxist, positivist and cultural geographers are indicative of much more fundamental differences in ways of thinking.

This presents special problems for the student of human geography. Think again about learning a language, like English. Where this is your native language, the process of learning has been a process of socialisation into a language-speaking community, hearing and reading the language, making attempts to communicate and being guided and corrected when you go wrong. Much the same process of socialisation happens during your undergraduate education, only in this case you are being guided to master particular ways of thinking. In the process, you absorb the unwritten 'rules' of the practice of human geography. Without labouring the point, it seems evident that even those who feel that they can speak with authority about the nature of these 'rules' do not speak with one voice. This lack of consensus is at once daunting and exciting, for it is the source of lively debate. Some might go so far as to say that contemporary human geographers speak not one language but many, that each approach to the practice of human geography has developed its own set of 'rules' and that these are incompatible one with another. Students of human geography, on this account, are being asked to master not just one language, but many, and all at once. (No wonder confusion often ensues!) To express this idea in a more philosophical way, we might suggest that there are many kinds of contemporary human geography, with each school of thought playing its own *language game* and each thus being *incommensurable* with the others. This way of characterising human geography owes much to the writings of one of the most prominent philosophers of the twentieth century, Ludwig Wittgenstein. It is no surprise, then, that some recent writings on the methodology of human geography have turned to the works of Wittgenstein for illumination and, sometimes, appealed to his work as a justification for a particular view (Curry, 1991 and 2000).

You will note that I have not committed myself on the question of whether or not contemporary human geography *is* characterised by a number of incommensurable language games. That, in itself, is a highly philosophical question and any attempt I might make to answer it would only divert attention from the task in hand. I hope that I have said enough, however, to convince all but the most determined anti-philosopher that the practice of research cannot be divorced from questions of philosophy. Methodological choices are, at heart, philosophical choices and since the diversity of contemporary human geography forces the former upon you, you cannot avoid the latter.

Before I proceed to examine some of the philosophical choices that will be forced upon you, I want to issue something of a health warning. This comes in two parts. The first concerns the language of methodological debate. This can be difficult and, although I will attempt to minimise the use of what may be unfamiliar terms, some, like 'epistemology', are so basic to a philosophical understanding that they cannot be avoided. If you have no idea of what such terms mean, you may have to resort to looking them up. *The Dictionary of Human Geography* (Johnston *et al.*, 2000) is very useful for this purpose. Secondly, you should be aware that, in writing about the philosophies that underlie human geography research, there is a constant danger of over-simplification. Most of the '-isms' which have become common currency in geography are labels which refer to a bundle of quite complex ideas. It simply will not be possible in the space of one short chapter to convey that complexity adequately, and the discussion that follows will only hint at it. Unless you are already well versed in the relevant literature, further reading will be crucial to the development of your own understanding of the issues. Guidance will therefore be given as to what you might most usefully read to extend your methodological understanding.

Finally, I want to emphasise that what follows is not an attempt either to cover all the philosophical choices which inform research in human geography or, indeed, to answer the philosophical questions (resolve the philosophical disputes) that I will mention. I do not intend to argue for a particular philosophical or methodological position. Rather, I want to *raise* questions in your mind and make you more aware of what you might otherwise take for granted. In sum, I want to encourage you to develop a deeper understanding of the *basis* of research in human geography. Doing a dissertation in human geography is about producing warranted knowledge and thinking about the issues discussed below will help you to know what to say if that warrant is ever called in.

Geographical research

One of your first tasks when preparing for your dissertation will be to identify a research topic that interests you. More specific guidance will be given in Chapter 3, but here I want to put that choice in a much more general context. Any research project will have a subject matter (*what* is being studied) and a method or approach (*how* it is being studied). These two dimensions define the basic methodological framework of the project but they are also interrelated. As students of geography, you are not only being socialised into the practice of research (the *how* dimension) when you attend lectures and practicals or read articles and books, you are also being taught about what counts as a valid or worthwhile research topic (the *what* dimension). This aspect of your education as a human geographer may rarely be made explicit and, in some respects, can only be learned by doing geography. It does, however, raise an interesting question. What makes a piece of research geographical? As Kirk (1963: 361) asked some time ago, 'Do we think and work geographically rather than think and work on geographical materials?' In other words, is it the 'how' or the 'what' dimension that makes our research geographical? Without being paranoid about protecting or delimiting academic turf, this is a question worth considering especially as it may be in the minds of some dissertation examiners.

First consider some standard answers to the question about what human geographers study: geographers study spatial relationships; geographers study the relationships between people and environments; geographers study landscapes; geographers study regions or localities. It is certainly possible to find evidence in the geographical literature that all these, and many more, have been objects of research. It is hard, however, to claim the exclusive academic ownership of such objects that would be needed to justify the claim that they are distinctively geographical. The 'how' dimension of research presents similar problems. Human geography is a social science which shades into both the natural sciences and the humanities. As such, it shares many of the approaches used by other social sciences (and faces similar philosophical choices). So difficult had it become to disentangle the methodologies of human geography from those of other social sciences that, nearly two decades ago, Eliot Hurst (1985) wanted human geographers to stop trying, to discard their empty geographical identities and throw in their lot with an undifferentiated social science. This has not happened and (perhaps paradoxically in the light of postmodernist claims that we are now in a post-disciplinary era) there has been a slow revival of the idea that there may be a distinctively geographical way of thinking.

This is best captured by the term 'geographical imagination', used by Harvey in *Social justice and the city* (1973: 24):

> *This imagination enables the individual to recognise the role of space and place in his* (sic) *own biography, to relate to the spaces he sees around him, and to recognise how transactions between individuals and between organizations are affected by the space that separates them.*

More recently, Gregory (1994b) has developed a similar idea in his vast book *Geographical imaginations*. Not that either is concerned to define geography, but both do draw attention to the importance of spatiality as part of a geographical imagination. Perhaps this is the key to identifying geographical thinking? Whether you agree or not, it is what may look like a small difference between the two conceptions that I want to draw attention to here. Harvey employs the term in the singular, whereas Gregory uses the plural. For Gregory there is not one geographical imagination, but many. This marks a shift in thinking over the twenty years between the publication of the two books, as Gregory, like many other human geographers today, has grown suspicious of the totalising discourses of the past. A totalising discourse is a way of thinking, writing and speaking which incorporates claims to be *the only valid* way of thinking, writing

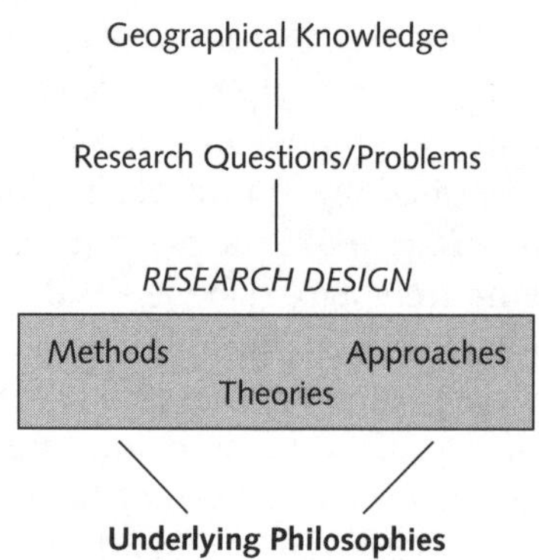

Figure 2.1 Philosophy, research design and geographical knowledge

and speaking; in other words, a discourse which tries to corner the market in warranted knowledge. The problem with exclusive claims to validity is that they have to be defended against competitors and, if you employ a totalising discourse in your research project, you need to know how to mount such a defence. Neither your empirical evidence nor your research design will help you here because both will be part of the research strategy you may be called on to defend. Once again you will need to appeal to the philosophical underpinnings of your work. If all this seems too daunting, you might decide to be less imperial and avoid claiming a monopoly over warranted knowledge. Like Hubbard *et al.* (2002: 237), you could insist that 'there is no right or wrong way of thinking geographically'. This should not be seen as a strategy for sidestepping the philosophical issues, however, as this itself is a philosophical position which is open to challenge.

Figure 2.1 illustrates the links between philosophy, research design and geographical knowledge. Philosophies underlie the design of any piece of research, which, in turn, must be appropriate to the questions or problems that prompt the research enterprise. The answers to your research questions, if they are warranted, constitute (geographical) knowledge. The relationships between philosophies and the theories, approaches and methods that shape empirical research are complex, however, and choices must be made at each stage. No one choice necessitates another but, since I have represented philosophical choices as basic to the research enterprise, I want first to examine some philosophical debates.

Some philosophical debates

Most textbooks designed to introduce students to the philosophies underlying human geography research resort at some point to organising their discussions in terms of a catalogue of different approaches used in geography, including all the major '-isms'. In addition, these approaches are often given a loose chronological sequence for the good reason that some took hold and were developed in human geography as responses to perceived deficiencies in what had gone before. Unwin (1992) is a good example of such a treatment and provides quite detailed accounts of the strengths and weaknesses of a variety of philosophical positions. Johnston (1986) is also useful if you want to gain a basic sense of what all the '-isms' mean. Cloke *et al.* (1991) cover the main approaches that have developed since the 1960s, in what is sometimes called human geography's post-positivist era. Their emphasis is on changing approaches, on how recent human geography is characterised by a state of constant revision and is still being refashioned. This is an accessible account of developments in the 1980s and well worth a read but, in starting from the rejection of positivism, the authors may leave the impression that positivism in human geography is dead. This would be unfortunate as the influences of this particular way of thinking still linger, especially, perhaps, in undergraduate dissertations. Hubbard *et al.* (2002) provide a more recent account of approaches to geographical thinking but they too concentrate on 'new' human geographies. If you want (and you should) an overview of the current methodological state of human geography, you must read one or more of these texts. It is not my intention here to duplicate either their historical treatments or their organisational frameworks. Indeed I intend to avoid running through a list of different 'types' of geography, not just because I would be repeating what has already been done elsewhere but also because of another sort of danger. Presented with a list, unwary readers might suppose themselves to be in some methodological supermarket where they can browse the shelves and pick according to whatever catches their eye. Deciding on your research design is not like selecting a tin of beans, however. In order to make an informed choice, you need to understand the philosophical basis of that choice. Let me illustrate this by exploring *some* of the major philosophical debates that underlie the choice of research methodologies in human geography.

Naturalism and anti-naturalism

One of the most fundamental choices that any social scientist has to make is that between the philosophical positions of *naturalism* and *anti-naturalism*. In a nutshell, naturalism involves the claim that research in the social sciences is essentially the same as that in the natural sciences. Thus naturalism is committed to methods adapted from natural sciences like physics. Anti-naturalism, as the term suggests, opposes such a view and maintains that, in some fundamental way or ways, social sciences are not like physics. The dimensions of this debate can be examined by considering some of the ways in which human geography has been refashioned since the 1960s.

In the past, when human geography tried to establish itself as a spatial science, naturalism held sway. This is evident both in terms of the approaches and methods adopted and the language that researchers used. Take, for example, the study of migration. Migration researchers borrowed directly from physics in their use of the gravity model to explain (and predict) the spatial movements of people. In fact this was part of an earlier movement known as *social physics*. Those who adopt such an approach are often impressed by the progress made in the natural sciences since the nineteenth century and argue that social scientists can only achieve similar progress if they copy the methods of physics and chemistry. There are two key elements to note in relation to what has become known as *the scientific method*: first, the centrality of empirical evidence and secondly, the importance science attaches to laws. Each has a crucial place in our conception of the natural sciences and I will deal with them in turn.

Modern science has always emphasised the importance of observational evidence as the basis of scientific knowledge. When a scientist claims to know something, that claim can be justified, or warranted, in terms of observation, experiment and the collection of data. Some philosophers of science (and perhaps some scientists) think that empirical evidence is the *only* valid evidence that can be used in the certification of scientific claims as knowledge. This sort of thinking is known as *empiricism*. Although empiricism can take several different forms, they all give epistemological primacy to evidence from experience; that is to say they base a whole theory of knowledge on the unique importance of this type of evidence. The appeal of such a view lies in the thought that we can only be sure of what we can carefully observe. We know that the Earth is round because we have seen images of it taken from space but we can say nothing about the shape of a leprechaun because no one has ever seen one!

There is more to science than the collection and checking of careful observation. Both physicists and chemists aspire to explain things and the kinds of explanations scientists seek are *causal*. In order to construct causal explanations, however, scientists develop theories (like the theory of gravity) which describe how things are related. Embedded within scientific theories are general laws. In the ideal case, scientific laws take the form 'All gases expand when heated'. This is a universal statement (independent of both space and time) about the behaviour of gases which summarises an observed regularity or constant conjunction. Scientists rely upon laws of this sort when explaining individual events (like a gas explosion).

One philosophical problem faced by the natural sciences is the basis on which they can establish their laws and theories as warranted knowledge. Most would appeal to the philosophies of *positivism*. Again there are several varieties of positivism but one of the most influential has been *logical positivism*, a type of empiricism developed by a group of scholars in Vienna in the early part of the twentieth century. These scholars were concerned to place all knowledge on a scientific footing by adopting the verifiability principle of meaning. Roughly, this is the principle that 'claims to know something can only be justified (are only meaningful) if they are open to empirical testing and verification'. The only exceptions allowed are the tautologies of mathematics or logic, which have no empirical content. Thus testing laws and theories against the carefully observed empirical evidence was seen not only as the hallmark of the natural sciences but as the *only* route to knowledge outside formal mathematics and logic. This is certainly one of the totalising discourses mentioned earlier.

Since the time of the Vienna Circle, some of the ideas of logical positivism have been discredited. The work of the philosopher Karl Popper (1968, 1969) has, for example, led to the abandonment of 'verification' and the substitution of 'falsification' as the rationale for testing the hypotheses of science. What distinguishes science from non-science, according to Popper, is that scientific knowledge claims are, in principle, empirically *refutable*. Since no amount of empirical evidence can confirm, in advance of the event, that the next gas we heat will expand, it must be the case that what scientists are really doing when testing their laws or theories is looking for counter-evidence to see whether or not they are false. Added to this modification of logical positivism is a growing appreciation that falsification, even in the natural sciences, is not a straightforward matter. Scientific theories are related to empirical evidence in complex ways such that empirical tests, even when the results are negative, do not alone settle the matter of whether or not the theory must be abandoned. Scientific theories, as philosophers would say, are underdetermined by empirical evidence. Despite these qualifications, the natural sciences still have elements of both empiricism and a more general form of positivism as their underlying philosophies. The formulation and testing (based on empirical evidence) of universal laws and theories continue to form the core of the scientific method. The question for social scientists is whether their studies can also be built upon this philosophical base.

Without settling the question, let me point out that there are some social scientists, including human geographers, who explicitly adopt a positivist philosophy to inform their research practice and who thus hold that the form of knowledge (epistemology) in the social sciences is fundamentally the same as that in the natural sciences. Others, perhaps many others, if pushed to articulate their philosophical choices, would turn out to be positivists, also espousing naturalism. Three articles by geographers, who all come down on the side of naturalism, illustrate the influence of positivist ideas in human geography. Hay (1985) is concerned to rebut some of the criticisms of scientific method and declares his belief that scientific methods will be retained in both human and physical geography. He makes it clear, however, that he does not suppose that this will preclude other approaches. (This is a philosophically problematic supposition that I will come back to later.) Macmillan (1989b) is, if anything, more anxious to promote naturalism in his discussion of the role of models in quantitative theory construction in human geography, although he does so less explicitly. His discussion of the nature of theory, for example, has both empiricist and positivist underpinnings. Lastly, Flowerdew explicitly declares himself a positivist in what he calls the 'broad sense' and embraces naturalism when he remarks, 'it is not clear to me that a different epistemology is needed to study human behaviour' (1998: 295). His discussion focuses on quantitative methods, however, and the apparent assumption that epistemology (positivism) is necessarily linked to a particular method (quantification) is problematic. As will be seen below, philosophical positions do not dictate research method. Hence a quantitative human geography need not have naturalism as its underlying philosophy, despite the historical coincidence between the two during the heyday of spatial science.

Because fundamental philosophical choices are rarely made apparent by those writing human geography (as distinct from those writing *about* human geography), they can be difficult to spot. Not all naturalist human geographers reduce human behaviour to the automatic responses of social physics. Indeed, many would argue that human behaviour itself can be subject to scientific scrutiny, with empirical evidence being used to construct and test general theories.

As a very general rule of thumb, human geographers who speak of models as theories, who construct formal (often statistical) tests for their hypotheses and who cite Harvey (1969) with approval are most likely to be naturalists believing that human geography is fundamentally the same as the natural sciences.

The alternative to naturalism is anti-naturalism and the last three decades have seen the mushrooming of anti-naturalism in the social sciences, including human geography. The basic claim is that the social sciences are *not* like physics, but that claim too, in order to be warranted, has to be supported by 'evidence'. There are many varieties of anti-naturalism but most have their roots in the *hermeneutical* or interpretative movement in philosophy. The term 'hermeneutics' comes from the name of the messenger of the gods in Greek mythology, Hermes. His task was to communicate (or interpret) the desires of the gods and convey these to mortals. In current usage, hermeneutics refers to the study of interpretation and understanding. Consider the following argument. The social sciences involve the study of the causes or the consequences of human action (or more colloquially, human activity). No social science can avoid this because it is part of their definition. When we think of explaining our own actions, like going to the cinema or to church, we do so in terms of our desires and beliefs. (We want to see a particular film, or we hold a certain set of religious beliefs.) Natural scientists, however, do not have to deal with desires and beliefs (it makes no sense to say that gases *want* to expand when they are heated), ergo, social science is fundamentally different to natural science. This, though a highly condensed version, is the sort of argument that supports anti-naturalism. Note the key role given to the mental qualities of human beings. Human beings can intend to do things, have reasons for doing them and can reflect upon what they are doing. Thus *intentionality*, *rationality* and *reflexivity* are fundamental characteristics of human action, the subject matter of the social sciences, whereas the subject matter of the natural sciences, so the argument goes, responds only to the forces of nature without any of these human qualities. Glaciers do not think! Phillips provides a useful summary (1987: 105):

> *The underlying point in all this is that the physical sciences, with their emphasis on uncovering the causes that produce effects, are not a relevant model for the social sciences for a simple reason: people act because they are swayed by reasons, or because they decide to follow rules,* not *because their actions are causally determined by forces.*

If you find this at all persuasive, then you must consider adopting an approach to your research which is both anti-naturalist and grounded in the hermeneutical philosophies. The use of the plural is important here as there is not only one road to hermeneutical understanding. All do, however, have one feature in common. They all take *meaning* seriously: that is to say, they are all persuaded that human action is essentially meaningful behaviour. The implication of this is that social scientists can only seek an understanding of human behaviour by rendering it intelligible in terms of its meaning or significance. This in turn involves providing an understanding through the interpretation of such behaviour and interpretation requires quite different research strategies to those employed by natural scientists when they set up and test scientific hypotheses.

Choosing a hermeneutical research strategy is far from simple as there are a number of ways of interpreting human action. Idealism, phenomenology, postmodernism, post-structuralism and even critical theory are all variations on a theme which focuses on meanings and rejects naturalistic approaches to human behaviour. All have advocates in the social sciences and in human geography. Understanding these variations and thus the choices open to you will involve delving further into the relevant literature (see Hubbard *et al.*, 2002, Chapters 2 and 3 for an introduction). The most I can do here is to tempt you to do so by pointing out how influential anti-naturalism is in contemporary human geography.

Humanistic geographers were amongst the first to take a strongly anti-naturalistic line in their insistence that human geography should be about human *experience* (Ley, 1982). The more human geographers began to explore this idea, however, the more it became evident to some that the challenge of understanding that experience involved understanding the language or languages in which it was expressed. Hence recent writings on human geography have taken a linguistic turn. This is evident in, for example, the 'new' cultural geography and one of its concerns – landscape. Landscapes can be defined and interpreted in different ways (Duncan, 1995). Landscape can be construed as a way of seeing, read as a text or even interpreted as theatre. Barnes and Duncan (1992) organise their collection, *Writing worlds*, around three concepts: text, discourse and metaphor. Their interest is in landscape writing and representation and they start from the rejection of the 'naive view' that words represent thoughts or objects in a straightforward and unproblematic manner. As Rosenberg (1988: 94) puts it, 'For these social scientists the philosophy of language is as important as the study of differential equations is for physics'. When you are reading the literature of human geography and come across terms like 'experiential place', 'text' or 'discourse', it is highly probable that you are reading a work which has its philosophical foundations in anti-naturalism and hermeneutics. Further, the extent to which such terms have permeated recent geographical writings can be taken as an indicator that human geography has entered a post-positivist era.

Realism and anti-realism

There is one other philosophical debate which requires a mention but which, again, is not easy to summarise in a few short paragraphs. Since, however, the attendant philosophical choices have been influential in the recent development of different schools of thought in human geography, here goes! In the most general terms, the debate is between the ontological positions of *realism* and *anti-realism*. This simply means that these positions involve opposing claims about the basis of existence (i.e. ontology), or what it means to say that something exists. Realism can be thought of as a common sense philosophy which maintains that there is a 'real' world, out there, independent of our conceptions of it. Thus when we study, say, some aspect of the human landscape, we are studying something that has a material existence beyond our perceptions of it. If you like, it is 'really' there. Anti-realists, on the other hand, deny this and argue that there is no possible justification for believing in a reality other than that constituted by the human mind. In a very fundamental sense we, as thinking human beings, *make* our world in our minds. This view is sometimes known as (metaphysical) idealism.

I do not wish to diminish in any way the importance of such ontological questions (which have exercised the minds of some of the greatest philosophical thinkers) but nor do I wish to pursue this debate here. More important in the present context are its implications for how we can gain knowledge of the world, for it is these epistemological implications which have the greatest impact on how we do research. To over-simplify for a moment, the epistemological debate revolves around the question of what role mental constructions (representations) play in our knowledge of the world. We might, for example, remain agnostic in relation to whether or not a world exists beyond our mental constructions of it but nevertheless argue that all we can come to know are these mental constructions. This does not dissolve the question of what can be said to exist but it does put it to one side. It also raises another question: 'What sorts of things can we come to know?' Is it the ways in which different people construct their worlds, with all the plurality of ways of seeing which this implies? Or is there a deeper world, behind the experiences of individuals, which is real just in the sense that we can come to know it? In human geography, humanistic geographers have been prominent amongst those who affirm the former although, most recently, a heightened concern with language and text bespeaks the development of different ways of coming to know human 'realities'. Others, however, lose patience with what can seem rather effete or even elitist concerns and wish to affirm that there is a deeper reality which human geographers can come to know. Most notable amongst the latter are those who argue for a version of *scientific or critical realism* as the philosophical basis of social science and thus human geography (Sayer, 1985, 1992). If it is not too much of a parody, an impatience with multiple realities often stems from their potential to trivialise 'real' social problems. Poverty may be a human construct but to emphasise its mental basis is to run the risk that it be treated as 'mere imagination' (see the debate between Yapa, 1996, 1997 and Shrestha, 1997).

All this is an over-simplification. There is much more to scientific realism (including its conception of causality) which the informed human geographer should know about. Equally, those anti-realists who, for want of a better term, we can call epistemological idealists, do not speak with one voice. Despite the different strands of realism and anti-realism, however, I hope I have conveyed something of each. Where the simplification becomes over-simplification is not in the lack of detail in which these philosophical ideas are described but in the oppositional way in which the philosophical choice is presented: realism *or* anti-realism. The opposition is clear, I think, in the ontological question about what can be said to exist (see Ladyman, 2002). No one can be both a realist and an anti-realist. Yet, when realism and anti-realism (idealism) are viewed epistemologically, the dichotomy becomes blurred, at least at the edges. (Walton, 1995, provides a discussion of this.)

I want to pick up the question of what sort of things we can come to know and how we can come to know them in the following sections of the chapter and in a rather more concrete way. First, here is a very brief review of the discussion so far. Social scientists, and therefore human geographers, are faced with a fundamental choice between naturalism (which would model human geography on the natural sciences) and anti-naturalism (which argues that human geography is quite unlike physics because it deals with human action and experience). On either side of this divide there are secondary choices to be made but ultimately – though not an uncontentious point – you cannot be both a naturalist and an anti-naturalist. Note, however, that any choice between the two will depend as much on your conception of the central characteristics of the natural sciences as it will on your idea of the nature of social science.

A further dimension of philosophical choice is introduced by the distinction between realism and anti-realism (idealism). The choice here is not only complicated by variations within each camp but also by the fact that, in some versions, they may appear to overlap. Nor, and this is most important, does the realism/idealism distinction map neatly on to the naturalism/anti-naturalism distinction, despite the fact that I have identified idealism as a hermeneutical (and hence anti-naturalist) philosophy above. Realism can be positivist (and hence naturalist) in some of its versions, but positivism is not always realist (*pace* what some geographers appear to believe). Equally, scientific realism, as a recently developed philosophy, does not ignore hermeneutics and thus sides with the anti-naturalists rather than the naturalists whilst, at the same time, leaving open the possibility of naturalism. If all this seems confusing, it merely reflects, I am afraid, the complexity of the matters involved and the philosophical choices open to contemporary human geographers. We are, however, only part way through our methodological journey and I now want to push further through the undergrowth, in a slightly different direction.

Human geography and social theory

Central to human geography is some concern with human agency and, unsurprisingly, there is much debate about how that concern should be conceptualised. What is the nature of human agency and what differences do various ways of thinking about human agency make to the conduct of research? Human beings, for example, are not isolated individuals but interact with each other as social beings. Further, their interactions are not haphazard or random but appear to be ordered in lots of different ways. Just as the task of the natural scientist is to find order in the physical world, so the task of the social scientist can be thought of as finding order in the social world. This raises another fundamental question: where do we look for that order? Or to put this in another way, who or what creates that order? This is not centrally an empirical question, although neither can it ignore observational evidence about how humans interact. Rather, it is a theoretical question in the sense of requiring us to formulate some theory about the nature of the social world. As with all theories, any particular formulation will be underdetermined by the empirical evidence at our disposal, which makes judging between competing theories especially problematic. Nevertheless, it is possible to detect some dimensions of choice which, because they underpin the detail of specific theoretical constructs, provide grounds for accepting or rejecting certain ways of conceptualising social interaction. These dimensions of choice provide the theme for the following discussion, which will be organised in terms of what is known as the structure–agency debate (Chouinard, 1997).

Marxism, feminism and social structures

It is only in the last decade or so that a concern with social theory has flourished in human geography. The story of why geography came belatedly to social theory has been told elsewhere (Gregory, 1989). Two points are of interest here as they link this part of the discussion both to what has gone before and to what is yet to come. First, the naturalism of the 1960s sent human geographers off on what is now widely recognised as a wild-goose chase, looking for laws of spatial pattern to establish geography's scientific credentials. In the push to make geography a morphological science, people often disappeared from geographical writings or were incorporated into abstract mechanistic models in ways which denied not only their humanity but the social character of their interactions. Even the advent of a self-consciously behavioural approach, which took human behaviour as the object of study, failed to promote an engagement with the wider literature of social theory. It was not, therefore, until human geography was released from the stranglehold of positivism (and its scientific method) by the critiques of both Marxists and humanists that thinking about 'society' became respectable. Secondly, and as a reflection on the history of geography, the *situatedness* of warranted knowledge becomes apparent. Intellectual inquiry is itself part of social life and therefore has a history (and a geography). The character of what is accepted as warranted knowledge can and does change over time and between places. This has led some to argue that all knowledge is relative, being unavoidably the product of the spatial and temporal circumstances of its creation. Such relativism is worrying for those involved in the production of knowledge (including those doing undergraduate dissertations), as we shall see. For the moment I simply want to underline that human geography has begun its engagement with social theory, that this has led to a rich and difficult literature and that I will not be able to do justice either to the richness or to the complexity here. All I can hope to achieve is to provide a few pointers which may help to further your own thinking and understanding.

Marxism is a particular way of interpreting the world which is often called historical materialism. Based on the writings of Karl Marx, it takes as its starting point the material conditions of human existence. A central concept of Marxian social theory is the 'mode of production' which characterises a society. This describes the (organised) means that people employ to sustain and perpetuate themselves. Capitalism, for example, is the mode of production of most contemporary Western societies and involves, according to Marx, a specific set of relationships between individuals, as well as certain attitudes towards nature. Such relations form the economic base or substructure of any society and all other dimensions of society (the legal, political and religious, for example) are parts of the superstructure, the nature of which ultimately depends on the economic base. Social organisation is, at root, economic. Further, there is a historical imperative in classical Marxism which posits a succession of modes of production through time as a society progresses towards communism. Capitalism is seen as inherently unstable and thus as inevitably bringing about its own demise. In a famous passage Marx (1974: 715) declares,

> *Centralization of the means of production and socialization of labour at last reach a point where they become incompatible with their capitalist integument. Thus integument is burst asunder. The knell of capitalist private property sounds. The expropriators are expropriated.*

Marxism, like positivism, is not a static set of ideas and many different types of Marxism have emerged as the classical works have been reinterpreted and modified (see Peet and Thrift, 1989 and Peet, 1998). Some of these have been developed specifically to circumvent perceived problems in the classical formulation but all share this as a common point of origin. There are some general features of these Marxisms which have attracted the attention of critics and which you need to think about when choosing the theoretical framework for your dissertation because they bring a Marxist interpretation of society into conflict with other social theories. I will outline these briefly and introduce some of the choices involved.

Classical Marxism is *historicist*, which is to say that it claims that, in order to understand the current state of any society (and predict future states), we need detailed knowledge of the past stages of development of that society. The present can only be understood in terms of the past. Further, it is the transformations in modes of production which are the key to understanding all other aspects of social change and these transformations or revolutions result from internal contradictions in the economic base as the struggle between classes in a society (the proletariat and capitalists under capitalism) is played out. Capitalism is thus – at least in some of Marx's writings – seen as ultimately a self-destructing economic system. These aspects of Marxism raise three related questions:

1. What is the validity of assuming that all societies undergo, or will undergo, the same historical sequence of development? What reasons are there for accepting such a grand theory, or metanarrative?
2. What is the validity of viewing social change as being determined by a society's economic base? Is there a conflict between the substructural determinism of Marx's theory and our understanding of the nature of human agency?
3. Is the relationship between economic classes in a society the most basic of all social relationships such that all other features (the superstructure) constitute an ideology or form of consciousness which arises from and legitimises the economic base? Are explanations of social institutions and relations among people to be found in facts about the means of production?

Consider the last of the three questions. (I will come back to the others shortly.) The problem with trying to answer this question is knowing how a positive answer might be warranted. What evidence is there to support such a claim? If you are tempted to give epistemological primacy to empirical or observational evidence as the positivists do, you may be forced to reject a Marxist social theory because the structures which are central to it cannot be observed. This would be equally true of the central concept of the theory of gravity in physics, of course, but the difference is (or so say the positivist critics of Marxism) that the indirect effects of gravity can be 'tested' whereas it is impossible to test the indirect effects of Marxist structures in an equally rigorous manner (a criticism of structural Marxism discussed at some length by Duncan and Ley, 1982). On the other hand, if you want to follow a hermeneutical philosophy which vests explanatory power in the interpretation of meanings, then Marxist theory becomes problematic on another count for it sees meanings in events which, it argues, are often hidden from the participants themselves. Individuals are said to be imbued with a false consciousness as the beliefs, roles and rules which influence the conduct of their everyday lives are part of an ideological superstructure. There may be ways out of this impasse but it explains why some philosophers, such as Karl Popper (1945), have reacted strongly against Marxism.

There is, however, another body of social theory which challenges all versions of Marxism in rather a different way. This is *feminism* and, since feminist thinking is influential within contemporary geography (see, for example, Bowlby *et al.*, 1989; Monk, 1994; Women and Geography Study Group, 1997), it is important to appreciate the nature of that challenge as well as the similarities and differences between feminism and Marxism. Again, I will only be able to provide a rough and ready sketch map, which will ignore much of the detail of feminist thought. At the core of feminist thinking is the concept of *patriarchy*, which sees an understanding of the historical relationships between men and women as basic to any analysis of how societies are ordered. In most (perhaps all) societies, power is unequally divided between men and women such that the male or masculine dominates. In Western capitalist societies there has been an ideological association between women and home-based reproductive labour (having children and nurturing the family) which relegates women to the least powerful positions in society and allows men, in all sorts of ways, to assume power over their lives. If the feminists are right then, at the very least, Marxist theory must be modified so that gender relations and their interplay with economic relations are seen to form the substructure of society (see Gibson-Graham, 1996). As a social theory, however, feminism is not dissimilar to Marxism in that it is a *holistic* theory which explains individual conduct by appealing to much larger scale social facts. Like Marxism, classical feminism must move beyond the meanings attached to events by individuals (women) who may not be aware of the ideological basis of their oppression. Only once they are brought to see it, is emancipation within their grasp. Thus some versions of feminism are equally problematic in respect to the warranting of claims to knowledge. Their meta-theoretical nature means that the relationship between theory and empirical evidence is not a straightforward one and any reliance on the idea of 'false consciousness' produces special problems for hermeneutical interpretation. Whether you side with naturalism or anti-naturalism, both classical Marxism and feminism are problematic when it comes to justifying their claims to know.

It would be wrong, I think, to leave the discussion there, for feminism offers us one way out of the apparent dilemma, which I will simplify in order to articulate it as clearly as possible. *If* the claims of feminism are right, then the conditions of patriarchy will pervade all aspects of our lives, including the everyday activities of academic work. Working from a feminist perspective, Rose (1993), for example, provides an enticing account of the ways in which the academic discipline of geography is dominated by men. She argues that this has serious consequences both for who can produce geographical knowledge and for what is taken to count as legitimate knowledge. To accept this argument, or one like it, is to open up the possibility that the major philosophical choices I have outlined above are a distinctively masculine way of representing the problems of legitimising or warranting knowledge. Rose herself attacks definitions of 'rational knowledge' which ignore the emotions, values and past of the knower. Perhaps there are other, even feminine, ways of justifying claims to know. I have to admit a personal struggle with this possibility in relation to feminist theory. As a female academic, I find that I am drawn to some aspects of feminism because they seem to accord with my own personal experiences. Yet, at the same time, I am not convinced that my emotional receptivity to feminist thinking is enough to establish feminist theory as warranted knowledge.

Humanism and human agency

The second question about Marxism, raised above, concerns its treatment of human agency. Several versions of Marxism rely, at least implicitly, on a mechanistic model of society where individual human beings are represented as so many cogs in a larger wheel. Capitalism is seen as a functioning system with its own internal logic, and individuals are reduced to passive agents in the production and reproduction of the system. This conception is hotly contested by those whose view of the individual in society is informed by the ideas of *humanism*. Humanism is not itself a social theory but rather a diverse set of ideas which have in common an emphasis on the humanity of individuals. People are capable of being creative (or destructive), reflective (or not) and, above all, they are moral beings, which is to say that there is a moral dimension to their actions. Any social theory which denies these characteristics of what it is to be human must, on the humanist view, be flawed because it fails to recognise the fundamental nature of its subject matter. Abstract models of 'man', whether Marxist or positivist, are out and full attention is paid to the lived experience of human beings.

Starting from this position, humanist social science assumes many guises but all show some concern for the mental processes of human beings. *Consciousness* and *intentionality* are central concepts and human meanings and values are emphasised. This is the active view of human agency which rejects any deterministic interpretation of human behaviour. Its moral dimension can best be illustrated by contrasting it with the passive interpretations of human agency found in the more structural versions of Marxism. The argument (and hence choice) here centres on the conceptualisation of structures, on what is meant by 'structure'. Accept for a moment the Marxist claims that the relationships

LIVERPOOL JOHN MOORES UNIVERSITY
LEARNING SERVICES

between people in a society are not what they seem to be 'on the surface'; that to understand these relationships we must dig deeper and come to appreciate the economic order which underlies these relationships and that this economic order, under capitalism, has a certain internal logic. It then becomes difficult to resist a conceptualisation of structure along the following lines (Harvey, 1973: 289):

> *The totality seeks to shape the parts so that each part functions to preserve the existence and general structure of the whole. Capitalism, for example, seeks to shape the elements and relationships within itself in such a way that capitalism is reproduced as an ongoing system.*

In such a scheme, individuals become no more than (unwitting) 'bearers' of structures (Althusser, 1972) and it makes little sense to blame individual capitalists for their exploitation of workers because they are merely responding to the demands of the system, the totality (Sayer's realism incorporates a similar, though not identical, view). Both responsibility and 'will' are removed from individual human beings and conferred on a supra-individual entity, the capitalist system. Now you are confronted with a dilemma, and a moral one at that. I will present it starkly so that you can see the conflict between what are in fact two extremes.

In our lived experience, we make all sorts of distinctions between good and bad practices, even in the business world. Our whole legal system rests on the general assumption that humans are morally responsible beings who have the choice of whether or not to act in particular ways. We talk of people intending to do things, making decisions, having reasons for their actions, and we expect them (sometimes force them) to take responsibility for what they have done. Can we accept a social theory that appears to deny these taken-for-granted aspects of lived experience? Can we confer moral blame on something as abstract as a system? The answer of the humanist would be 'No'. We cannot ignore the beliefs, intentions and desires of individuals because this would be to remove from them their subjectivity, to dehumanise them. Thus a humanist social science must take as its starting point an active view of human agency which allows that individuals are able to exercise their will.

The choice which I have sketched is essentially a choice between determinism and voluntarism and, if you are tempted to choose the former you might want to ask yourself what 'determines' that choice! A more usual reaction is to try to avoid the choice altogether by building on certain criticisms of both in order to reach a better understanding of the relationship between structure and agency. Among those in the social sciences, including human geography, who are most in sympathy with the idea that everyday social practices are influenced by underlying structures, one influential restatement of the relationship between structure and agency is the theory of *structuration* developed by Anthony Giddens (1984).

Giddens emphasises the need to recognise the duality of structure in which the relationship between individual and structure is taken to be reciprocal. As he puts it, '. . . according to the notion of the duality of structure, the structural properties of social systems are both medium and outcome of the practices they recursively organize' (Giddens, 1984: 25). Human agents are seen both as operating within a specific social context (structure) and as active in determining the precise outcome of their social interactions. According to Giddens, the analysis of society involves three 'levels': structures, institutions and agents. Agency and structure 'interpenetrate' in complex ways. Structures, through a host of institutional arrangements, both constrain and enable human action, while human agents, by their behaviour, reconstitute and may change both institutional arrangements and structures. Moreover, Giddens extends the notion of structure beyond the economic, political and legal to include the communicative structures of language. To this extent he introduces hermeneutical concerns to his structuration theory, which he terms a 'post-Marxist' theory.

The confirmed humanist, whose sympathies lie on the voluntaristic side of the divide, tries to colonise the middle ground in rather a different way. Broadly speaking, Giddens's attempt to dismantle the dualism between agency and structure involves modifying a Marxian idea of structure to give more power to people. Approaching from the opposite direction where people have all the power, modification means giving more credence to structure. Unsurprisingly, because of their views on what it means to be human, many humanist social scientists reach a somewhat different middle ground. Duncan and Ley (1982), for example, declare that they are not against macro-structures and recognise that individual actions cannot be fully explained without reference to the contexts in which individuals act. They continue (1982: 32, my emphasis):

> *The relations among the many actors may become patterned or structured; they may become institutionalized and taken for granted. Individuals may be unaware of all the causes of their behaviour, and intentions may form only a portion of the explanation of action. However, we suggest that macro-scale social structures should be precisely defined, that they do not have autonomy or an existence that is not* ultimately reducible to cumulative human actions and interactions, *and that the processes linking structure or context and individual or social action need to be carefully specified.*

Ultimately power remains with people. You must judge for yourself whether this constitutes a difference in substance or merely in emphasis between humanist and post-Marxist views. What you do need to recognise is that there are different ways in which both human agency and structure can be represented and that, in doing a dissertation in human geography, you will almost certainly be making assumptions about both.

Postmodernism and the death of grand theory

There is one further set of theoretical ideas concerning the nature of contemporary society that should not escape your attention because it is widely referred to in the geographical literature. These are the ideas of *postmodernism*. As the label suggests these ideas arise from a critique of modernism. This brings us back to the first question about Marxist theory raised above. Marx provided a grand theory or metanarrative describing the way in which societies are transformed by the inner logic of capitalism. His mission was to uncover what he calls 'the law of motion of capitalist society'. To this extent, his was a naturalistic account which attempted to establish social laws of much the same kind as the natural laws of the 'hard' sciences. Postmodernism, however, rejects the notion of a single general theory of society and sees social order as much more partial and even contradictory. As Cloke *et al.* (1991) point out, postmodernism is infuriatingly difficult to define and I will not attempt to do so here. There are, however, two aspects of postmodernism I want to mention: postmodernism as a condition of society and postmodernism as a method for the social sciences.

On the one hand, postmodernism can be taken to refer to an historical era in which the 'postmodern condition' (see Harvey, 1989) predominates. There is no simple way of summarising the character of that condition but it has a cultural expression in art, architecture and literature where it has come to be associated with a particular postmodern style (Graham *et al.*, 1992). Of special interest to geographers is another aspect of postmodernity which Harvey (1989) calls 'time–space compression'. This describes a shrinking world in which the friction of space has increasingly been overcome to accommodate social interaction. Hägerstrand's (1975) time–space geography has thus been revitalised as a framework for studying these new relations, while others have sought to explore the manifestations of postmodernism in the built environment of the city (see Soja, 1995, for an accessible account of the 'six geographies' of Los Angeles). Whether or not the conditions of postmodernity are sufficiently different (to those of modernity) to mark a new era is still much debated. However, the spatial component of the suggested transformations has, understandably, excited many geographers for they see here a key role for geography. Space and place become important concepts in the understanding of societies (Soja, 1989; Dear and Flusty, 2002). Geography matters! This has led Dear (1988), amongst others, to claim that postmodernism presents a special challenge and opportunity for human geographers. By taking up the postmodern challenge, Dear argues, we can reconstruct human geography and place it '*at the very centre of a newly defined paradigm of human inquiry*' (Dear, 1988: 267). At the heart of this paradigm lies a particular characterisation of society, for society is to be viewed as a 'time–space fabric' with the details of all aspects of human life inscribed upon it (Graham, 1995). Thus geography, or more appropriately, geograph*ies* become central concerns of all social sciences.

Of greater interest here, however, is the question of how the postmodern condition might be studied, for postmodernism can also be interpreted as method. In eschewing metanarrative, postmodernism allows for multiple voices. No theory, particular aspect of society or 'voice' is privileged (that is, given priority) over others as postmodernism celebrates difference. Attention to language is an important part of postmodern method and postmodern social scientists have turned to *deconstruction* as a way of analysing and understanding 'texts'. Their aim is to disentangle, or deconstruct, the different discourses that inhabit the text (Graham *et al.*, 1992). Thus knowledge is seen as multiple and situated (or positional). This promotes sensitivity to different ways of 'reading' social relationships and their human geographies and favours minor theories over the totalising metanarratives of modernism. Thus feminist geographers, for example, are increasingly becoming aware of the problems and dangers of first-world geographers writing about third-world women as if there is one feminism rather than many (Radcliffe, 1994; Bondi and Davidson, 2003). And cultural geographers now offer landscape interpretations that are grounded in a concern for the (multiple) discourses through which culture and nature are imagined and experienced (Cosgrove, 2003).

If the situated knowledges and multiple 'truths' of postmodernism challenge exclusionary theories of the nature of the social world, its close relation, post-structuralism, further destabilises understandings of knowledge itself. Post-structuralist thinkers such as Jacques Derrida elaborate a method of deconstruction that is designed to expose the problematic relations among text, language and meaning. Language is viewed as 'impure', in the sense both of being imbued with cultural prejudice and of imposing an illusory order on the world. According to post-structuralists, what we regard as knowledge is often built upon categorisation and opposition (as in the identification of 'self' and 'other', or 'structure' and 'agency'), and thus serves to 'repress, divide and segregate' (Hubbard *et al.*, 2002: 86). This critique of modernist thinking raises fundamental philosophical questions about the nature of language and of knowledge. Even discussing these questions becomes problematic since post-structuralism profoundly disturbs deeply entrenched forms of academic enquiry. It is worth noting, for example, that the either/or choices I have outlined above would be inimical to post-structuralist ways of thinking. However, to reject the necessity of making choices between the various philosophical positions as I have presented them is itself a choice that requires some kind of justification or warranting. This, then, is not an easy option because, like postmodernism, post-structuralism has attracted criticism from both within and beyond geography (see, for example, Peet 1998).

Neither postmodernism nor post-structuralism provide a recipe for doing human geography, or any other social science. Rather, they embrace a complexity of ideas both about the nature of contemporary Western society and appropriate ways of thinking about and studying that society. The acceptance of these ideas may obviate the need to choose between, say, Marxism and humanism because both can be rejected as totalising discourses. (See Gregory, 1989, for further discussion of 'the crisis of modernity'.) Yet the very toleration of different approaches which the celebrants of postmodernism welcome is also, according to its critics, its greatest weakness. The danger of rejoicing in the many is that 'anything goes'; where all interpretations are equally valued, this removes the grounds for judging between different accounts and makes the warranting of knowledge especially problematic. Postmodernism has been roundly criticised for this *relativism* (Sayer, 1993), an issue that links debates about postmodernism and post-structuralism to more fundamental ontological questions (i.e. questions about what can be said to exist) in philosophy.

Geography in practice: approaches and methods

I have sketched certain choices that any human geographer makes when carrying out research and have suggested that you are better to be aware of them than not. Underlying your dissertation research will be certain philosophical assumptions (those of naturalism or anti-naturalism, for example) and, probably, a particular conception of the nature of relationships in society. Ignorance of your philosophical roots implies a less than full understanding of what you are doing. However, the practice of geography, its approaches and methods, does not map neatly onto the philosophical choices you might make. Although philosophical problems and answers are implicated in choice of method, theories and approaches, particular philosophical positions do not *necessitate* particular research designs. As one writer on philosophy and social inquiry comments, '. . . a positivist, qua positivist, is not committed to any particular research design. There is nothing in the doctrines of positivism that necessitates a love of statistics or a distaste for case studies' (Phillips, 1987: 96). Sayer, an advocate of realist geography, makes much the same point when discussing the links between radical or Marxist research and realism (1985: 161):

> *. . . this association of realism and radical geography is not a necessary one: some radical work has been done using a nomothetic deductive method . . . and acceptance of realist philosophy does not entail acceptance of a radical theory of society – the latter must be justified by other means.*

It is impossible, then, to specify in advance which methods are most appropriate to the philosophical and theoretical choices you might make. Nevertheless, I do want to offer some comments on approaches since, as a matter of historical record, some ways of doing geography have been strongly linked to certain philosophical and theoretical positions.

Any research project starts from a question or problem that defines the topic, or *what* is to be studied. It may seem quite straightforward, then, to assume that the approaches and methods used to tackle the research problem (i.e. *how* the study is to be conducted) should be appropriate to the question posed. If your aim is to investigate the relationship between deprivation and health across the UK, then quantitative modelling appears to be the most suitable method whereas, if your intention is to understand the (multiple) meanings of a particular place for different ethnic groups, then qualitative methods involving in-depth interviews seem much more appropriate. Notice that both of these aims are rather generally specified and a good research question requires much greater focus (see Chapter 3); nevertheless, they serve to illustrate two very important points.

First, the way in which the aim of a piece of research is conceptualised and expressed contains within it assumptions that tend to direct researchers towards certain methods. For instance, the assumption that there is a singular relationship between deprivation and health across space encourages a research design that 'tests' this relationship and thus uses methods similar to those found in the natural sciences. In contrast, attention to multiple meanings is part of a postmodern anti-foundationalism that emphasises difference and the role of language in representations of the world. This requires sensitivity to what people say; hence the use of qualitative methods designed to elicit their discursive constructions. In both cases, the way in which the research is 'framed' – its aim conceptualised – contains within it a methodological direction. Thus deciding *what* to study and *how* you are going to study it are not separate dimensions of research design but are closely intertwined. This means that in formulating your research aim (or aims) you are to some extent prejudging the issue of what methods are appropriate for achieving your objective/s. A much more fundamental question, then, is whether your research problem itself is appropriately formulated.

Secondly, it is the framing of the research problem that links epistemology and social theory to method. An interest in deprivation and health could be conceptualised quite differently to the example above if expressed in terms of experiences of deprivation and multiple understandings of 'health', leading to the use of qualitative rather than quantitative methods. Framing the research problem in this way would ground it in theories of knowledge and society at odds with the positivism implicit in the original conceptualisation. Yet theory and method are not correlated in the uncomplicated manner this might seem to suggest. McKendrick's (1999) tabulation of the relationships between research traditions in geography and a variety of methods is imperfect (Graham, 1999) but his main message is sound: epistemology informs, rather than precludes, methodological strategy. Innovative research might use methods familiar within one research tradition to investigate research questions grounded in a different philosophical or theoretical position. The challenge for the researcher is to adopt a critical attitude to research design and the key to this lies in the initial formulation of the research problem.

A good research project requires a well-formulated research problem. There are numerous ways in which research questions can be conceptualised but a critical attitude to research design requires an understanding of the philosophical underpinnings implicit in the way a research problem is 'framed'. It is this framing that informs the choice of research method/s, and that choice is itself open to critical interrogation (Hoggart *et al.*, 2002). Since epistemology does not prescribe research method, there can be no recipe book for bridging the gap between theory and research practice. The translation of critical realist philosophy into an empirical research strategy, for example, is not unproblematic (Pratt, 1995; Yeung, 1997). The same can be said of Giddens's theory of structuration, which Gregson (1987) calls a second order theory, removed from events or contingencies of particular periods and places (i.e. from the domain of empirical enquiry). The difficulties of extending such abstract ideas and applying them in empirical research are considerable and go some way to explaining why the philosophical discussions of human geographers can seem so divorced from the business of doing research. Nevertheless, it is only through a dialogue between theory and practice that knowledge and understanding will be advanced. This is especially challenging when you are embarking on your first independent research project.

Finally, I want to raise one further issue. Earlier I used the term 'warranted knowledge' to indicate that answers to research questions need justification. It should now be evident that philosophical and theoretical choices form part of that justification, along with the details of empirical evidence (or data) collected. I also suggested that the purpose of an undergraduate dissertation, or any piece of human geography research, is to add to our stock of knowledge.

It is important to recognise that this 'stock' is in a constant state of flux as research findings which are at first accepted may later be modified or rejected. In addition, rejected knowledge claims may be reinstated at a later date as more empirical evidence becomes available or new ways of thinking confer a new significance on them. Knowledge, then, is cumulative, not in the sense of adding to some otherwise fixed pile but rather because it accumulates through a process in which current researchers assume a critical attitude towards past research.

For some, this critical attitude has a moral dimension and is applied not only to the work of other researchers but also to the human world being researched. The point of research, on this view, is not simply to add to knowledge but to change the world for the better. Thus critical researchers who adopt this attitude aim to do more than analyse social relationships or human landscapes, for their purpose is to change society itself. Habermas (1974, 1978), a social philosopher and theorist, sees this as a third type of science, *critical* science, and his ideas have been developed in human geography especially by Gregory (1989, 1994a), who has written extensively on critical social theory. Gregory is quite clear about his own position when he says in the introduction to *Geographical imaginations* (1994b: 10):

> *My particular concern is with the multiple discourses of* critical theory*: discourses that seek not only to make social life intelligible but also to make it* better.

A critical human geography would produce action-oriented knowledge and it may be that you choose to adopt a similar aim in your dissertation research. If so, you need to think about the normative element of your project both when formulating your research design and when presenting your findings, for there are a number of different ways of arriving at conclusions about how society *ought to be*. Further, there is a tension between postmodern ideas of differences and diversity and the aspirations of a critical human geography (McDowell, 2002). If there are many 'voices' and we privilege none of them, what grounds would we have for telling others how things ought to be?

This chapter has sought to raise a whole series of questions about how knowledge can be justified, about philosophical and theoretical choices and about the ultimate aims of research. All these issues remain contentious in human geography and it is up to you to join the fray and participate in the debates. With these debates in mind, you can move on to what may seem to you more practical matters of how to choose a research topic and decide upon data sources and methods. Remember, however, that philosophical choices underlie these decisions. If you set up a project that attempts to replicate the methods of the natural sciences, you are probably assuming the positivist aim of uncovering laws of behaviour. If, on the other hand, you ignore social structures and focus on the intentionality and motivations of individual human beings in their lived experience, you are assuming a humanist interpretation of the nature of human agency. The choices are up to you but consistency and a deeper understanding of the basis of your research design demands that you take philosophy seriously.

BOX 2.1

Further reading

Hubbard, P, R Kitchin, B Bartley and D Fuller 2002 *Thinking geographically: space, theory and contemporary human geography*, London: Continuum

A good introduction to theoretical thinking in human geography, with a focus on post-positivist approaches. It touches upon some of the debates outlined more starkly in this chapter and provides an extended discussion of the practice of human geography organised around themes of the body, text, money, governance and globalisation.

Gregory, D, R Martin and G Smith (eds) 1994 *Human geography: society, space and social science*, London: Macmillan

A collection of essays by various authors which seeks to emphasise the relationship between geography and other social sciences. The chapter by Derek Gregory on 'Social theory and human geography' is particularly useful for its discussion of discursive spaces and multiple voices, outlining some of the influential developments in the understanding of both social structures and human agency.

Phillips, D C 1987 *Philosophy, science and social inquiry*, Oxford: Pergamon Press

For the more adventurous, this provides a philosophical treatment of debates about the nature of science and social enquiry. Although no mention is made of geography, it is written in a style accessible to non-philosophers and its 'mild Popperianism and naturalism' should be particularly challenging to human geographers who favour post-positivist, hermeneutical approaches.

Section A Preparing for the research project

For many students, the business of actually starting a research project is one of the most difficult stages of all. This is in part because the deadline for the project initially seems such a long way off and there are always so many more urgent things to be done, but it is also due to the apparent immensity of the task, and a lack of experience in organising such a large piece of work. The difficulty really begins with the decision as to what is to be studied. Some geography departments will offer students a list of approved topics from which to choose, but many will give a fairly free choice, subject to a tutor's approval. When a topic has been tentatively identified, it is essential to build up a broad picture of the work that has been undertaken in that area before. This helps to avoid the obvious danger of doing something which has already been done, but is also important in identifying outstanding research questions and appropriate methodology for that particular topic. This section of the book is intended to help you get started, and it will probably be the part that you are most likely to read first. Chapters 3 and 4 will be relevant to every project, and Chapter 5 is likely to apply to most.

In Chapter 3 Tony Gatrell and Robin Flowerdew deal with the vexed question of choosing a topic. By using some real-life examples, they demonstrate one of the most important messages of this book, namely that there is no 'magic formula' for doing a good dissertation. The chapter shows how successful projects can be undertaken by those who begin with only a general area of interest, as well as those who have a clearly defined objective from the start. There are many possible sources of ideas which can be developed into good research questions, and the relative merits of investigating an issue of local interest, working with an external organisation, taking a topic from one of your taught courses or exploiting your own outside interests are each discussed. There are also some useful hints on the kinds of things to avoid. Gatrell and Flowerdew offer some helpful comments on how to decide whether it is really research, whether it is really geography, and whether it is practical, although you should be aware that different academic geographers are likely to give different advice here! Everything you read here should be weighed against the advice you receive from staff at your university, and it is important that supervisors are involved from as early as possible. Their support should be secured before you actually begin the work. At the end of this stage you will hopefully have been able to identify some fairly clearly defined research questions which summarise what you are trying to find out. These are the kind of questions to which everything else in the project must relate – the sort of questions you should be able to pin to the wall over your desk as a reminder of where you are going. It will soon become apparent that undertaking research without clearly defined objectives is a fascinating experience, but is unlikely to result in any very clear conclusions!

As a potential project topic begins to emerge, it will become necessary to investigate previous work more systematically in order to discover what has been done in the same area. This will sometimes be an iterative process, in which initial research questions are found to be impractical or already answered. Both of the following chapters provide advice on how best to make use of online resources. While many key research resources are most easily accessed online, indiscriminate use of the web at this stage can potentially lead you astray. In Chapter 4 Robin Flowerdew provides guidelines for finding previous work. All too often, students at this stage rely on lecture notes or a couple of the most obvious text books in the library, but, although useful, these will not always be the best sources. Student textbooks generally have to cover a lot of material in order to accompany a taught course and secure a wide audience. This can be very useful in providing a general overview of a field, but means that there is rarely space for the coverage of detailed issues in great depth. Research texts and papers in journals are more likely to deal with specific topics, and the challenge here is in locating the publications that are most relevant to your chosen topic. An effective literature review needs to take advantage of the various indexing systems available, both paper and computer-based. Having got this far, it may be necessary to establish what information is available to help investigate your research questions. The information that you obtain at this stage is most likely to be secondary data, i.e. data that someone else has already collected.

Gordon Clark addresses the wide range of secondary data sources that may be relevant in Chapter 5. Secondary data sources may be important in two ways. Sometimes they will provide the main source of information for the project work. It is important to recognise that this is usually quite acceptable, although you should check whether your department has a specific requirement that projects must involve an element of primary data collection (or alternatively that it should be entirely library based!). For example, there is an enormous body of research, some of it very influential, which has been undertaken using the results of censuses of population as the sole data source. Many good projects are based on the analysis of historical data sources, and it is important to realise that 'data' in this context does not necessarily imply numeric information. For some projects secondary data sources will provide important contextual information for the primary work to be undertaken later on. Possible approaches include updating an existing source to examine how things have changed, or using published statistics in order to identify the most appropriate target groups or sampling frames for your own primary work. It may even be helpful in framing interview questions.

Secondary data analysis has a number of advantages and disadvantages in its own right, associated with the separation between the researcher and the data collection process. The advantages are generally in terms of access and resources (someone has already done an expensive and time-consuming data collection task, and made the results available to you). Disadvantages tend to involve the different purposes for which you want to use the data, and the inability to control the ways in which the data were collected and presented. The chapter contains lots of helpful information on data sources that should point you towards the information most relevant to your chosen project.

Having read the chapters in this section and made some initial decisions concerning a research topic in the light of the relevant literature, it would be wise to go back to Chapter 2 and consider carefully some of the philosophical implications of the research you are going to undertake. You may also want to look forward to some of the later chapters and think through the methodology that you are going to use. In the later sections it will be possible to pick and choose to some degree if you are using the book to help you undertake a specific project, but there is no effective alternative for good preparation. Although there are many possible approaches, the advice offered in this section should aid you in identifying a workable topic and getting the necessary groundwork done.

3 Choosing a topic

Anthony C Gatrell and Robin Flowerdew

Synopsis

We offer in this chapter some practical guidance on topic selection. Most of the chapter focuses on selecting topics for dissertations, certainly the major project you will undertake as an undergraduate. But much of what we have to say will also help in sharpening your ideas for more modest pieces of coursework. We talk about strategies for generating research ideas, and consider ways of ensuring that your ideas meet the criteria of being both 'geographical' and 'research', as well as leading to a piece of work that is capable of being completed. We include a range of examples, since we want to address all students doing human geography.

A tale of two students

Rather than beginning by giving you some general principles, it might be most helpful to tell you about a couple of undergraduate students whose dissertations illustrate contrasting approaches to choosing a topic. Let us introduce these students. One is called Emma Gaffney, the other is Mark Smith. We have chosen to tell you about these two – and these are their real names – for two reasons. First, both produced dissertations of some quality. Secondly, and more important for our purposes, each adopted a rather different strategy in selecting their topic. We use their real names so that you can, hopefully, empathise with them.

Emma came to see one of us (Tony Gatrell) with the intention of doing some research in the area of medical geography or the geography of health, nothing more specific than this. After discussing a number of possible research topics, it was clear that some reading was called for and it seemed that it might be sensible, in terms of subsequent career opportunities, to think about something that was on the agenda of current public health research. Health organisations in England and Wales are required by law to produce an Annual Report into the state of public health in their area. Armed with a couple of copies of such reports, Emma returned saying that she thought there might be some work to be done locally (in the Morecambe Bay area) on the geography of low birthweight.

She arranged to meet with a consultant in public health medicine, who could not have been more helpful, providing her with individual-level data once Emma had decided that this was the topic for her. This is not the place to describe her project in detail; suffice it to say that it involved producing some statistical descriptions of the incidence of low birthweight, relating this to well-known indicators of social deprivation (derived from census data) and also interviewing a small group of women about their experiences of antenatal care. Her earlier reading also provided a more general context within which to set her own empirical research. You should note that this kind of project generally requires approval from a local medical ethics committee because it involves the use of personal medical data, and consequently involves careful planning and plenty of preparation time.

Mark Smith arrived with a very specific idea for dissertation research. A keen walker in the Lake District of north-west England, Mark had for some time been fascinated by the detailed guides to walking in this area produced by Alfred Wainwright. To those who know little of the area this name will mean nothing; but to those who have a strong attachment to the area, and especially those who share Mark's love of fell-walking, Wainwright is well known as the author of many pocket guidebooks to different parts of the Lake District. These contain some wonderfully detailed sketches and maps, coloured by personal comment about the landscapes. Mark's idea was quite specific: what use do people make of these guides and to what extent do they help communicate a sense of place? Does this vary with social group, among other things? Clearly, Mark's topic was already pretty well defined. What it lacked was the background to place it in a more general context of studies of landscape and attachment to place, an area of research that writers such as Steve Daniels and Denis Cosgrove (see Daniels 1993, for example) have done so much to promote. In initial supervisory meetings, Mark learned about this research tradition and, with further reading, was able to modify his original idea so that it could be linked to a wider set of geographical literature.

We hope it is clear from these personalised accounts how different students might adopt different strategies in seeking a topic. Emma began with little more than an expressed interest in the geography of health, and with some guidance from both her supervisor and a sympathetic external person, narrowed this down to a specific topic. Mark, on the other hand, had a very specific topic up his sleeve. The help he required was in setting this idea in a more general context; in seeing how it related to existing geographical concerns. Neither is necessarily a better strategy than the other, but the important point to grasp is that both ended up with a clear statement of a research problem or question, and both succeeded in setting this within a more general context. Barbara Kennedy (1992) has similar points to make and is well worth reading; strategies for choosing a topic are also discussed by Hoggart *et al.* (2002: 49).

Generating a project

We will now consider some specific advice that can be offered to you by way of strategy for topic selection, since we imagine that many will be closer to Emma's position than to Mark's. Where can you turn for sources of inspiration? Let us suggest five possible sources, and give you some examples to illustrate what we mean.

At the risk of stating the obvious, our first suggestion for inspiration is to invite you to consider a course, or set of lectures, or maybe just one class, that has interested you. Has the tutor (lecturer, instructor, call him or her what you will) touched on a particular theme or idea that has intrigued you? Here is an example from one of the authors (Tony Gatrell) to help make the point. He had been attending a series of lectures on transport networks, given by Peter Haggett. (If you do not know who Haggett is, look at the collection of essays written in his honour: Cliff *et al.*, 1995). Haggett spent part of a lecture telling us about some work done by the American geographer Howard Gauthier on the relationship between transport improvements and economic development, work that linked to the arguments of economic historians such as Rostow and Fogel about whether improvements in transport infrastructure were a precondition for, or followed, urban and economic growth. Gauthier had suggested a particular approach and had conducted his own research in Brazil. Adapting his ideas to another context seemed to be a possible dissertation topic, and New South Wales, for reasons no longer clear, seemed a good place to choose. Resources did not allow for primary fieldwork in Australia, but data could be collected during many happy hours in the Map Room of the Royal Geographical Society, extracting information from old maps, and in the Library of Australia House, collecting data from historical censuses.

So, while taking notes, or simply taking part in a classroom discussion, why not be alert to the possibilities of research; be aware that, if something intrigues or interests you, it might just be possible to follow it up in your own research.

You are used, as a student, to being given reading lists and no doubt you are already used to chasing up books and articles in an effort to add colour and depth to what has been said in class. You may be doing this because it will help you produce better examination answers at the end of the year (or term, or semester). That is a laudable reason. But – and this applies more to articles in serials (journals) than to books – the author(s) will usually end with a paragraph or more that calls for further research or identifies issues that have been raised in the body of the text. No doubt some of these are issues which he or she is already working on! If the article has interested you, however, it may be that one of these is something you can work on yourself. Think about it, and discuss it with others. You might even try contacting the author for advice, something which has happened to us on several occasions. Get a postal (better still, an electronic mail) address and see what response you get; but do ask specific questions rather than expecting someone to come up with a topic for you!

Reading is important. You need to read material closely in order to understand the argument and findings. But there is much to be gained from browsing the library shelves. By this we mean skimming the contents of recent issues of journals. Look through them to see if any of the titles appeal. Maybe you have already narrowed down your interests to, say, economic geography, in which case certain journals (for example, *Environment and Planning A*, *Economic Geography*, or *Antipode*) will suggest themselves. Such browsing will not, of course, substitute for a proper bibliographic search when you have chosen your topic (see Chapter 4 for more details) but it may prove useful at an early stage of idea generation. Internet searches can fulfil the same role.

Related to this is the browsing of newspapers, of magazines, or of television and radio programmes. We challenge you to find any copy of a national or local newspaper that does not contain something of geographical interest. It may well be that you lack the resources to pursue the topic, something we return to later, but, particularly if the newspaper is a local one, there may be names of people you could contact for further information or discussion. Here are just two examples. A local paper in Lancashire carried articles for a number of weeks on two prominent local stories. One concerned the closure of a former long-stay hospital, a process that owed much to the development of 'care in the community'. The health authorities wished to sell this site to developers in order to build up to 1000 new houses on it. The site was adjacent to a large village. Some local residents were none too happy with this proposal and a local group emerged to fight it (in the event, they lost the battle and the development took place). Here is an issue full of geographical implications – the use of land, the politics of land use decisions, the ways in which different social groups band together to form protest groups, the possible visual intrusiveness of developments, and the other environmental consequences of a major development in a county (Lancashire) that prides itself as the leader of 'green' local government. Had this issue arisen in your area, you could have made a start by contacting the local district planning officer, or the secretary of the action group. The other big local issue that had been important for over ten years concerned the operation of a cement-making firm in a nearby town, Clitheroe. Local residents had for years been worried by emissions from the kilns, particularly as the company had been burning a fuel derived from liquid wastes such as solvents. They claimed excesses of asthma and other respiratory symptoms among children living nearby. What can be more geographical than a study of the health consequences of such emissions and the attempt to detect clustering of ill health around a suspected point source of pollution? In Bailey and Gatrell (1995, Chapter 4) one of us tried to suggest possible approaches to problems such as these, but a student could pursue this in the first instance by seeking a meeting with someone from the local department of public health.

This leads us to suggest a fourth strategy, closely related to Emma's approach that we talked about earlier. It does rather assume that you have already settled on a broad topic area (e.g. cultural geography, social geography, or possibly something a little more focused such as the geography of crime, health or unemployment). In any event, you have a broad idea of your area of interest; now you seek a specific research problem. Why not try to 'solicit' one from an organisation or individual working for such? Your department may already encourage this; for example, Lancaster University developed the idea of 'enterprise dissertations' (Clark, 1995), whereby the student works on a problem suggested by national or local organisations. Here are some examples, along with the collaborating body: forestry privatisation in a national park (a national park authority); land management of a bird reserve (an ornithological organisation); quarry traffic and the environment (a parish council); deprivation and health on a housing estate (a health authority); and tourism development appraisal (a district planning department). As Clark (1995) notes, such projects help develop an awareness and experience of the business or enterprise environment.

A fifth strategy is to seek to develop one of your own outside interests into a research topic. For example, another student, Ewan Brown, was a regular attender of one of the large rock festivals; as a result, he sought to do some research on the economic and social impact of the Reading Rock Festival (see also chapter 10) on the local community. The research involved survey work of local residents and business people. Another example is work by Richard Middleton, who had a keen interest in sport, notably football. He conducted a study of the multiple criteria involved in selecting a site for a new football stadium and used digital data in a Geographical Information Systems (GIS) framework in an attempt to suggest a set of possible locations. Other examples we have supervised include geographical studies of silver bands, fox hunting, family history, and Pentecostal churches. Other students have chosen topics related to their family background (see Shah, 1999) or to medical conditions they have suffered from, such as Crohn's disease or ME. So, ask yourself if it is possible to fit the research to your own interests. One advantage of this is that the research should at least keep you well motivated! On the other hand, you must force yourself to pay attention to the intellectual issues you are addressing, however fascinated you are by the minutiae of the activity from which the dissertation has developed.

As a last resort, you can, of course, always talk to one of your lecturers! Or, perhaps a postgraduate student or research assistant in the department. This might be a supervisor to whom you have already been assigned, or it may be someone you regard as approachable. Here's an example. Carolyn Cadman had already decided on a possible topic. But at an initial supervisory meeting it was pretty clear that her heart was not in the subject. So it seemed worthwhile to explore other options and the discussion turned to her plans for the summer. Apparently, part of this involved travelling up and down the British motorway network on her boyfriend's motorbike (yes, we know, it is not our idea of the perfect summer either). A conversation developed about the merits and otherwise of motorway service stations and what sorts of 'places' they were like; how distinctive they were, what people's experiences were of these places, how such experiences perhaps varied according to time of day, how they were regarded as very transient places by many but as places of work for others. John Urry's fascinating work on *Consuming Places* (Urry, 1995) seemed relevant to these ideas, and she decided to try to develop her dissertation along these lines, which she did successfully.

What should you avoid when generating ideas? In some instances it is a good idea to think of previous projects you have worked on, maybe at school or on field classes, for example. One of these might well have fired your imagination, and you would like to modify or extend it in some way. But be just a little wary of this. One dissertation some time ago (where the student sought virtually no supervisory help) was on central places in Cheshire. It was pretty uninspiring and seemed to be just the kind of thing the student had worked up as an A-level (senior high school) project. At university level, we are looking for more than this. Similarly, try to avoid anything that smacks of a study of your 'home town': in other words, a general geographical description without any sense of research problem. Again, Barbara Kennedy (1992) has some good comments to make on this.

Try also to avoid the 'technique in search of a problem'. What do we mean by this? For example, lecturers who teach GIS may find that some students approach them asking if they can 'do something on GIS'. It may well be that there is an interesting technical issue about GIS into which they can conduct research. More commonly, however, they have got 'hooked' on GIS and want to use it. But this is the wrong way to go about geographical research. First find a research problem, then worry about what kind of approach, or method, or technique, is appropriate. It may well be that GIS – or, in a very different context, qualitative research methods – are useful; but do not be driven by technique.

Is my project research?

We hope it is clear from what we have just said that you should always be looking for a research *problem*: a question, or set of questions that are worth asking, an issue that merits attention or requires solving. So what criteria will be used to judge whether your project is deemed to be 'research'? Probably the main criterion is originality. Do not, please, imagine this means coming up with an entirely novel theory or completely new idea; but it does mean avoiding the application of old methods to old problems, or the umpteenth replication of an idea that might have been fashionable thirty years ago. What it could mean is asking if there is a geographical dimension to something not previously thought of as 'geographical', or asking questions of data that have yet to be explored much by geographers. Carolyn Cadman's work on the local cultures of motorway service stations is perhaps a good example of the former. As for the latter, among many examples that could be drawn from the literature, look at David Ley's work on graffiti in Philadelphia (Ley and Cybriwsky, 1974), or Wilbur Zelinsky's use of interstate variations in subscriptions to magazines as a way of characterising cultural variability in the United States (Zelinsky, 1974). Both of these might well require resources beyond the means of the solitary undergraduate, but they do illustrate the point, and as 'classics' in the field are worth reading anyway.

You should also be aware that 'originality' does not necessarily mean the collection of data first-hand. If you can get access to a large data set and do some novel analyses using it, this will serve you better than relying on a modest set of data you might yourself have collected. If you want evidence that this is a viable strategy, take a look at some work by the editors, each of whom has built a considerable reputation partly on the strength of analysing secondary (such as Census and health), rather than primary, data (Flowerdew, 1993; Martin *et al.*, 1994). This principle may distinguish you from your physical geography colleagues, since they will usually be expected to collect primary data in the field, albeit perhaps as part of a larger expedition.

Another criterion worth thinking about is topicality: is the subject of contemporary interest, either to academics or the public, or both? This links back to our earlier remark about ransacking newspapers for ideas, since anything appearing in the current papers is, by definition, 'topical'. We mentioned a couple of local issues earlier. And, of course, select a topic that will keep your interest over several months. It must capture, and fire, your own imagination (as well as, hopefully, the examiners'). The topic that is feasible but dull is hardly guaranteed to keep you motivated or impress an examiner.

Is my project geographical?

You may well be contemplating a topic for study, but be worrying whether it is 'geographical' (cf. Shah, 1999). This is certainly something that will concern your tutor, though we would wager that if you asked three tutors for their definitions of geography you would get four or five answers that did not necessarily coincide! There are good grounds for arguing, as did a recent newspaper article, that geography is the archetypal postmodern science; in other words, anything goes. You have only to skim the contents of contemporary journals to confirm this. The conference programme for the Annual Meeting of the Royal Geographical Society with the Institute of British Geographers in London, September 2003, featured titles such as: 'The spaces of Joyce's *Ulysses*', 'Trance and visibility at dawn: racial dynamics in Goa's rave scene', 'Café encounters', 'The topography of anxiety' and 'Tattoos in prison: pictures of masculinities at the verge of society'. These were not on your school syllabus! Yet despite the rich diversity of subject matter we can probably pick out some common themes: a concern with place, with landscape, with space, and location. These remain core concerns, even though the methods used to explore them, and the scales at which they are investigated, may differ substantially.

Let us, the authors, commit ourselves and suggest what we take geography to be about. The fundamental question is: does space or place or location 'matter', or make a difference? Is there place-to-place variation in economy, in culture, or society? Think about it: if there was no spatial variation there would be no geography. If, to pick up our own area of interest, there were no variations in health experience, in access to health care, in disease risk, and so on, there would be no 'geography of health'. If rates of urban growth in the nineteenth century had been uniform there might be no urban historical geography. If women had equal access to spaces, places and resources, there would be no need for a feminist geography. Regardless of what sort of geographer you are, it is a fundamental concern with place-to-place variation that motivates us all.

Having said all this, of course, your tutors will want to satisfy themselves that the topic meets with their approval. Hopefully they will give you a fairly free rein, but listen to what they say; they are themselves conducting research in some branch of human geography and probably know what they are talking about! In summary, you should be able to find a 'geographical' angle to most topics, but if it is at all non-standard, you should be prepared to defend its geographical character.

Is my project practical?

Finally, when looking for a topic you will need to consider some practical constraints, revolving around time, cost, and personnel. You may well wish to study the harvesting of non-timber forest products in the high mountains of Nepal (another paper in the RGS-IBG Annual Meeting in 2003) but unless you are a lottery winner, or part of a wider expedition, this may well be impractical. A major constraint will be time. If you are working on an undergraduate dissertation you will probably have about six months from conception to completion, of which at most a third will be devoted to data collection. You need to allocate plenty of time to reading background literature and to analysing whatever data you collect. You need to ensure that your required data actually exist or are capable of being collected in the time available. Have you obtained any permission or clearance necessary to collect the data? Are ethical issues involved? Some of these questions, partly of a practical nature, are addressed in later chapters. But if your project calls for a substantial piece of empirical research, and the data just are not available or possible to collect in the time you have, you may well have to reconsider your plans. Do not spend so long contemplating data availability – or indeed, choosing a topic at all! – that you never actually make a start. And once you have started, do not give up on it but persevere. You will encounter problems – all of us engaged in the research business do – and some of the other chapters in this book are designed to help you avoid, or deal with, such problems. But there are dangers in abandoning a topic late in the day; commit yourself to it once you have started.

Other constraints will be financial. Your department may be prepared to make some contribution towards expenses, or you may be working collaboratively with an outside organisation, as we suggested earlier; this organisation may provide you with limited resources to help you conduct the research. But do be aware that any research costs money; for travel, subsistence, accommodation, photocopying, and so on.

Please be sensitive too to issues of safety. This applies as much to the student wandering around the streets observing the urban landscape as to your friend collecting stream sediment samples on some isolated upland catchment. Your department should have a safety policy that addresses issues of fieldwork. Read it soon, and do not pick a topic that puts you in any danger.

Summary

If you have been struggling to find a good topic we hope what we have had to say has been useful, practical advice. To recap, our five suggested strategies were as follows (Box 3.1). First, to consider whether a theme or idea addressed in a lecture, a class, or series of classes was worth following up. Secondly, to engage in some reading arising from a class, paying particular attention to whether the author was giving an indication of where future research might be carried out. Thirdly, to get into the habit of browsing, either journal articles or newspapers and magazines, or possibly radio and television programmes, in an effort to see

BOX 3.1

Top tips: five strategies for generating research ideas

1. Follow up an idea that arose in class.
2. Read an article, or section in a book, on a topic that interests you.
3. Browse! Articles, newspapers, radio, television, etc.
4. Talk to outsiders about possible research ideas.
5. What outside interests do you have that might generate a research topic?

whether some idea gets triggered off. Fourthly, to consider soliciting a topic from some outside body or organisation. Finally, to ask yourself if your own outside interests might suggest a topic for investigation. We also suggested you think about sitting down with a member of staff, or maybe a postgraduate student or research assistant, in an attempt to tease out some possibilities.

Some of these suggestions involve no more than what your tutors themselves do when looking for research topics. For some, their research is mapped out at an early age, as they develop an idea first explored in a Ph. D. dissertation; the rest of their career is an endeavour to expand this research area. This is rare, however. Others, like us, get ideas from hearing other people at conferences, or reading what they write. Or we may get ideas from our own students: good students teach their teachers. Often, we develop research in collaboration with others; a colleague comes to see us with a problem or an idea, or a funding agency invites research proposals and this triggers off some discussions and submissions of grant applications. By no means all research is generated by virtue of the inherent intellectual interest of the topic; often, nowadays, it is commissioned by a funding body and has to be seen to be 'useful'. One clear advantage of the student dissertation or project is that it does not have to be useful to anyone. This may be a criterion that some of you will wish to use in defining your topic, but it certainly need not drive your search for a topic: intellectual curiosity will do nicely! See Hampson (1994) for some reinforcement of this point.

We have also sought in this chapter to suggest that you should aim to be original, but cautioned that this did not necessarily mean you had to devise your own theory or method. We invited you to ask whether your idea was 'geographical' and gave you some indication of what we, at least, would take this to mean. Lastly, we touched on some practical issues and constraints, of which you need to be aware in conducting the research.

So, where do you go now? Hopefully you are now starting to get some ideas, and the main business of the research can begin. Above all, enjoy it! You are (probably) on your own. You are out of the classroom and working for yourself. You may be collecting data that nobody has ever collected before. You may develop a 'taste' for research and want to do more. Make the most of the opportunity, because whether you realise it or not you are picking up all sorts of skills (organisational, management, survey, analytical, and so on) that you will find useful in later life.

We hope it goes well.

BOX 3.2

Further reading

Clark, G 1995 Enterprise dissertations revisited, *Journal of Geography in Higher Education* 19, 207–211

This focuses specifically on projects that are 'linked' to an external organisation.

Gatrell, A C (1991) Teaching students to select topics for undergraduate dissertations in geography, *Journal of Geography in Higher Education* 15, 15–20

This is similar to the present chapter, though written more from the perspective of the teacher than the student.

Hampson, L 1994 *How's your dissertation going?* Unit for Innovation in Higher Education, Lancaster University

A very useful little booklet that covers many aspects of project and dissertation work, not only those concerned with topic selection.

Kennedy, B 1992 First catch your hare . . . Research designs for individual projects. In A Rogers, H Viles and A Goudie (eds) *The student's companion to geography*, 1st edn, Oxford: Blackwell, pp. 128–134

4 Finding previous work on the topic

Robin Flowerdew

Synopsis

An important part of a dissertation or project is the literature review, in which the writer shows his or her knowledge of earlier work on the topic. A good knowledge of the literature is also useful in designing the research, thinking about specific points to investigate, and avoiding problems that others have already encountered. The chapter is mainly devoted to discussing methods of finding relevant literature, from fairly obvious sources like your supervisor, the Internet and the library catalogue to abstracting services, bibliographies, and citation indexes. Many of these sources can now be consulted most easily using computer technology, but there are techniques that can be learnt for these systems to be used most effectively.

Introduction

Once you have decided what the general topic of your dissertation will be, perhaps even before you have decided exactly what you are going to do, it is important to find out what other work has been done on the topic. There are three good reasons and one bad one why this is important.

The first and best reason is that you are a university student and acquiring an education involves increasing your knowledge by finding out about your chosen topic. You are going to become a bit of an expert in the field of your dissertation, and it is very important to become well informed about the whole field in order for your own contribution to be taken seriously. Also presumably you are interested in the topic!

Secondly, other people working in the area will have thought about it and come up with some interesting ideas and findings. These can give you good ideas about your project, and can help you by suggesting interesting new angles on it, providing background information, telling you if some of your initial ideas seem to hold good or not, and giving you some suggestions about methods to use.

Thirdly, and more pragmatically, people expect a 'literature review' as a component of a dissertation. It is something that you can get on with at an early stage of the project (though it will require revision later on); it is very comforting to have a substantial section of the dissertation drafted before you even start the fieldwork! People marking the dissertation will expect you to have read the other relevant studies. They will be impressed if you have found substantial amounts of relevant previous work, and dismayed if you do not seem to know about the major work in your field.

The bad reason which students sometimes are concerned about is the fear that somebody else will already have had your idea – it has been done already! You should not worry about this. What you say about it will be different from what your predecessors will have said; you actually have the major advantage of being able to read their work and to design yours to avoid their mistakes and to follow up issues which they have not concentrated on. In general, a previous study on your topic should be something to build on. Even if it is good and your initial feeling is that you will not be able to add anything, detailed study is bound to reveal issues which you disagree with or where you can add an extra element to the study – at worst, you can update the findings. Probably the only situation in which a previous study might make your own study less promising is if the first study involved spending substantial amounts of time talking to exactly the same people that you will want to interview – they may be fed up with being studied.

Cast the net widely

Do not think of your topic in narrow terms when you are doing your literature review. If you are studying, for example, the geography of the sports footwear industry in North Lancashire, please do not conclude that there is no relevant literature to discuss simply because you cannot find anything on this specific topic. The literature review should certainly refer to anything you can find on the geography of sports footwear, or of sports goods or footwear more generally, regardless of location; it should also include other material on industry in North Lancashire and beyond. More importantly, there may be review or theoretical articles which provide very interesting and relevant perspectives on aspects of industrial geography which can be applied or discussed with respect to your work, or even ideas of even more general application. Some of the best dissertations and projects apply current new methods or theoretical perspectives to a new topic.

Making a start

Perhaps the first and perhaps the best approach to finding relevant literature on your topic is to talk to your supervisor and to other people in your department; do not confine yourself to the academic staff – graduate students may be knowledgeable and pleased to be asked. There may even be a final-year undergraduate who has been through exactly the same process as you the year before. Especially if they helped you develop the idea for your project, they will probably know some of the background work you will need to consult. Do not expect too much, however; you may be given precise and accurate citations, but you may be told something like: 'I think there was something on your topic in the *Transactions* two or three years ago by Willis – or was her name Wilkins?' It may take a bit of detective work before you eventually find the 2002 *Progress in Human Geography* paper by Smith and Watkinson!

Staff may also know people in other universities who have worked on related topics, or who they think would be interested in what you are doing. It may be worth contacting such people for ideas or information, though bear in mind that they may have their own students to advise, and may not have time to be of much help. Generally, in helping you identify relevant literature as in many other respects, people are more likely to respond positively if you approach them politely and if you come across as knowledgeable and enthusiastic about your subject. Do not be too downcast if they are too busy or self-important to help.

Particularly for the theoretical or methodological aspects of your project, you will probably have come across books and articles that are highly relevant to the more general issues underlying your project and to the methods you will use to work on it. An up-to-date textbook on the relevant part of geography should at worst provide some general context and may well contain references to highly relevant studies. I hope that some of the other chapters in this book will be helpful too, both for the points they make directly, and for the references they provide to other work on their theme.

If your work involves study of a particular group or organisation, they may be aware of other studies that may be relevant. A local authority's planning department, for example, may well have conducted relevant work themselves, and will probably be aware of any major academic studies that impact on their domain.

Subject catalogues

Another obvious place to look is the subject catalogue of your library. Of course, this will only be helpful in finding books, not journal articles, and will be constrained by how the cataloguing has been done. Geography is notoriously poorly served by the most widely used library cataloguing systems. Sometimes it can be difficult to know what to look up – the best advice here is to think laterally and try several alternatives. It does not do any harm to find lots of things that are not relevant, but it would be a shame to miss out on a few that would be really helpful.

I also recommend going to the library shelves and browsing. You may notice things that you had not spotted in the catalogue, and you can find out much more about a book by looking at it than you can from a catalogue entry, from things like style and readability to the specific contents and emphasis. It may be particularly useful for checking out edited volumes that may or may not contain chapters of direct relevance to your needs. Lastly, you may see other books which might not be obviously relevant to your topic at all, but look intriguing and could contain an interesting idea that you can apply to your own research topic. Sometimes these serendipitous discoveries can lead to the most interesting new ideas.

It may also be useful to consult catalogues for other libraries. This is often possible to do on-line. The Consortium of University Research Libraries (CURL) has set up a system for doing this, called COPAC (the CURL Online Public Access Catalogue), which gives access to twenty-six research libraries. It is also possible to search the British Library Online Catalogue and other sites connected to other public library catalogues in Europe, the United States and elsewhere. These sites may tell you about other books relevant to your project which are not available in your own library. You may be able to use another library at a nearby institution that has books or journals that your own library does not take. There may be a library near your home or near your field work site that you can use during vacations. A quick look at their catalogue can tell you whether key references will be available there, and also if there are other useful books and articles not available at your university. Most libraries will permit students at other institutions to consult their collections on a 'reference only' basis, though it may be useful to take a letter from your dissertation supervisor or your university librarian to persuade them you are a genuine student.

Bibliographies

There may also be bibliographies available in a variety of forms. These can be in book form or may be available in a variety of different formats, and are obviously excellent as a source of references, though sometimes frustrating if they contain lots of material which is not available in your library. Most bibliographies are arranged according to subject matter, so again you are dependent on how well the organisational structure adopted accords with your needs – you may well have to look in several subsections to find what you are looking for. It may be rather a hit-or-miss affair whether anybody has produced a useful bibliography for your topic, but it is worth keeping an eye open for something of the sort. Perhaps the most useful bibliographies are those which are annotated – i.e. where comments have been written about the references cited. Annotations can be very useful in helping you decide whether something is worth following up or not, though they may not always be helpful as the person doing the annotation may have been looking for very different things from those which concern you.

Abstracts

Almost all journal articles are accompanied by a short abstract, usually printed at the start of the article, which is intended to give a digest of the contents of the article, so that somebody can find out quickly and easily whether they want to read the whole thing. These abstracts are often collected together in order to help people find journal literature on a specific topic – exactly what you will need to do at the outset of your dissertation research. The problem is that it may be difficult to find your way around a set of abstracts to pick out those that are of interest to you and of relevance to your specific research project. This makes you rather dependent on how well the abstracts are organised and indexed. Using information of this kind has become easier now that abstracting services have mostly become computerised, as it is possible to search on a computer for words in the titles or abstracts of articles.

A series called *Geographical Abstracts*, covering both human and physical geography, was started in the 1960s; as the amount of periodical literature in geography increased, it split into more and more separate series, covering different parts of the discipline. It changed its name to *GeoAbstracts*, and is now fully computerised. Data from *GeoAbstracts* from as far back as 1980, combined with other abstracting services mainly in the environmental field, are available under the name Geobase in many university libraries. Geobase is accompanied by a set of instructions and should be usable by geography students without requiring any special expertise.

Human geographers may also find abstracting services from some other social sciences useful. Staff in your library should be able to tell you what the main sources are, and how they are most easily accessible.

Reference lists and citation indexes

Often the hardest part of researching a new topic is the beginning. Once you have found one article, it is usually easier to find more. This is because of the academic habit (which you should adopt) of giving a list of references at the end of a paper (or sometimes at the foot of each page) which will probably cover most of the major sources consulted by the author. Of course, not everything cited in a relevant article will also be relevant to your interests, but probably some will be, and others may give some useful general context. When you look up the articles cited, they will in turn cite more articles, and so on, perhaps going back to some influential early papers on the subject. Not only does this process help you to identify relevant literature, it also gives you some idea of how different writers on your topic have influenced each other, and of how current thinking on your topic has developed. You may find disagreements or controversies between different writers; perhaps your study will allow you to take sides, or even to suggest that they are all wrong!

Reference lists are very useful for tracing earlier articles. They are no good, however, for finding the most up-to-date literature on a topic, for the simple reason that an article written several years ago cannot refer to anything that had not yet been done! A common criticism of students' literature reviews is that they have failed to take note of recent developments in their subject – which can be a natural problem to encounter if you have not found any very recent work on your topic. However, the problem can be avoided by using a citation index.

A citation index works like a reference list in reverse. If you have found a relevant article, you can use it to look up other articles that have cited it. It is quite likely that other people working in your area will have read the article and included it in their reference list. It will then be included in the citation index. This allows you to work forward from a relevant article you have identified, and to find recent work on your topic. Of course, once you have found a recent article on your topic, you can consult its reference list and find other older ones. Alternatively you can look for citations for the new article, and perhaps find still more recent work.

Citation indexes are available as part of a useful general service called the Web of Knowledge. To use this in the UK, you will need to register with the ATHENS system. The relevant part of this is called the Web of Science, which gives you access to any or all of three citation indexes, the Science Citation Index, the Social Science Citation Index, and the Arts and Humanities Citation Index. In human geography, the Social Science Citation Index is likely to be the most appropriate. It is a multidisciplinary index, with searchable author abstracts, covering the journal literature of the social sciences. It indexes more than 1725 journals spanning 50 disciplines, as well as covering individually selected relevant items from over 3300 of the world's leading scientific and technical journals. The citation indexes used to be available in print, with a series of thick volumes appearing each year in rather unmanageable format. Nowadays they are usually consulted online. It is then possible to type in a reference (to an article of interest to you) and find out how many subsequent articles have cited it. Details of these articles can then be displayed, including the list of references they themselves cite.

Although citation indexes make finding up-to-date references much easier, there are still problems involved in their use. First of all, it is much easier to locate references where the author has an unusual name. If you are looking for references to an article by Robin Flowerdew, you can safely assume that any references to 'Flowerdew, R' are to the right person. If you look up David Martin, however, references to articles by the co-editor of this book will be mixed in with references to the work of other David Martins and indeed to that of Daniel, Denise, Derek, Dianne and Didier Martin – only initials, not first names, are used, and you cannot rely on middle initials being used in a consistent way. A second problem arises because references are often incorrect or given in different format: this means that separate listings may be given for references if the journal name is abbreviated in different ways.

Computerised databases

Finding relevant literature has become far easier in the last few years with the growth of computerisation and the development of methods of information retrieval. It has become quick and easy to search large databases for articles containing certain key words and to retrieve either the articles themselves or bibliographic aids to locate the articles. Some contexts in which this can be done have already been mentioned, but there are many others in addition. This can be done in systems like Geobase or the Web of Science already mentioned or in other systems such as ArticleFirst (indexing journal articles) or WorldCat (indexing books). There are also systems relating to specific themes (like Geobase), which may be relevant to certain dissertation topics. These include EconLit (economics), ERIC (education), GPO (US government documents) and MEDLINE (medicine and health).

Generally one has to type in a word or phrase and a computerised system searches the database and finds how many times the word or phrase occurs. It is often difficult to predict what this number will be, and it may well turn out to be dauntingly large (or sometimes disappointingly small). In this situation, most computerised systems allow you to restrict or extend your search by specifying other key words. For example, if you start by specifying 'tourism' you may be told there are several thousand references ('hits'); you can narrow your search down, perhaps by asking for references including both 'tourism' and 'ecological', or 'tourism' and 'Guernsey', or 'tourism' and 'postmodern' as appropriate. This will certainly produce a more manageable list; if it is still too long to be usable, you can try combining with a third key word. Alternatively, you may find there are no references combining your chosen key words, or you may be worried that you are missing important references. It is then possible to broaden your search – for example by combining 'tourism' and 'ecological' or 'environmental', or 'tourism' and 'Guernsey' or 'Jersey'. You have to think logically about exactly what you are asking for – it is easy to get confused over exactly what 'and' and 'or' mean in this rather technical context.

Another facility which is often useful is that many systems allow you to specify any word beginning with a particular combination of letters: thus articles about tourism may actually use 'tourist' or 'tourists' in their title or key word list rather than 'tourism'. Rather than searching for all these independently, it may be possible to search for any words beginning 'touris' or 'tour'. This approach can give you more than you bargained for however – the prefix 'eco' for example will generate references using various forms of the word 'ecology' such as 'ecologist', 'ecological' and even 'eco-friendly', but it will also pick out all the words derived from 'economics', not to mention references to the Italian writer Umberto Eco or to French articles about Scotland (Écosse).

However, like most computer systems, these bibliographic search systems do exactly what they are asked to do and have no ability to apply any common sense. They will thus retrieve items containing the key words you specify when the context is completely different from the one you are interested in; a student studying the geographical mobility of schoolteachers specified the combination of 'teacher' and 'mobility' and retrieved a set of references on the use of dancing in physical therapy. Similarly, these systems will miss items, which are precisely on your topic, if the exact key words you specify are not used. The only solution to this is to think broadly and to keep trying all the synonyms you can think of.

Overall, searching computerised databases is more of an art than a science. You can learn how best to do it but there are no sure-fire rules which will guarantee that you will locate everything that is relevant. Unfortunately different systems work in different ways and you really do need to consult the help system in order to learn how to combine search terms, how to specify prefixes, and how to display the references you have found. Nevertheless it is well worth doing, both as a means of locating relevant work which may help your own study, and to construct a long and impressive reference list to impress your teachers.

The Internet

If you have used the Internet, you will be aware of the vast quantity of material held in different parts of the system and the vast and amorphous set of links allowing you to access them. If your use of the Internet is to be at all systematic, you will need to become familiar with some of the major facilities and with the search engines used to find your way around (see also Chapter 5). The systems already described, such as the Web of Knowledge, ArticleFirst and Geobase, are available via the Internet, and there are many other sites offering bibliographic information as well as everything else to those who discover them. With a little work, you will discover many such resources.

Using a search engine on the web may be useful for finding previous literature as well as for other things. Your search may lead you to sites containing academic references that can be followed up, or to stories featuring research by specific individuals, who can in turn become the subject of searches. Most academics will have an up-to-date list of their publications on a website that can be found from the staff list on their institution's website.

Inter-library loans

When you undertake a literature search, you will almost certainly find that some, perhaps many or most, of the books or articles you most want to follow up are not in your university library. As suggested above, you may be able to consult the catalogues of other libraries to which you can get access, and this may solve some of the problems. You may be able to persuade your supervisor that a book is so important that it should be ordered for the library. Most of the time, however, the Inter-library Loans service is the best way forward. Inter-library loans in the United Kingdom allow users of any library in the system to obtain any book or journal article they wish to consult which is not in their own library, provided it can be located elsewhere. Books are sent to the library that requested the loan, to be consulted and returned in a few weeks' time. A journal article is photocopied (subject to certain copyright restrictions) and the photocopy is sent to the library, and does not need to be returned. The service is generally quick (materials usually arrive within two weeks or less of the request being made, though this may depend on how easy they are to find). The service is, however, fairly costly, and many libraries are reluctant to let students make too much use of it. You may therefore have to get your application countersigned by your supervisor, and you may be rationed to a small number of requests. Nevertheless the service exists to be used, and you should certainly take advantage of it to obtain material important to your project.

Conclusion

Finding relevant literature is an important activity to get on with as soon as possible. It can help you design your study, suggest new lines of investigation, warn you of dead ends, and put your study into context. It is also a worthwhile ability in its own right, potentially useful in further study and in many jobs. It is also, almost invariably, something you will be given credit for when your work is being marked. However good your study is, an inadequate literature review will be a major negative feature; and even if your empirical work does not achieve its objectives, the markers will still be impressed if you demonstrate a good knowledge of the literature. Chapter 17 gives some advice on how to construct your literature review; see also Hart (1998).

The chapter has outlined several approaches to finding references on your subject. In practice, most people will find plenty of material without using all the methods outlined above (providing of course you take a sufficiently wide interpretation of your topic). Sometimes people may aim for completeness – to include everything that has been written on your subject – but this is probably a foolish aim. You can never be sure that there is not some obscure article or pamphlet that your search has failed to unearth. Indeed, if you did find everything that was remotely relevant, you would not have space to write about it all. The best advice is to be pragmatic – do your best in the time available; do not be dismayed if you cannot find much that is exactly on your subject; and keep your own project firmly in mind as you write it up.

BOX 4.1

Further reading

Material on this topic, especially on computer-related bibliography, is likely to date very quickly. There may also be differences in what can be accessed at different places. However, a book-length study of literature searching has been produced:

Hart C 2001 *Doing a literature search: a comprehensive guide for the social sciences*, London: Sage

Another useful source, written specifically for geographers undertaking a research project, is:

Healey M 2003 How to conduct a literature search. In N J Clifford and G Valentine (eds) *Key methods in geography*, London: Sage

5 Secondary data

Gordon Clark

Synopsis

This chapter reviews why secondary data are important for dissertations and projects. It considers their strengths and weaknesses. Suggestions are made about where to get secondary data and the range of material that is available. New developments in the online provision of secondary data are discussed. Finally, some problems that can arise when using and interpreting secondary data are discussed, with suggestions about how to cope with them.

Secondary data – definitions and functions

What are secondary data?

Secondary data means information that has already been collected by someone else and which is available for you, the researcher, to use. The distinction is with *primary data* which you gather for yourself and which forms the subject of Section B of this book. The secondary data sources relevant for dissertations comprise existing information which is publicly available even if not already published. So, tax returns are confidential and are therefore not available secondary data, whereas planning records are, because they are open to view. Frequently the data are official in the sense of having been collected by public bodies (e.g. the census of population). In the case of census data, the individual records are confidential but much aggregated information is available. Private firms also collect information and this too could be helpful, but commercial confidentiality and cost often limit its usefulness. The official status of much secondary data (perhaps collected under statutory powers) gives them an authoritative quality that can be both reassuring and beguiling.

The best-known types of secondary data are the major government surveys such as the censuses of population, agriculture and production, and the statistics on health, crime, environment, housing, transport and employment. Such data are found in most countries, they are available for past decades, and they can usually be disaggregated to quite small areas such as administrative and political divisions. They are therefore rightly popular with geographers. However, such massive statistical resources should not blind us to the wealth of other secondary data that can be found. In addition to other censuses and surveys, there are administrative records of all kinds, business records (e.g. from firms) as well as personal papers and diaries. Newspapers and publications such as company reports and planning documents would also fall into the category of secondary data. In addition to printed material (often statistical in character), secondary data includes maps, aerial and other photographs, remotely sensed data, video and sound recordings (e.g. radio and television programmes and oral histories). Some of these data sources are discussed further in later chapters.

In summary, secondary data are likely to be an important element in nearly all human geography dissertations and so you need to be aware of the wide range of secondary data which exists – the second part of this chapter deals with their availability. In the final part we shall guide you through the pitfalls in using secondary data and explain their potential. Useful books containing further discussion of the issues raised by secondary data include those by Dale *et al.* (1988), Irvine *et al.* (1979), Fielding (2000), Dorling and Simpson (1999), Healey (1991) and Walford (2002). White (2003) also discusses some aspects of using secondary data from the viewpoint of an undergraduate research project.

The uses of secondary data

Although the majority of human geography dissertations are based on primary data, it is acceptable in most institutions to base your dissertation entirely on secondary data: you should check the details of your local regulations. There are many opportunities for original analysis or interpretation of such data, although sometimes more may be expected of students who have not had to tackle primary data collection themselves. Even if you are collecting primary data,

BOX 5.1

A summary of the features of secondary data

The strengths of secondary data:

- It exists already (and so is cheaper and quicker to obtain than primary data)
- It provides the researcher with contextual material for his/her primary research
- It is usually of proven quality and reliability
- A very wide range of secondary material is available

The weaknesses of secondary data:

- Its inflexibility (you cannot customise it to your needs)
- Its quality is unverifiable since it is not replicable (you have to trust it)
- It may cost you money and take you time to obtain
- Secondary data is a cultural artefact, produced for administrators with priorities and ways of seeing the world that may be very different from those underpinning your dissertation

there are three reasons why you should make the best use you can of the secondary data which are available on your dissertation topic. First, they are a vital guide to the geography of your topic and your area as these are currently understood. They tell you what the area and its people are like now and what they were like in the past. Secondary data can be analysed or reanalysed to demonstrate actual or potential relationships between variables – between environmental conditions and health, for example. The linkages, which may be purely statistical at this stage, may become the starting point for your detailed investigations of people and places later in your dissertation. The secondary data sketch out the issue or question you will pursue in your own primary research.

The second reason for the importance of secondary data is to provide a context for the primary data you will subsequently collect if your project involves original research on a case study. Secondary data provide three overlapping types of context – geographical, historical and socio-economic. They allow you to compare your case-study material with other areas at the same scale or with larger areas. Your village or neighbourhood can be compared with others, with the wider region and with the rest of the country, none of which you have the time to study directly yourself. The historical dimension is provided when you use the secondary data to create time series extending back from your present-day observations. You can research the past less fully than the present. Where your dissertation is a study of a social or occupational group such as farmers, disabled people or immigrants, then the secondary data can allow you to compare them with other groups or the population at large. This comparative element is important because only in this way can you identify spatial and social differences and historical trends. Arguably, without such contextual material there can be no geographical understanding of the processes of change.

A third reason to use secondary data is that a demonstrable ability to collect, manipulate, interpret and present secondary data is a very useful skill in its own right for your *curriculum vitae*. It helps you show potential employers that you can be relied on to find out what is already known about any topic.

A warning

Secondary data are not static. New sources are being created and older ones are destroyed by time or accident or they cease to be collected. Historical manuscripts are an obvious case of such losses, but even new technologies are subject to obsolescence; early computer formats such as magnetic tape and punched cards may be virtually unreadable today. Some data become available in new formats (e.g. on DVD or online) and others are now available in a wide range of media (e.g. newspapers as paper copies, microfilm, CD-ROM and online). Sometimes changes are introduced as agencies seek to minimise their costs of statistical data collection or to increase their availability to potential users. The reorganisation of agencies (e.g. the reform of UK local government in the 1970s) can lead to a loss of data if files are lost or thrown out.

Less obviously, there can be changes in the organisations that collect data and how they may be contacted. This may lead to alterations in when and how data are collected and the spatial units for which they are published. Even more

Table 5.1 Online sources of secondary data

UK-based archives (many not restricted to UK data sets)	
http://www.bl.uk	British Library
http://zetoc.mimas.ac.uk	British Library's Electronic Table of Contents of Current Journals and Conference Proceedings
http://www.cadensa.bl.uk	British Library Sound Archive
http://www.britishpathe.com	British Pathé Film Archive
http://www.bufvc.ac.uk	British Universities Film and Video Council
http://www.mimas.ac.uk/experian	Experian geodemographic data (requires registration)
http://edina.ac.uk/digimap	Digimap Ordnance Survey map data for academic use (requires registration)
http://www.old-maps.co.uk	Old maps of the UK – scanned historical mapping
http://edina.ac.uk/StatAcc	*Statistical Accounts* of Scotland of the 1790s and 1830–40s (some services require registration)
http://www.edina.ac.uk/ukborders/	UKBORDERS service providing many census and administrative boundary data sets (requires registration)
http://www.ukop.co.uk/ukop	UK Official Publications since 1980
UK agencies and government departments	
http://www.defra.gov.uk	Department for Environment, Food and Rural Affairs
http://landreg.gov.uk/propertyprice/	Land Registry property price data
http://www.magic.gov.uk	Multi-Agency Geographic Information for the Countryside
http://www.statistics.gov.uk/	National Statistics online
http://www.direct.gov.uk	Direct Government website
http://www.ordsvy.gov.uk/	Ordnance Survey – Britain's national mapping agency
http://neighbourhood.statistics.gov.uk	Neighbourhood Statistics Service
http://www.cabinet-office.gov.uk/agencies-publicbodies	UK public bodies and agencies
http://www.pti.org.uk/	UK Public Transport Information
http://www.census.pro.gov.uk	UK Population Census 1901 from the Public Records Office
http://www.upmystreet.com/	Up My Street – compendium of local information indexed by UK postcode
International archives	
http://countries.eea.eu.int/SERIS	European Environment Agency
http://europa.eu.int/documents/index_en.htm	European Union publications
http://europa.eu.int/comm/eurostat/	Eurostat site – source for European statistics
http://www.census.gov/	US Census Bureau – including census statistics and mapping
http://www.census.gov/main/www/stat_int.html	US Census Bureau's listing of international statistical agencies
http://www.who.int/	World Health Organisation – including extensive comparative international health data
http://www.newsbank.com	Newsbank newspaper archive
Academic portal sites	
http://census.ac.uk/	ESRC/JISC Census Programme, with links to other UK Census resources
http://www.britac.ac.uk/portal	PORTAL – British Academy directory of online resources in the humanities and social sciences
http://www.copac.ac.uk/copac	COPAC – consortium of UK university reference library catalogues
http://www.sosig.ac.uk/	SOSIG – Social Science Information Gateway
http://www.Gesource.ac.uk	GEsource – geography and environment gateway

subtle is the redefinition of key terms such as 'farmland' and 'unemployed' which can render the unwary liable to serious misinterpretation of secondary data.

Consequently this chapter can be only a snapshot of the situation in 2004. Therefore it concentrates on general features of secondary data which are likely to retain their validity for a long period. Most of the examples cited directly are drawn from the UK although the general principles hold internationally and new online resources are continually being developed. Many of the secondary data sets noted here are available via the Internet but are therefore prone to periodic changes in URLs. 'Deep links' (i.e. those that point to specific files deep within the structure of a website) are most likely to be altered. Students wishing to pursue specific online resources from this chapter will find the relevant URLs collated in Table 5.1, and these will also be maintained on the book's companion website (*www.pearsoned.co.uk/flowerdew*).

Availability of secondary data

Range and major sources of secondary data

For human geographers perhaps the single most widely used source of secondary data is the Population Census (produced in the UK every ten years by the Office for National Statistics (ONS); the General Register Office (Scotland) and the Northern Ireland Statistics and Research Agency). Not only does it provide demographic information but also details on education, transport, work, housing and many other aspects. The census of population is unique in its near-complete population coverage with availability down to small geographical areas. Additionally, nearly every sector of human geography has at least one major source of data that anyone doing a dissertation should track down – for farming it would be the agricultural censuses produced annually by the UK Agricultural Departments. In the UK the main source for manufacturing is the Annual Business Inquiry; for employment the Department of Employment's Labour Force Survey and NOMIS; for public attitudes the British Social Attitudes Survey; for tourism the UK Tourism Survey and International Passenger Survey; for personal expenditure the Family Expenditure Survey (published as *Family Spending*) and the General Household Survey; for crime Criminal Statistics England and Wales and the British Crime Survey (the Home Office); property prices from the Land Registry, and so on. Most of these are at least annual publications, although the interval between population censuses varies widely and is most commonly decennial. You need to check which areas the data cover – some British statistical publications deal with the United Kingdom, others with Great Britain, while some refer only to England and Wales or just to England, with separate publications for Scotland, Wales and Northern Ireland. Dorling and Simpson (1999) give some excellent reviews of sources, topic by topic. Students outside the UK will usually find that similar sectoral data are available for most developed countries, although the amount, nature and arrangements for access may vary greatly.

A further important development in England and Wales is the Neighbourhood Statistics Service, which brings together area-based data aggregated from the records held by many different government departments. Most of these sources are cross-sectional; they report on conditions at one point in time. You can create a time series by compiling a list of a given statistic over a sequence of years, which will show you how conditions overall have changed in an area, although many such data series are difficult to interpret due to changes in the geographical areas for which they have been compiled. You cannot usually use such cross-sectional data *longitudinally* to trace the history of individual people or firms. A few sources do allow this – in the UK, the British Household Panel Study (from the Data Archive at the University of Essex) has started to track people and families, while the New Earnings Survey Panel Dataset tries to follow employees year by year. In addition, the Office for National Statistics Longitudinal Study links census and vital statistics data for a sample of people from 1971 to the present, and a series of birth cohort studies record information for a sample of people all born at the same time and interviewed every few years. None of these longitudinal sources is ideal – they suffer from respondents dropping out and access to them is more complex. A few of the cross-sectional sources (e.g. *Regional Trends*, *Social Trends*, Transport Statistics Great Britain and the Annual Abstract of Statistics) do at least provide ready-compiled time series of their key data.

How to find out

Perhaps you do not know what secondary data are available for your topic or you suspect there is more than you know about. You may want brief details of what specific sources can provide and where to get hold of them. The best starting points are the websites of national and international statistical organisations and directories of official publications, which in the UK would include the following:

- Office for National Statistics (2000) *Guide to Official Statistics*, London: HMSO
- The home page of the Office for National Statistics – *http://www.statistics.gov.uk/*
- The home page of Eurostat – *http://europa.eu.int/comm/eurostat/*
- UK Official Publications since 1980 – *http://www.ukop.co.uk/ukop*

The Office for National Statistics indexes are particularly useful since they include an alphabetic keyword list of topics in the paper version and a keyword search facility in the online version. Statistical agencies in other countries will have analogous publications and websites; White (2003) lists web addresses for a selection of twenty-three countries.

Cross-cutting sources

The sources described at the start of this section are each arranged around a theme – population, agriculture, health, and so on. However, if you wanted to compile a profile of several different aspects of an area, you would have to examine each of these sources to find the regional tables. There is one UK source (*Regional Trends*, HMSO, annual) which pulls together a large amount of data using a standard regional framework; this can save you a lot of work. *Regional*

Trends does not cover all possible variables and its spatial resolution is still fairly crude but it is a good starting point. An analogous publication is *Social Trends* (HMSO, annual) which provides a different set of data tabulated by social or socio-economic class. Again, it brings together within one volume data selected from a wide variety of sources. Unfortunately there are many other cross-cutting perspectives which are not well served – age, religion and sexual orientation, for example. For areas such as local authority districts and counties a good starting point is the planning reports for the area; often these will give a profile of the area derived from many sources. If searching by postcode or place name in the UK, the Neighbourhood Statistics Service gathers together census and other data for small areas.

International secondary data

So far we have considered the availability of data for one country, with special reference to the UK. If, however, you want data on several countries, then you need to find sources that have pulled together such information. Trying to access individually the statistical sources for ten or twenty countries would be far too time-consuming for most undergraduate dissertations. The European Union, through Eurostat, provides a great deal of data which allows you to compare conditions across member states. Other international organisations also produce annual compendia. The United Nations has its *Statistical Yearbook*, the International Labour Organisation produces a *Yearbook of Labour Statistics*, and the World Health Organisation publishes a *Statistical Annual*. The European Environment Agency's *Europe's Environment: The Second Assessment* (1998) is another pioneering attempt at international comparisons. The EEA's website brings together national online environmental reports while several national statistical organisations maintain useful listings of other such agencies, allowing you to track censuses and official statistics for the country in which you are interested.

Of course, there are serious problems with making international comparisons once you have got the data. We shall return to this issue in the final section of this chapter.

Sources of qualitative secondary data

So far this chapter has focused on traditional quantitative and statistical sources since these are relevant for most dissertations and are still the most frequently used types of secondary data. However, we also need to consider qualitative and non-statistical sources. These include newspapers, photographs, sound recordings and diaries (see also the chapters by Aitken, and Aitken and Craine later in this volume for a discussion of how such sources might be used).

Newspapers can be an important source of information. How they define and treat 'news' is an important topic in itself; see for example Burgess and Gold (1985) and Hay and Israel (2001). Newspapers cover local and national scales and comparisons between them are interesting. They can be used qualitatively (how they report events) or quantitatively (the number of stories). The advent of online and CD-ROM versions of newspapers has made searching them much quicker. Yet these new formats are often incomplete, lacking photographs, advertisements and, not being facsimiles, typefaces and layout. They have also been available for only a few years so you may still need to have recourse to paper or microfilmed copies, especially for local newspapers. The indexing of paper copies of newspapers is a problem. If you do not know exactly when an event happened, you need a subject index. The only UK newspaper which has a full published index over a very long period is *The Times*. Another useful source is Keesing's *Record of World Events* (Cartermill, Cambridge, monthly, started 1931). Online and CD-ROM versions of newspapers with a search facility partially overcome this problem with printed papers.

Photographs tend to be less well indexed. Collections are scattered, and it can be unclear to the outsider what they contain since their catalogues are less often accessible online. Photographic libraries vary greatly in their internal arrangements – some are well indexed and specific images can be identified and retrieved quickly; others are less easy to use or browse through. There will also be some cost if you want copies of photographs. Television companies catalogue their output well, but access and reproduction rights may be problematic and/or expensive. Film libraries in the United States tend to be very much larger and better endowed than those in the UK such as the British Film Institute and the regional film archives (e.g. in the North West and Scotland). The online British Pathé Film Archive allows you to search their news films for 1896–1970 and, for a fee, to download clips or stills.

In most countries sound recordings, apart from radio broadcasts, are a less well developed resource. Ethnographic and oral history collections do exist but at least in the UK are limited especially for speech as opposed to music. The main British repository is the British Library Sound Archive and from this website you can search their holdings. Diaries are particularly difficult to identify and track down. Some may still be with the authors or their families, whereas others may be in university libraries or record offices.

With all these qualitative sources there is a serious issue which might be called 'representativeness'. It is well known that newspapers often put a distinctive slant on the items they cover and how they deal with them. Similarly, diaries are personal products that may be written simply to record events or they may have an element of self-justification or 'setting the record straight'. Photographs and film (even those labelled 'documentary') may have been staged for effect or scenes selected because of their interest or picturesque quality. The question to be asked is always why a photograph or film was taken in a particular way at that time by that person. Careful interpretation is every bit as necessary with qualitative sources as with statistical data. Indeed, how people, places or environments are presented may be an interesting dissertation topic in itself.

Historical records

Historical records pose some particular problems for dissertations. They are less likely to be available in electronic form and so you may have to travel to where the records are stored. The main repositories are university libraries, county and national record offices, local authorities (e.g. in planning departments), company headquarters and government offices. Modern styles of records (those during the last 50–100 years) are likely to be similar to current ones – though be very wary of historical comparability problems (see below). Earlier records may run into problems of different languages or units of measurement, different spatial units, and patchiness of coverage due to the loss, deterioration or physical destruction of records. This loss of data is very prevalent when organisations are restructured or taken over. The opportunity to throw out records may be irresistible.

Online secondary data

The Internet has rapidly evolved from a rather specialist research tool to a mainstream online source of information for everyone. It is helpful to think of the Internet as providing different sorts of information and sites.

Single sites
- Information formerly provided by a single agency in other forms (e.g. paper publications, books of statistical data, newspapers) may now be available from a website.

Maps
- Maps as digitised images or as digital data files can be downloaded and printed.

Catalogues
- Access is given remotely to catalogues formerly available only in paper form or in a single building.

Compendia
- Place-based compendia of data (e.g. bringing together information generated by different agencies for a given place such as a postcode).

Portals/gateways
- Bringing together in one website similar types of information formerly scattered across different physical and/or electronic locations.

Single sites – access online to material published on paper

Your search for official statistics online should start from the home page of the national statistical organisation of the country in which you are interested. These sites generally show the range of data available, although not all of it may be directly downloadable from the web. It is, however, still the case (though decreasingly so) that online statistical sources are less detailed than the traditional ones. National statistical organisation's sites also tend to offer less detailed information than some of the key government department sites. So the place to go for agricultural data and sustainable development in the UK, for example, is still the website of the Department for Environment, Food and Rural Affairs (DEFRA). The Direct Government website (*http://www.direct.gov.uk*) gives you access to all the ministries and agencies and their statistical sections and local government websites. Some statistical data covering all the member states of the European Union are available from Eurostat.

Newspapers are still published on paper but searching them for items on a given topic can be slow, and local papers may not be available outside their area. Hence web-based access to international, national and local newspapers has made them a much more usable research tool. Newsbank is an online collection of major national newspapers. Of course newspapers demonstrate the general point that web-accessible material tends to be modern material. Few newspapers are available on this medium before 1990. For earlier periods you need to use microfilm or microfiche versions of newspapers or the actual paper copies. In general most historical sources are less likely to be online. Two interesting recent exceptions are the UK Population Census of 1901 and the *Statistical Accounts* of Scotland of the 1790s and 1830–40s.

Most of the websites listed here are free, at least to academic users, although you may be required to register, using information provided by your university's library or computing services. There is a charge for using the 1901 Population Census site.

Maps

Traditionally geographers wanted maps for the areas they studied (see Dunn's chapter later in this book) and have traditionally looked to paper maps to provide that information. The cheapness and ready availability of paper maps is obvious. However, the inflexibility of published paper maps is notable – one has to accept the size and scale of maps provided by the mapping agency. The advent of online maps has altered all that. Numerous websites provide online mapping and these are addressed in more detail in Chapters 16 and 19. Students should be aware that copies printed without the use of a colour printer may result in confusing images; also there may be variations in which of these services are available to you, depending on the country within which you are working and whether your university is registered with specific service providers such as Digimap in the UK.

For historical maps back into the nineteenth century one can buy a scanned version on CD-ROM; photocopies of paper maps in your library may be easier and quicker to organise. The Old-Maps website provides a wide range of UK maps from the mid-nineteenth century onwards.

The Multi-Agency Geographic Information for the Countryside provides many useful base maps for rural topics in the UK while the EDINA site provides a very wide range of boundary maps for administrative and electoral areas.

Catalogues

While catalogues may not be the most exciting things to read, they are a vital tool for finding out what you can read and where to get hold of it. The web has made catalogue searching much quicker. For example, the catalogue of the British Library lists nearly all material published in the UK. However, although their catalogue is available online you may have to go to London to read the material or order it from Boston Spa via your university library's inter-lending service. More practical is COPAC, which is the merged catalogues of twenty-two of the largest university libraries in the UK and Ireland.

More specialised catalogues are also becoming web-accessible. Examples include the following: the British Universities Film and Video Council at *http://www.bufvc.ac.uk*; UK Official Publications since 1980 at *http://www.ukop.co.uk/*

ukop; the British Library's Electronic Table of Contents of Current Journals and Conference Proceedings at *http://zetoc.mimas.ac.uk*; European Union publications at *http://europa.eu.int/documents/index_en.htm*; British Library National Sound Archive at *http://www.cadensa.bl.uk*.

Compendia

While the sites above are good for large areas, if you want a range of data for small areas, then various academic and commercial sites may be of interest. In the UK these include Experian (for which registration is required for academic use) and more informal sources such as Up My Street. Care must be taken for these and other sites to make sure you know whether data presented are based on observation or are estimates based on possibly inaccurate assumptions.

Portals

A portal is a website that does not provide information itself, but is a list of hyperlinks to other websites usually having something in common. You can expect more of these to be developed, so it is worth seeking specialist portals relevant to your area of study.

Results of UK censuses of population from 1981 to 2001 are available online from *http://census.ac.uk*. There are several different types of census output, including aggregate statistics and their associated geographical boundaries, Samples of Anonymised Records and interaction data. All are accessible from this entry point although user registration is required for access to much of the data holdings. Another example is the US Census Bureau's listing of international statistical organisations, which provides a route into official statistics for many countries. Similarly, the UK Public Transport Information site brings together a large number of websites for separate transport undertakings in one place.

Perhaps the single most useful such portal for UK dissertation students is the one which gives access to all national and local government departments at *http://www.direct.gov.uk*, while *http://www.cabinet-office.gov.uk/agencies-publicbodies* does the same for public bodies such as hospital trusts and development agencies.

Multi-disciplinary academic portals include PORTAL which is a directory of online resources for the humanities (including social anthropology and geography) run by the British Academy and the Social Science Information Gateway which covers the social sciences including economic and political geography. A specifically geographical site that gathers together many geographical and environmental sources is GEsource. Portals are invaluable as a first base for searches for information. They are, however, vulnerable to obsolescence if they are not maintained to ensure that all the hyperlinks still work.

Web-wise

The Internet (or web) has the ability to bring you existing information more quickly than conventional methods. Examples include online newspapers and government statistical data. The web speeds access from any networked computer (which increasingly could be at home) and at any time of the day or night that suits you. The web may also allow you to search large data sets and catalogues of information by keyword much more quickly than browsing the shelves or turning the pages. Some of these data sets (e.g. the full results of censuses of population or large-scale maps of a whole country) may be so voluminous that many libraries will be unable to hold them all, so the web provides the only practical solution to accessing so much information. Similarly the fact that the web brings many sources of information together in one place, as it were, is a major saving on travelling time from library to library. The rapidity of publication on the web is another of its clear advantages, so is the ability to print out the information, and sometimes to download data to common office software for your own analysis.

Information on the web must however be treated with some caution. The range of statistical material may be less detailed than that available in other forms. In some cases only the headline data or that for large spatial units may be made available on the website. There is a tendency to show only *current* data, and for data for previous years to disappear or be replaced. You may have to revert to more traditional sources for some of the information you need. This is particularly likely to be the case for information from small agencies and private firms.

Sourcing information from the web has two dangers. The first is that whereas you can see how big the book is and can spot the footnotes and other explanatory material which tell you what the data mean, on the web you get the information a screenful at a time. All the other things you need to know in order to interpret the data properly are less visible, being on other parts of the website. You need to remember that you will have to read more 'screenfuls' than just the one with the information you want, if you are to understand fully data on the web.

The second danger derives from the essence of the web – that it is largely unregulated. Anyone (for example, lobbying groups) can put any information they like on the web without it being checked, whereas most libraries buy only items that the academic staff feel are reputable. The web contains the reliable and the unreliable, and it is up to you to decide which is which. So when using the web and particularly search engines which can take you anywhere in cyberspace, you need to be very cautious about what you are reading. Is it from a reliable source? Is it authoritative? From which standpoint has this information been produced? Is there enough background information to allow you to judge whether you are reading an opinion, evidence or the results of careful research? Be cautious with the web.

Search engines

Yahoo!, Altavista, Lycos and Google are all well-known search engines which may help you find the website with the information you need for your dissertation. 'May help you find' is the operative phrase. The danger is that you search for, say, 'urban transport' information and get a million references, all of which seem totally irrelevant. Here are some tips for using search engines productively.

- Make your search term as precise as you can (e.g. 'Chicago' rather than 'urbanisation').
- Narrow your searching by limiting it to material in a particular language.
- Narrow the spatial reach of your search using AND and NOT (e.g. railways AND UK; gangs NOT United States), though not all search engines support such Boolean searches.
- Use search terms in double quotation marks to get matches with those exact phrases (e.g. "tourism information centres" – individually each of these three terms will yield a vast number of references).
- Watch out for ambiguous search terms – 'port' could be a place for ships, a drink or the left-hand side of a ship; 'ford' could be a river crossing, car manufacturer, a prison or a former US President).
- Use an asterisk to cover related words and spellings (e.g. 'defen*e' will pick up the UK and US spellings of defence; 'Americ*' will pick up 'America' and 'American(s)').
- Use OR to cover synonyms such as 'organic farming OR agriculture biologique'.
- As a rule of thumb, if the item you want is not in the first 50 references the search gives you, then you are unlikely to find it. Start a new search with different terms. Alternatively try and identify which organisation has the information you need, get their web address (many can be guessed and most can be found easily with a search engine) and go direct to their website.

Problems with secondary data

Secondary data as cultural products

Secondary data reflect the aims and attitudes of the people and organisations that collected the data. No better example of this can be found than the CORINE environmental database run by the EU. It was originally intended to be a comprehensive set of environmental variables, available at large and small scales, and collected on a comparable basis in each member state. In practice it is still a partial view of environmental quality (only what was easily collectable was collected), it has gaps in some countries which choose not to collect certain variables, and even apparently similar data are not strictly equivalent in different member states. Different countries collect data on those aspects of the environment they consider to be important and, it is argued, they do not collect what might be politically embarrassing. From this perspective, the lack of comparability of data (which conveniently precludes international 'league tables' of environmental quality) is as ecologically unfortunate as it is politically expedient.

This cultural perspective on secondary data is particularly relevant to historical data. Units of measurement may represent different physical quantities in different places and periods (e.g. the acre and the bushel). Phrases may be used literally or as legal clichés; technical terms (e.g. tenant, borough, estate) may have had local or general meanings which are dissimilar to current usage. Even with contemporary data, its interpretation may require careful handling and some understanding of the mindset and administrative concerns of the department collecting the data. A case in point is crime statistics, which cannot be taken at face value (see Pantazis and Gordon, 1999). Internationally, terms may be defined in different ways. There has been some standardisation in how unemployment, for example, is defined and measured in different countries, but this is still an exception.

Availability of secondary data

Whereas much of this chapter has stressed the wide range of secondary data that awaits the dissertation student, there are several reasons why secondary data might not be available. The simplest is that no one has thought it worth the administrative cost of collecting it. The environmental effects of farming may now seem to us a self-evidently important topic, but those who run agricultural censuses may still confine their surveys to data on food production. As departmental budgets shrink, one of the casualties may be the programme of data collection. Annual surveys may become less frequent; censuses may be replaced by sampling; fewer questions may be asked; and less material presented to the public, and in a simpler form with fewer cross-tabulations and analyses.

Even if the data source you want has survived, it may be difficult for you to get access to it. Records relating to individual people or firms will probably be confidential (in the UK for 100 years) and data which are so disaggregated that individuals or businesses could possibly be identified may also be suppressed (witness the removal of parish-level agricultural data in the UK).

A final hurdle regarding statistical data is cost. Government agencies responsible for their own budgets may charge students for accessing data that are in the public domain but have not already been published. They may plead that they do not have the time or staff to look out material for you. They may claim copyright over the information and seek to charge a fee or control how you use it. This can apply not only to data but also to photographs and copies of manuscripts and maps.

The best advice would be to rely on sources which are already in the public domain (e.g. in libraries or online). If you have to approach an organisation for data, stress your student status and consequent poverty; this might persuade them to waive charges. Perhaps your supervisor will be able to help you make your case.

Comparability of data

At the start of this chapter we noted that one of the principal functions of secondary data was to provide a context for more detailed case-study work which you might conduct yourself. Our model here is that you might want to compare your primary information with someone else's secondary information, and perhaps with secondary information for different areas and periods. A precondition for these comparisons is that any differences you observe are due to the different conditions on the ground and not to differences in the structure of the primary and secondary sources of information. You need to be able to argue, for example, that the high unemployment and poor health record on the housing estate you are studying are due to, respectively, fewer jobs and more illness there, rather than because you measured unemployment and health differently to the government. So, whether secondary data are actually comparable (with other secondary data and with primary data) is a crucial element. Note for example that unemployment data from British official sources are not comparable with unemployment data from the census, which are self-reported.

There are a number of common problems that make comparability difficult. The first, particularly relevant for geographers, is the spatial units for which data are provided. These may change over time as local government is reorganised; levels of data provision may be withdrawn; boundaries may change (e.g. wards and enumeration districts). They may also differ between countries; what is available at one spatial level in one country may not be available at that level in another country. The pressure is therefore to move to larger spatial units (e.g. regions) where the boundaries are less changeable – but that gives you a less detailed geography.

The second element that can frustrate comparability is the date to which data refer. The data you collect may well refer to the present; secondary data may refer to last year or several years ago (the latest census data available can be as much as twelve years old in the UK or USA). Are the differences between your data and the official data due to the time gap between them or because your study area really is distinctive compared with the rest of the country? Probably both factors will be at work; the balance between them may be difficult to determine.

A third element concerns methodology. Ideally your data and the secondary data should have been collected using the same methods: other things being equal, it is a good idea to word questions in a questionnaire in the same way as in your secondary data source. In practice the lone student will be hard pressed to achieve exact comparability, though you should try as best you can. You will not have any statutory authority for your survey, so your response rate may be lower. Alternatively your lack of official standing may be an advantage in gaining the cooperation of those groups who look sceptically on officialdom. Time constraints may force you to limit yourself to a sample (rather than a census) so your results will have a margin of sampling error that the official data will not have. Time may also limit your work to a smaller area than any for which official data exist; so an exact comparison for your area may be impossible. Overall you will find several obstacles to comparability due to the impossibility of replicating official data-collection procedures. Only in certain cases will you be able to turn this to your advantage by providing information at a level of detail previously unexplored; by trading off depth of study against lack of coverage; and by gaining the trust of those who shun official surveys.

Where you are using imagery (photographs, aerial photographs and remote-sensing data) there are other problems. You probably cannot fly an aircraft or space mission to get your own images but you could try to replicate old ground-level photographs. The problems here are numerous. You need to be able to date the old photograph, but often this can only be estimated from visual clues (such as the styles of cars and clothes). Comparability of images means you have to go to the same place in comparable weather and season, using a similar camera lens. This will need considerable preparation to achieve.

An analogous exercise is the updating of maps for changes to the man-made environment. Again, you are not going to build a complete new map and dating should be easier since most maps have revision details. In practice, the latest revision date may not apply to all the features on the map. You need to check with the mapping agency as to what was revised and when. You will also need to know how they defined the features on their maps. How many trees need to be how close to each other for it to be mapped as continuous woodland rather than scattered trees? Only when you know the answers to that type of question will you be able to replicate and update the surveyor's work with confidence that the changes you record are real ones.

The final problem for comparability leads on from this cartographic example. In brief, you need to be speaking the same language as the compiler of the secondary data. It may be obvious to you what a farm is, but in the English and Welsh Agricultural Census an 'agricultural holding' has a precise and not wholly intuitive definition (which has changed over time and which is slightly different in Scotland and Northern Ireland). 'Unemployment' has officially been redefined thirty times in the UK between 1979 and 1989 (Thomas, 1999) and UK figures cannot be directly compared with European ones since each country defines unemployment differently. Similarly 'crime' as perceived by the public is not the same as the crime which is recorded by the police. There are, of course, interesting studies to be done into these differences between the perceived and official worlds. Many of the chapters in Dorling and Simpson (1999) give some useful examples.

The inflexibility of secondary data

It is an obvious point that secondary data have to be accepted in the form in which they are presented to you. Unlike any data you collect yourself, you cannot plan what to collect or how to present it. So if your secondary data source says nothing about, say, people's ethnic origin, there is nothing you can do to rectify this. This is a problem that is likely to affect everyone doing a dissertation; you have to make the best of what you can get.

However, this inflexibility is a particular concern for those researching in relatively new areas such as environmental, cultural or gender studies. These have become more common subjects for academic study but the inertia (administrative and conceptual) among those who provide the secondary data is such that environmental and gender questions and disaggregations may not be provided for some time. So the inflexibility of secondary sources may be more of a problem for some dissertation topics than for others.

Conclusion: potential and pitfalls

It is hard to imagine an undergraduate dissertation that could not benefit from using secondary data to some extent. Every human geographer has to be aware of their potential and of the pitfalls, which manifest themselves in two ways – in their availability and in their use and interpretation.

The range of secondary data is vast; some of it is becoming more easily available, especially online and on CD-ROM; new sources are continually emerging. Potentially, it can greatly enrich our view of the world and place our fieldwork in proper perspective.

Yet some of that potential availability and usefulness may be illusory in practice when data are withheld or are too expensive to acquire. You can do nothing to plug the gaps in what others have collected. Above all, you have to be very careful about how you interpret secondary data and compare it between different areas and periods. This means, at its simplest, reading very carefully the explanatory notes and the critical reviews of the data source. Do not just unthinkingly extract the numbers, quotations, photographs or text from your source. Ensure that you know as precisely as you can what they meant to the original author, compiler or respondent, and consider what interpretation you can reasonably put on this information. Remember to record your ideas in your dissertation so that the examiners can see your train of thought and assess its validity.

You must of course acknowledge the source of the secondary data you use. If it has been taken from the Internet, make sure you provide the full URL and the date you accessed it. Remember to make any necessary copyright declarations (e.g. Crown Copyright) and ensure that you have permission to use the data.

Happy hunting!

BOX 5.2

Further reading

Campbell, D and M Campbell 1995 *Doing research on the Internet*, Reading, MA.: Addison-Wesley

Dale, A, S Arber and M Proctor 1988 *Doing secondary analysis*, London: Unwin Hyman

Dorling, D and S Simpson (eds) 1999 *Statistics in society: the arithmetic of politics*, London: Arnold

Fielding, J L 2000 *Understanding social statistics*, London: Sage

Irvine, J, I Miles and J Evans (eds) 1979 *Demystifying social statistics*, London: Pluto Press

White, P 2003 Making use of secondary data. In N J Clifford and G Valentine (eds) *Key methods in geography*, London: Sage, pp. 67–85

Section

B Collecting primary data

It is tempting to see primary data collection as the major component of your project, and in many cases it will certainly take up a considerable proportion of the time. However, it must be set in a context which ensures that relevant questions are asked and that the information gained is relevant to the original research questions. Primary data collection may take various forms, and the chapters in this section address questionnaire design and sampling, interviewing techniques, focus groups, 'participatory' research methods and participant observation. Again it is necessary to stress that the word 'data' here is not intended to imply that the information being collected has to be numerical.

It is common for research projects to be timed such that students find it necessary to undertake their primary data collection over the long summer vacation. This has the advantage that there is a reasonably long period of time available for the work, but means that students are likely to be conducting a significant part of the work without any direct contact with their supervisor. (You should find out before the start of the vacation whether your supervisor is willing to be contacted and provide dissertation advice by e-mail.) It is therefore important to plan the primary data collection exercise very carefully in advance, including such tasks as the establishment of necessary contacts with key individuals and the piloting of questionnaires. There is little point turning up to interview some of your key respondents just after they have gone on their summer holidays, or in printing 200 questionnaires only to discover that most people misinterpret one of the central questions. A common failing is to see the primary data collection as the starting point for a research project, and to conduct it without adequate background reading or planning. Those who adopt this approach may subsequently find that their main results were already clearly established in the literature, but that there were other interesting angles to the problem that they did not think to address. To be used effectively, primary data collection must be part of an integrated process that begins with the underlying research questions, is informed by an understanding of previous work, and which is designed with specific plans in mind for analysing the data. Of course, involvement with the real world will mean that you make new discoveries as you go along, and may need to adapt your original strategy, but ideally this should be the exception rather than the rule.

Questionnaires have undoubtedly been one of the most widely used data collection tools in human geography projects and in Chapter 6 Julian Parfitt presents an extensive discussion of the issues surrounding their design and conduct. Using some real examples, he demonstrates the stages in questionnaire surveying, emphasising the importance of piloting and revising questionnaires before undertaking the main data collection task. It will usually be the case that your first attempt to draw up a list of potential questions will be too long. However, it is possible to hone the content to ensure that questions will really elicit the relevant information, and that the format, sequencing and wording are as effective as possible. It is necessary to have a clear view of the types of error that can affect the reliability and validity of results, and to minimise these wherever possible. Different strategies are presented for obtaining attitudinal information, although if you are primarily interested in people's attitudes, you should also carefully consider the potential advantages of the other approaches introduced in the following chapters. An important issue in the implementation of questionnaires is the selection of an appropriate sample of respondents, and consideration is given to the different sampling strategies that can be adopted.

It will become immediately apparent that there are limitations to the kinds of information that can be obtained using conventional questionnaires, and in Chapter 7 Gill Valentine discusses the merits of conversational interviews as a means for data collection. In-depth interviews provide an alternative means for exploring issues in more depth than is generally possible using questionnaires. Such interviews are commonly tape-recorded for subsequent analysis. The kinds of interview being discussed here are generally unstructured or semi-structured, in contrast to the highly structured information gained from questionnaires. Primary data collection through the use of in-depth interviews will inevitably involve fewer respondents than questionnaire-based data collection as the interviews themselves and the subsequent analysis will take much longer. The aim here, however, is not to be representative, but rather to give the researcher deeper insight into respondents' feelings and attitudes. The opening section of the chapter contains a fuller case for the use of interviews as a research tool, and the reader will again perceive one of those areas in which different researchers tend to favour different research traditions. As elsewhere, these approaches can be seen to have different strengths and weaknesses, and wise readers will think carefully about how these apply to their own research questions, and may even contemplate some form of multi-method research. The rest of the chapter addresses many of the practicalities of conducting interviews including decisions about who to talk to and where; tape-recording versus note-taking; how to conduct the conversation; and how to practice your skills. Those contemplating this approach should note again the exhortation to brief themselves very carefully on the research topic beforehand, in order to maximise the potential benefits of their chosen method.

One of the obvious problems of in-depth interviews is that you can only talk to one person at a time. The alternative of conducting focus groups can be used to obtain opinions and experiences from a group of people interacting with each other, in a more natural setting than the rather artificial setting of conducting one-on-one sessions in questionnaire or interview research. David Conradson discusses the strengths and weaknesses of focus groups as a research method, with examples from previous undergraduate dissertations. The use of focus groups in human geography research has grown considerably over the past five years and, although an advocate of the method, Conradson identifies various potential problems as well as stressing the advantages. The chapter covers all aspects of the process from selection and recruitment of an appropriate group through to methods for the analysis of transcripts from focus-group discussion.

In the next chapter, Mike Kesby, Sara Kindon and Rachel Pain discuss the use of 'participatory' methods in human geography research. They distinguish between two uses of the term 'participatory'. First, research may (or perhaps should) be conducted not by an outsider working alone, but with the active and informed participation of the people 'being researched'. Secondly, participatory methods involve activities designed to help members of the group being studied to express their views and experiences, and perhaps to convey a message to decision-makers as well as to the academic community. Techniques like 'participatory diagramming' can clearly be a lot of fun, both for the researchers and the researched. Each of the three authors has used these methods in different contexts, and much of the chapter is made up of three case studies drawn from their research experiences.

The last chapter of this section deals with ethnographic research. Ian Cook explains how this approach has developed out of a desire to understand groups of people 'from inside'. The chapter provides an introduction to the methods of participant observation, in which the researcher deliberately becomes involved in the activities of the people being studied, recording events and impressions. The chapter begins by exploring the kinds of research questions to which ethnographic research may be particularly well suited, including the issue of gaining access to the particular community to be studied. A further important issue concerns the role that researchers adopt, and whether they identify themselves as researchers to the group being studied. The chapter also addresses the ways in which research 'data' can be constructed from observation of this kind, and how the findings should be presented. Such work might be complemented by other primary or secondary data, assembled using different approaches. The discussion of ethnographic research brings to the fore the many ethical, practical and personal issues which surround the organisation of primary data collection – issues to which again there are no clear-cut right answers. It is important to realise that to a large degree these issues are present in all approaches to data collection, although they are not always adequately addressed.

Once you have read the relevant chapters for the data collection methods that you intend to employ, you should turn to the following section and think about the ways in which you will need to analyse the results. Whichever research methods you choose to use, much effort can be saved by thinking ahead and recording data in a format that will be suitable for your subsequent analysis.

6 Questionnaire design and sampling

Julian Parfitt

Synopsis

Questionnaires are frequently used as tools for collecting data in human geography and related areas of research. With reference to three extended examples, this chapter discusses the processes of designing and conducting a questionnaire survey. After a discussion of types of data and error encountered in survey research, there are sections on questionnaire design, on sampling and on the execution of questionnaire survey fieldwork. Throughout the chapter, the emphasis is on practical suggestions for good practice, and on pitfalls of which students should be aware.

Introduction

This chapter covers the arduous process of designing and conducting questionnaire surveys. In the context of human geography, the questionnaire survey is an indispensable tool when primary data are required about people, their behaviour, attitudes and opinions and their awareness of specific issues. An understanding of how such data are collected is important not just to the researcher wishing to conduct a survey of their own, but also to anyone using the vast amount of secondary data collected by means of questionnaires (e.g. the Census of Population, the British Social Attitudes Survey, the Family Expenditure Survey, national opinion polls).

The use of questionnaires is ubiquitous and hardly a day goes by without the media announcing the results of a survey on something. Market researchers, planners, government departments, academics of various types, opinion pollsters and sales people are all busy conducting surveys. It is therefore all the more important not to add further to this intrusion into people's lives unless you have very good reason. If you do have good reason, it is essential that your survey is thoroughly thought through before any fieldwork is undertaken, otherwise the respondent's time (usually given free of charge) will have been wasted and public tolerance of surveys will have been further eroded. Conducting a survey therefore carries with it a weighty responsibility.

In preparing for your dissertation you should have become aware of previous research in your chosen area (Chapter 4) and secondary data sources (Chapter 5). Has the local council recently conducted its own transportation study? Do official data exist on tourism or recreational activities within the National Park? From this desktop stage you should be able to assess whether or not a survey can be avoided. You may feel that you need to conduct your own survey because you are required to conduct original research. Check this with your department, but generally it will be recognised that there is little point in a single student trying to replicate a big official survey, and in these circumstances credit will be given for original analysis of secondary data.

The tone of this advice is designed not to discourage the use of questionnaire surveys, but to reinforce two important points about them:

- The need for a survey must arise from a carefully considered set of research objectives and not from a general desire to 'go out and do a survey on my topic'.
- It is often a surprise to those inexperienced in conducting questionnaire surveys how much work (intellectual and physical) they entail.

Nevertheless, questionnaire surveys are an important means of eliciting different sorts of information from a target population.

Different data and variable types

Survey data can be broadly classified into three types:

1. Data that *classify* people, their circumstances and their environment. Such items of data are sometimes referred to as *respondent variables* and include information such as age of respondent, income, housing tenure, household size, locational variables (such as distance from a planned major shopping centre), state of health and so on. Some characteristics are more easily established than others, but generally fewer problems are posed than with collecting other data categories.
2. Data that relate to the *behaviour* of people. For instance, where do they do their shopping? How often do they visit their local cinema? Where do they work and how do they travel there? Apart from data collected by direct observation, the hazard with such data is that behaviour expressed at interview can often differ from actual behaviour, depending on the nature of the subject of the question.
3. Data that relate to *attitudes, opinions and beliefs*. These tend to be the most difficult data category to collect. The problems encountered include patterned responses and insincerity (particularly the tendency of respondents to want to please) and the related problem of 'attitude forcing'. The latter is the process whereby the questionnaire device itself creates attitude data, either because of a sense of embarrassment in respondents without a particular opinion on the subject at hand, or as a function of the way in which responses are elicited. Furthermore, attitude questions are very susceptible to biased responses depending on how they are asked (choice of wording and tone

adopted by an interviewer) and may be greatly influenced by the nature of the preceding questions in an interview. How such *response errors* can be reduced or avoided is one of the primary objectives of careful and considered questionnaire design and interviewing technique.

Large-scale surveys, such as those conducted by national institutions or for opinion polls, are mainly concerned with counting numbers of people in particular categories and with particular characteristics, such as the proportion of the population with no access to a car or those who are uncertain how they might vote in an election. Population estimates derived from such sources need to be based on samples that are as representative as is practicable of the population as a whole and sample sizes are therefore large. This type of *descriptive survey* may be distinguished from the more *analytical survey* design that attempts to explore the more difficult 'why' questions.

Analytical surveys are more concerned with explanations and causality (or at least establishing associations between variables) and are therefore more frequently adopted by academic researchers. Questionnaire surveys based on analytic research designs are the main focus of attention here.

Oppenheim (1992) identifies four types of variable that should be considered in an analytic survey:

1. *Experimental* or *independent variables*. These include all variables that may be possible predictors of the main effects that are being studied. Does the likelihood of using the city's 'park-and-ride' scheme vary with the age of respondents or with whether or not they have young children? Does the public acceptability of a planned wind-energy scheme vary more with respondent's distance from site than with any measure of environmental beliefs? It is often the case that various selections and combinations of possible independent variables have to be examined as part of the survey design.
2. *Dependent variables*. These are the main variables for which explanations are being offered in terms of the way in which the independent variables influence them. The awareness and frequency of use of a 'park-and-ride' scheme amongst city-centre shoppers would represent dependent variables in a survey.
3. *Controlled variables*. In exploring the relationships between a dependent variable and a set of independent variables there will be a number of factors that need to be held constant in order that the relationships can be more clearly established. These are called the controlled variables. If, for instance, a comparative survey were to be conducted amongst those travelling on 'park-and-ride' buses originating from four different sites around a city, there would be a number of controlled variables in the fieldwork design. Time-of-day and day-of-the-week effects would have to be considered in a study conducted over a series of weeks within a given season. Given a sufficiently large team of interviewers, all interviews could be conducted on the same day to eliminate day-of-the-week effects. Alternatively, with fewer interviewers, fieldwork could be scheduled so that overall each site is similar in terms of the number of Mondays or the balance between morning and afternoon interviewing shifts.
4. *Uncontrolled variables*. All variables other than those specifically controlled for in the experimental design are uncontrolled variables. These can often result in the wrong inferences being made from the data. For instance, a study

which sets out to explore the relationship between exhaust emissions from vehicles and human health in a certain district might record the distance that the respondent lives from a motorway flyover. The hypothesis might be that those living closest to the flyover have the most health impairment due to greater exposure to airborne particulates. Important uncontrolled variables in this study would be the residents' exposure to airborne particulate matter from other sources, confounding socio-economic variables (the area near to the flyover may be poorer) and lifestyle variables affecting observed health outcomes (exercise, diet, alcohol consumption). Although uncontrolled variables can easily frustrate what might appear to be a compelling hypothesis, it is important that they should be fully discussed and not swept under the carpet.

The extent to which these four types of variable can be identified at the outset will depend on the researcher's knowledge of the topic. In setting up a research hypothesis it is rarely the case in human geography that a simple causal model is under consideration. More often, a complex of interrelated independent variables may influence the dependent variable and a multi-causal model must be considered.

Main stages of the survey process

The main stages of a survey are set out in Box 6.1. As with many research processes, the starting point is hard to define, and the research idea is forever

BOX 6.1 The stages of a questionnaire survey

General activity	*Specific tasks*
Initial research idea: refine and develop analytical design	Developing aims and research objectives. Literature review/secondary data sources. How much is known already?
Design of research	Hypotheses formation; basic research design. Consider dependent, independent, controlled variables. Choice of survey methodology: postal, telephone, Internet, personal interview? Drafting questionnaire.
Further refinement of the research instrument and sampling	Pilot work. Post-pilot revision of questionnaire. Sampling: sampling frame. Think about possible sampling biases. Consider systematic/purposive techniques (for a summary of sampling terms see Box 6.3).
Main fieldwork	Interviewer briefing (if appropriate). Assess response rates as questionnaires completed/returned.
Processing/analysis of data	Data processing control, manual edit checks, coding of data. Data transcription from questionnaire to computer. Machine edit checks. Statistical analysis, production of tabulations.
Results	Results, hypothesis testing. Research report: conclusions in relation to research hypotheses.

being modified as data are collected and further insights are obtained. Moreover, the success of the survey has more to do with the researcher's ability to think through and integrate the various stages of the process than with following a rigid set of rules.

One trick which forces careful thought at the outset is to think through the research process 'back to front': start with a sketch of the final analysis and work backwards to the original research objectives. Write down the main dependent variables (those central to the research hypotheses) and independent variables (such as those defining important population subgroups) in the form of dummy tables that anticipate those likely to be contained in the final analysis. Ensure that the questionnaire includes questions that will collect the data required to fill the dummy tables. Next examine the sampling procedure. Is the sample likely to be representative of the target population with respect to the population characteristics in the dummy tables? Is the sample size sufficiently large such that the main cells in the tables will have enough respondents in them to allow statistically reliable comparisons to be made? For most statistical tests each cell will need to contain at least thirty respondents. How do the elements in the tested hypotheses relate back to other data sources and the wider context of the research? Although thinking through the research involves much guesswork, at least it reduces the chances of committing one of the cardinal errors of questionnaire surveys: measuring something quite different to what was originally intended.

Every survey that collects questionnaire data has its own particular design challenges; no two surveys are alike and few are executed without hitches. Presentation of a pat description of how to do a good survey is therefore not of great practical benefit, although it is useful to consider some of the most common pitfalls. Three contrasting case studies have been adopted to illustrate different aspects of the survey process. The characteristics of these surveys are initially described in Box 6.2. Each one exhibits design strengths and weaknesses,

BOX 6.2

Three examples of questionnaire surveys

Survey 1: The influence of local factors on household waste recycling: a Norfolk case study (Powell *et al.*, 1996)

Techniques: interviewer-administered questionnaire survey, observational study, systematic random sample with variable sampling fraction.

District-level household waste recycling rates are regularly published for England and Wales, but little information is available on the sorts of factors that can influence recycling at the local level. A survey was designed which involved house-to-house interviewing in the vicinity of three mini-recycling centres in Norwich to which recyclable materials (primarily paper, glass and aluminium) were brought by the public (Powell *et al.*, 1996). A total sample of 543 addresses was drawn from the electoral register using a standard sampling interval (a systematic random sample). The sample was selected in order to obtain a target number of interviews at each site. A strict call-back régime was used to maximise the chances of an

Box 6.2 continued

interview at the addresses selected. A total of 417 interviews were completed – a response rate of 77%. In addition to the questionnaire survey an observational study was carried out at a selection of urban and rural recycling centres in and around Norwich in order to measure the quantities of material received and to record the distances travelled and mode of transport of all site users on particular days. In all, 533 observations were made over a total of 14 site days.

Survey 2: Public knowledge and attitudes in relation to HIV and AIDS: a case study of North Lancashire (Lovett and Bayman, 1993)

Techniques: postal and interviewer-administered questionnaire survey, 'snowball' method of recruitment within pre-set quota.

Public education campaigns are known to have improved knowledge about HIV and AIDS but have rarely led to changes in behaviour. One possibility for making such campaigns more effective is to target educational material more specifically at different social, religious and ethnic groups. Based on this premise, the main survey objective was to assess geographical variations in knowledge, attitude and practice with respect to HIV and AIDS within the study area of North Lancashire. A postal survey technique was rejected in favour of an interviewer-administered questionnaire as a result of an identified social class bias in responses to a pilot survey. Ward-level census data were used to classify the 119 wards within the study area in terms of their socioeconomic and demographic characteristics. Wards were then selected for fieldwork on the basis of a cluster analysis based on 35 variables from the 1981 Census. A total of 353 interviews were achieved in 7 wards using a quota system for respondent age and a 'snowball' method (a *purposive* technique) of recruiting participants.

Survey 3: Conservation advice and investment on farms: a study in three English counties (Clark, 1989)

Techniques: semi-structured interviews, questionnaire survey, qualitative analysis, purposive sample.

The main focus of the research was how the conservation advice network available to farmers in Suffolk, Hereford and Worcester, and Derbyshire influenced conservation and countryside management. As conservation advice to farmers is not a single agency function (23 different agencies were identified in the study areas), it was necessary to establish the nature of the advisory network within each area. This was done by use of secondary data sources and through a series of 'expert' interviews within organisations involved with conservation advice. These interviews were not questionnaire-based. A questionnaire survey of farmers was then conducted in each county. As the proportion of farms receiving conservation advice was thought to be low, two different sampling frames were required to recruit a 'conservation advised' sample and a control sample. A total of 172 interviews were achieved: 85 in the 'advised' sample and 87 in a control sample selected to represent the characteristics of the wider farming community in each study area.

which will be discussed at relevant points in the main text, as will all of the terminology used.

The three examples have been chosen because they exhibit most of the techniques discussed later in the chapter: observational study, interviewer-administered questionnaire, postal questionnaire, semi-structured interviewing. Although Surveys 1 and 2 were not conducted as part of student dissertations, similar sample sizes have been achieved within the constraints that you are likely to face. These two examples used highly structured questionnaires in which the main analyses were quantitative, indeed they contribute to the examples of categorical data analysis discussed in Chapter 12. The third example, also based on a structured questionnaire, had a much smaller sample size and the data were analysed in a more qualitative way. This example was taken from part of a Ph.D. thesis and fieldwork was undertaken in three widely scattered locations. A shorter dissertation with fewer resources would probably have required a study 'closer to home' and probably of just a single case-study area.

Error in survey research

Different types of error

At the most fundamental level researchers employing quantitative survey techniques must address the twin issues of *reliability* (can the results be replicated?) and *validity* (does the survey measure what it was intended to?).

Good survey design is partly achieved by attempting to anticipate and minimise various types of error that may ruin the reliability or validity of a questionnaire survey. Errors can be introduced at any stage: from poor hypothesis formation, through to data transfer errors at the data processing stage. The different sorts of random and systematic error which underpin these questions of validity and reliability can be subdivided into errors associated with how respondents have been selected: i.e. *sampling error*, and those introduced by questionnaire design biases of one sort or another i.e. *non-sampling error*.

Sampling errors

The simplest sort of error to be aware of is sampling error, although controlling it usually has resource implications. Take the case of simple random sampling in which respondents are randomly selected from the population, each having an equal probability of selection. If the selected sample is too small, less than 50 for instance, there might be a relatively high probability that the sample population will be atypical of the target population with respect to key characteristics. In other words, as a function of the small sample size, there is a high probability of chance differences occurring between the sample and the population from which it has been derived. This would not matter if the population were extremely homogeneous. We know from statistical theory, and from common sense, that these differences – largely attributable to sampling error – will be greater for a smaller sample than for a larger one.

If the information on which the sampling procedure is based (the sampling frame) is imperfect it is likely to introduce a sample bias. For instance, using a trade directory to sample businesses in an area might bias the sample against smaller firms, if proportionately fewer are listed compared with larger firms. A blatant example of sample bias could occur in the case of the recycling survey (Box 6.2). If it had been decided to question only people using the recycling sites, rather than a systematic random sample of householders, the sample would not then have been a valid measure of recycling attitudes and behaviour in the wider local population.

Response errors

It is convenient to consider non-sampling errors in terms of distortions introduced in the process of interviewing (*response error*) and errors that arise through biases in who did and did not respond (*non-response error*).

The process whereby ideas are exchanged and recorded during interviewing (whether questionnaire-led or otherwise) is subject to error. The questions asked may not be understood in the way intended, the respondent may feel pressured into agreeing with the researcher's own ideas, or other sorts of biases may enter into the responses given. These sorts of errors, called response errors, can be reduced through careful questionnaire design.

At one time it was thought that interviewers' opinions, if held strongly enough, would influence respondents, but research has shown this to be less important than originally supposed. However, this can happen with respect to the way in which interviewers treat responses that are somewhat ambiguous. Having built up a picture of the respondent from earlier questions, there is a tendency (albeit unintentional) to fit unclear responses into ones consistent with opinions expressed earlier in the interview. Alternatively an interviewer builds up a picture of the relationship between key independent variables (age, social status, sex) and the main types of responses to questions through the experience of earlier interviews. When the respondent is vague or ambiguous in a response, the interviewer then lends a helping hand (at least a mental one) by assuming the same relationship holds.

This type of 'expectational error' is particularly a danger in social research when the sole interviewer is also the researcher. Information that conforms to the hypothesis under investigation might be considered more attractive than that which contradicts. This is a part of the great temptation to equate the quality of research with whether or not something conclusive was found. If this type of error does occur to a significant extent, apart from being unscientific, it might reduce the chances of discovering the unexpected in the data.

Apart from response errors largely attributable to the interviewer, the respondent also introduces errors. Most of these respondent and interviewer errors can be attributed to poor questionnaire design, especially inappropriate question wording, which is discussed below. Inappropriate choice of survey technique (see below), poor interviewing technique (see below and Chapter 7) or more likely a combination of these factors, may also contribute.

Non-response errors

Non-response errors are those associated with refusals or non-contacts (people 'not at home' or 'moved away'). If a sample of 400 addresses has been drawn to represent a cross-section of the population, but only 150 questionnaires are successfully completed, then to what extent can it be maintained that the 150 are representative? The refusal rate may be influenced by the interviewer, the subject matter, or a combination of factors. It could also be that a certain subset of the population were less likely to be at home when the interviewer called: all of these have the potential to introduce 'non-response' bias. For instance, in the case of the recycling survey, response rates were lowest in the area that had the highest proportion of households which were economically active (i.e. the main wage earner was not retired or unemployed).

At the risk of stating the obvious, non-response bias is a problem if the survey population contributing to the completed interviews is significantly different in key characteristics compared with the sub-population of non-responders. The problem of bias will be greater as the proportion of non-response increases.

Although it might be possible to adjust the results through weighting to remove such non-response errors, the big problem is deciding what these errors might be in the first place. In many cases, key groups of people who might be less likely to give an interview can be identified in advance (perhaps as a result of previous research) or may be revealed as a result of a pilot survey. The latter was the case with the HIV/AIDS attitude survey: the pilot stage identified a social class bias in response rates.

Rather than adjusting for non-response bias, it is far better to ensure that it does not get out of hand in the first place. This is largely a function of choosing the right survey technique for the subject matter – a subject that will be considered after the challenges of questionnaire design and sampling have been discussed.

Questionnaire design

In terms of the stages of the survey process, the researcher will usually have a firm idea of what sort of survey technique is to be adopted – a decision which has important implications for all aspects of questionnaire design. Before assessing the 'pros and cons' of various techniques, it is necessary to consider questionnaire content, wording and format.

What to include?

The content of the questionnaire needs to be firmly rooted in the research question or hypothesis under investigation. If these research issues are confused or poorly formulated, the researcher cannot hope to make them meaningful to the respondent. There is no salvation in thoughtful questionnaire design if the questions do not relate to the central ideas or hypotheses under investigation.

To leave out data crucial to the research hypothesis would be negligent; equally, if a lot of extraneous material has crept in, the quality of the research will suffer. All questions included in the questionnaire must be really needed: none of them should be there because it was felt that 'they *might* come in useful'. The original research hypothesis and subsequent list of different variables should be referred to, as well as the preliminary ideas drawn up for the main analyses. Sometimes it is not easy for the researcher to identify redundancy or omission in his or her own questionnaire, which makes it all the more important that a pilot survey is conducted.

What to include is also a function of what questionnaire length can be sustained for a given research topic. If the length of the questionnaire is excessive and the topic less engaging, data quality will suffer. Generally, if an interviewer-administered questionnaire exceeds half an hour in length, the later questions will suffer from a 'fatigue' bias. Beyond a certain limit there is a chance that respondents will refuse to answer any more questions. Long questionnaires also take far longer to analyse, and the main objectives of the research may easily become obscured by superfluous questions.

The questionnaire content must also set out to measure what is practicable and relevant to respondents and to give them the maximum opportunity to respond. It is essential that the content of the questionnaire must be readily understood and be about something that respondents are likely to have opinions on or that will be meaningful to them. No questions should rely on recall of events or behaviour too distant in the past, nor should a formal questionnaire be used to tackle a topic that is likely to be so emotionally charged that valid responses cannot be expected.

Questionnaire format, sequence and wording

Following on from the general comments about content, there are a number of points to do with question wording, language and questionnaire 'flow' which can help to reduce response biases.

Successful questionnaire design has much to do with the ability of the researcher to empathise with the prospective respondents. The language and tone of the questions must not put the respondent out of his or her depth. Conversely, oversimplified 'Play School' type language will also put respondents off. Striking the balance is to some extent a matter of judgement, but it also requires the questionnaire designer to have had some prior contact with the survey target group. Although the researcher will be familiar with the language and ideas of the research topic, the same will probably not be true of the respondent, so avoid unnecessary jargon. The questions might also be making assumptions about the meanings of terms that might differ from the respondent's frame of reference. For instance, when researching attitudes towards conservation advice given to farmers, the terms 'conservation' or 'advice' might mean something different 'on the ground' compared with the researcher's use of the words. Establishing whether there is any mismatch between respondent's and researcher's definitions is yet another good reason for doing a pilot survey.

Just as conversations that jump erratically from topic to topic tend to be irritating and tiring, the same is true for questionnaires. If a questionnaire gives respondents the feeling that it has been anarchically designed, many will refuse to complete the interview. It helps if related subjects are put together in a block of questions. Continuity statements should be used when moving from one subject to another: e.g. *'I would now like to ask you some questions about personal transport . . .'*

There is also a need for a standardised introductory statement to attempt to put the respondent at ease, such as: *'Hello, I'm . . . (*give name, show ID Card*) . . . , from . . . (*Institution, show letter of explanation*) . . . I'm carrying out a survey about . . . and I would be very grateful if you would answer a few questions. Any information which you provide will be kept strictly confidential and only used for statistical purposes'*. The introductory questions which follow should be easy, so as to act as a 'warming up' exercise for the respondent. In the HIV/AIDS survey (Box 6.2), for instance, the questionnaire begins with questions to do with where information is obtained about HIV/AIDS (leaflets, TV, newspapers etc.). Any riskier questions should ideally be kept towards the end of the questionnaire, so as to minimise information lost if the respondent refuses to continue. A few easy questions at the end of an interview are a good idea – to provide some tension release for the respondent. An open-format question (discussed later) at the end is sometimes used for this purpose.

Questionnaire design is a creative process and there is no single recipe for a good set of questions. It is easier to pin down the features of bad question phrasing. The following are some of the most frequent mistakes.

Double-barrelled questions

Such questions are a very common design error: Question: *'Have you travelled on business by train or by air within the last 3 months?'* This is two questions in one: they should be asked separately. Never use such a question, as you cannot be sure which part of it the respondent is replying to.

Negatives and double negatives

Be careful when wording all questions containing negatives and in particular avoid use of double negatives. The responses from the following question cannot be reliably interpreted: Question: *'Would you rather not use non-nuclear sources of electricity?'*

Recurrent or habitual behaviour

Questions relating to such behaviour should be as precisely worded as possible. Words such as 'usually' or 'recently' present difficulties in interpretation and should be replaced by a particular time frame such as 'in the last month'. Remember that memory can be unreliable, so do not expect accurate recollection of distant events. For example, in the recycling survey questionnaire, respondents were not asked 'what quantities do you usually recycle?', or 'how much do you recycle within a typical month?', but about the most recent visit they had made to the recycling centre.

Leading or loaded questions

A leading question gives a hint of the researcher's own point of view and will therefore induce a response bias. A loaded question is one that is emotionally charged so that a respondent is not so much reacting to the issue posed by the question as to the loaded phrase itself. Question: *'Don't you think that burning less coal, oil and natural gas is crucial to the survival of the planet and for the health of future generations?'* This example is both leading ('don't you think?' leads to a positive answer) and loaded (the wording that follows is emotive).

Failure to state the alternatives

Leading questions are often inadvertently created by a failure to state the alternatives: Question: *'Do you think this upland area should be maintained as mixed farming?'* Compare this with *'Which of the following land uses do you consider to be the most appropriate for this upland area in the future:* (read out all categories) *be selectively reafforested, be maintained as mixed farming, be developed for leisure activities or some other option?'* (if 'other option', probe *'and what would this option be?').* The second question is more likely to get a valid response, particularly if the interviewer instructions are followed and information on other options is collected.

Inconsistency and suggestion

Nothing must be left to chance with instructions on a questionnaire, even if the instructions are written to yourself as the sole interviewer. You need to ensure that the way in which the questionnaire is administered is as standardised as possible. If the procedures used in asking questions are not formally written down, it is possible that the approach to interviewing may subtly change through time. Follow-up questions are particularly prone to this type of problem as they are often tagged onto main questions as a note to 'probe for main reasons'. In fact such questions require the same care in formulation as the parent question. Consider the question: *'How often do you use the car share scheme?'* followed by *'Why don't you use it more often?'* Here the follow-up has been poorly devised as it suggests that the respondent should use the scheme more often, depending on where the emphasis is put when reading out: a phenomenon called 'intonation bias'. It is better in this case to have two separate questions, in which the follow-up question is asked of a specified subset of respondents, following clearly stated instructions:

> ***Interviewer instruction:*** If respondent uses a car share *'less than once a month'*, ask Q12b.
> *Q12b: Please give the main reasons for not using the car share scheme more often.*

This is a more neutral type of follow-up question and it is less likely to be accidentally left or misinterpreted.

Potentially embarrassing questions

There are many reasons for questions being embarrassing; a question might deal with very private matters, or ask about socially disapproved attitudes or behaviour. People are often loath to admit their own anti-social tendencies. One tactic is to word questions in such a way as to suggest that the taboo subject or socially disapproved issue is accepted. Another would be not to ask questions in a personalised form. None of the questions on the AIDS/HIV questionnaire (Box 6.2) involved asking respondents about their private lives apart from a few basic questions at the end used for classification purposes. If any of the questions about AIDS/HIV had been of a more personal nature, the cooperation of respondents would have been more difficult to secure. The creation of a permissive atmosphere in the questionnaire, together with an absolute guarantee of anonymity, can help to overcome such obstacles.

Classification questions

Classification questions such as '*How old are you?*' or '*How much do you earn?*' can often cause problems. Rather than ask a direct question, creating categories for the respondent to choose from can be helpful. Make sure that categories, such as for age ranges, do not overlap: they must be mutually exclusive. Although many people tend to be less coy about their earnings than in the past, it is sensible to use a 'show card' for sensitive information, such as income. The respondent is shown a card (never hand over the questionnaire instead) on which income categories are listed and labelled by letters, e.g. A = £0–4999, B = £5000–9999 and so on. Remember that in creating categories to classify income there are implications for data analysis and which statistical tests are appropriate. You now no longer have an interval variable but a categorical variable (it is not known precisely within £5000–9999 the respondent's income lies). However, this may be preferable to the loss of information if respondents refuse to answer the question.

Different types of questions

A thoughtfully designed questionnaire is one that deploys a number of different question types in order to provide different 'shots' at finding something out which is relevant to the research. It would, after all, be a rather dismal and trivial research question if it could be answered simply by a 'yes' or 'no'. Choosing the array of question types that perform best for a given research topic is part and parcel of all of the other decisions that shape the research: in particular whether or not it will be interviewer-administered or a self-completion questionnaire.

The relative strengths and weaknesses of different questionnaire options at conveying different question types will be discussed later. First it is necessary to consider the main types of questions and the way in which they are handled at fieldwork, editing and analysis stages.

The most basic distinction that survey practitioners make is between open and closed questions. A closed question is one in which the questionnaire designer provides most, if not all, of the alternative answers: the question is said to be 'precoded'. For example:

Q2 Are you aware of any recycling sites in your area?

Yes	1	go to Q3
No	2	go to Q5a

This closed question has two precodes associated with it: '1' for 'Yes' and '2' for 'No'. Next to the codes are 'filter' instructions to the interviewer – Questions 3 and 5a are 'filtered' on the responses recorded at Question 2. Such filter instructions must be clearly stated on the questionnaire and must be checked for errors.

The main problem with closed questions is that they may force respondents into adopting false positions, particularly when asking questions that present an array of precoded responses. However, for the researcher, closed questions are far easier to analyse, as the answers need no further categorisation. Closed questions also encompass a range of more innovative question types, such as Likert scales and ranking questions (see the Attitude measurement section below).

In the words of Oppenheim (1992: 113), 'free-response questions are often easy to ask, difficult to answer, and still more difficult to analyze'. Open questions do not force respondents into giving particular answers. They are therefore more work for the interviewer as no 'precodes' are set out and space is provided on the questionnaire into which the answer is to be written. The comments are later read through and a coding sheet is drawn up which classifies the most frequent responses and creates codes for them. The output from open questions can be a considerable extra burden on the hard-pressed researcher and it is quite common for such questions to be asked but not to be fully analysed. Successful open questions are generally those which are not open-ended invitations for respondents to 'tell everything', but which direct the respondent in a more focused way. For instance, the recycling survey did not ask respondents to discuss the benefits of recycling in general; instead they were asked: '*What do you believe to be the main advantages to you of having a recycling centre in your neighbourhood?*' The code frame generated from this question is used later to illustrate the coding process.

Which type of question format is best: open or closed? The main advantage of open questions is that they allow for spontaneous responses unencumbered by the sorts of answers which the researcher regards as valid. They also add colour to the survey write-up: verbatim comments of what was actually said can bring to life otherwise turgid pages of cross-tabulations. A more qualitative survey, such as Survey 3 in Box 6.2, is bound to use more open questions. However, such questions are more demanding of respondents, particularly if they are not very articulate, and take more time to complete than closed questions.

Closed questions are therefore far easier to ask, answer and process. However, they can suffer from the bias of constraining responses to certain categories or 'putting words into people's mouths' by introducing them to responses that might not have occurred to them. Unless the questionnaire has been well piloted, the precodes may not be the most appropriate choice. Most questionnaires use a mixture of question types, sometimes deploying both to tackle the same issue.

Attitude measurement

Attitudes and opinions are generally considered to be the most difficult category of social survey data to collect. Many different types of scale have been devised to measure attitude, but there remains the big uncertainty: no matter how error-free the procedure for attitude measurement, there still remains the psychological uncertainty over what such scales actually represent. Designing question formats to measure attitudes and opinions is an industry in itself and much ingenuity has been applied to their design.

Attitude data are of interest to the researcher because of their potential to predict how people might behave in the future: how they might vote, make transport decisions, use their leisure time, manage the countryside and so on. Attitudes are by their nature very imprecise and prone to fluctuate, to be ill-formed, to endure for only a limited time or perhaps to last for a whole lifetime. They can be referred to as some measure of an individual's underlying state of mind on a particular aspect of the world. Their acquisition helps us to structure our responses to the external stimuli of the outside world through drawing on our own past experience or those of others. Attitudes are abstractions: they have a cognitive component, an emotional component and an 'action tendency component' (Oppenheim, 1992: 175). The various criteria that have been used to define 'attitude' are discussed by Allport (1954).

The link between attitudes and behaviour is often unclear and in some areas totally unreliable. Furthermore, external influences may intervene to modify any attitudes measured by a questionnaire and a proportion of respondents may not behave as expected. An attitude expressed at interview may not be a reliable guide to behaviour for other reasons – prestige bias (for instance, in Survey 1, people claiming during an interview to recycle more waste than they actually do), interviewer bias and other response errors. Attitude data derived from a questionnaire survey must therefore be used with caution and must not be pushed too far. A selection of the simpler and more useful means of attitude measurement is reviewed here.

The first (Figure 6.1) is taken from the postal pilot survey of attitudes towards HIV/AIDS. It uses a Likert-style format in which statements are provided and respondents are asked to indicate the extent to which they 'agree' or 'disagree'

Question 11: *Please indicate how you feel about each of the following people on the scales listed below.*

PERSON WITH CANCER
(PERSON WITH AIDS: same exercise repeated on a separate scale)

	Strongly agree		Neither agree nor disagree		Strongly disagree
Should be isolated from society	☐	☐	☐	☐	☐
Should change lifestyle	☐	☐	☐	☐	☐
Should be supported by others	☐	☐	☐	☐	☐
Should receive publicly funded care	☐	☐	☐	☐	☐
Family should be told	☐	☐	☐	☐	☐

Figure 6.1 Example of the 'attitude battery' approach to attitude measurement

using a scale with five positions. These positions have not been coded on the questionnaire, although it is quite common to see such a scale with positions numbered 1 to 5 if the questionnaire is interviewer-administered (the code values do not influence the respondent).

Such an assembly is known as an 'attitude battery'. The main advantage of such a device is that respondents can be led fairly quickly through a range of statements that explore different aspects of the topic without over-burdening them. The use of an extended scale rather than just 'agree', 'disagree' and 'don't know' allows for some measure of the strength of opinion. The attitude statements need to be carefully selected from a larger pool of possible statements. These statements should not be dominated by extreme views, nor should they be purely anodyne. The creation of an attitude battery in which a single topic is explored through different statements allows the individual measures to be combined into an aggregate attitude measure, in this example an overall score of respondent empathy towards cancer or AIDS/HIV victims. To achieve this, care must be taken with the coding of scores such that the 'more empathetic' end of each scale is given the same value. For instance, if a respondent disagrees strongly with 'should be isolated from society' and scores '5' and agrees strongly with 'should receive publicly funded care', then they should also score '5'.

Attitude scales composed around the 'agree–disagree' format are a very common device that should not be overused in any one questionnaire. It is therefore often important to use attitude choices as an alternative. For example, in the farm conservation questionnaire (Survey 3), farmers' attitudes towards conservation and farming were gauged using the following format (abridged from the original):

Of the statements below, which do you feel most typifies your approach towards conservation and farming:

- I have undertaken some conservation work when opportunities have permitted and have some understanding of the priorities or needs of my farm.
- I regard conservation as separate from my farming operation and have done very little as a result.
- My farm planning takes into account a wide range of objectives amongst which is conservation.
- I have a good understanding of conservation but have been unable to undertake such work as I do not have the resources or because I lack necessary advice.

Semantic differential scales are another useful device: they consist of bi-polar scales, usually of seven intervals, defined at each end by opposing pairs of adjectives or descriptors. The labels used must be truly opposed (light/dark, clean/dirty). Figure 6.2 shows the semantic differential scales that might be used to assess people's attitudes to a proposed wind farm development. The scales would be administered either on a self-completion questionnaire with the respondent selecting a position on each scale, or, if interviewer-administered, as pairs of descriptors printed on a card with the respondent being asked to choose a position on a scale of 1 to 7.

Question: *On a scale of 1 to 7, please tell me how you feel about the proposed wind farm for the Widgely Green site. Is it . . .*

(use SHOW CARD and read out items) . . .

	1	2	3	4	5	6	7	
beautiful		√						ugly
noisy					√			quiet
advanced technology			√					backward technology
unsafe							√	safe
reliable						√		unreliable

Figure 6.2 Example of semantic differential scales for attitude measurement

Notice that the 'favourable ends' of the bi-polar scales are not all on one side: it is always a good idea to mix the direction of the scales to reduce the patterning effect to which some respondents are susceptible. The aggregate score for the respondent's favourability towards the wind farm could therefore be created by assigning the score of 7 to each favourable scale end and summing the respondent's score: 6 + 5 + 5 + 7 + 2 = 25 out of a possible maximum of 35.

Problems with semantic differential scales are often caused by the chosen descriptors not being true opposites in the minds of some respondents. Also the use of overly terse descriptors can easily lead to ambiguous interpretation. What does 'reliable' mean in relation to the wind farm example? Will the respondent be more influenced by perceptions of how reliable a wind turbine is mechanically, or will they be wondering how reliably the wind blows across the Widgely Green site?

There are many other attitude measurement devices besides the relatively simple ones mentioned here. For further ideas, help should be sought from one of the recommended textbooks, each of which devotes separate chapters to this subject.

Sampling

Surveys are usually conducted by sampling from a population rather than contacting all of its members. Needless to say, there are few instances in which a census of a particular population would be feasible, unless it was small in size. Although sampling saves time and money, there is always a trade-off between decreasing sample size to reduce effort (and cost) and increasing sampling error. However, even with a relatively small sample, if great care is taken over the sampling methodology, it is still possible to draw some conclusions about a target population. Furthermore, if a probability sample has been drawn, it is possible to calculate statistical confidence intervals for the survey estimates.

Before dealing with the nitty-gritty of sampling methods, two basic questions must first be addressed and answered clearly by the researcher:

1. What are the appropriate units of study?
2. What is the target population?

The choice of appropriate units of study is an important part of research design from which other decisions will flow, not least the sampling strategy. For many research topics the individual will be the unit of study. This is likely to be true of any study in which the individual's decision-making is of prime interest: health and lifestyle issues, transport use, personal attitudes to a local development proposal and so on. Projects which look at different sorts of economic or institutional entities, such as 'small businesses', 'agencies which give conservation advice to farmers' or 'secondary schools in Suffolk' may well have 'the organisation' as the unit of study, with different people within each contributing data to a single case. Surveys may also have 'households' as the legitimate unit of study (we might want to know how different family members influence decision-making), though there are practical difficulties in recruiting whole households for interview and such research is more likely to take the form of qualitative 'group' discussions than a quantitative survey.

The unit of study is not necessarily the same as the sampling units. For instance, a survey of hazardous waste producers may have 'waste producing firms' as the sampling unit, but each questionnaire may generate information on a variable number of hazardous wastes of different types (asbestos, spent acid solutions, contaminated solvents and so on). The main unit of analysis may well be the waste stream types (and where they are dumped), rather than the companies themselves. It is not possible to use 'different types of hazardous waste' as the sampling unit as no dependable listing exists from which a sample could be drawn (i.e. there is no sampling frame).

With the issue of the unit of study sorted out, it is now time to consider the nature of the target population of the chosen units. The definition may include a number of different elements:

- *A geographical boundary*. Each of the three case studies in Box 6.2 has a different type of geographical unit: a local catchment area of a recycling centre in Survey 1, a formal administrative boundary in Survey 2, and something less geographically specific in Survey 3: farmers who have received conservation advice in three different case-study counties.
- *A temporal boundary*. This will be determined by the time frame in which the research is to be conducted. However, there may be important time-related factors to consider within the likely fieldwork period that will influence the nature of the target population. If an attitude survey is being conducted on something which varies greatly by season, such as tourism in a National Park, it will be important to consider factors such as whether or not the schools are on holiday, the proximity of proposed fieldwork dates to bank holidays, midweek versus weekend and so on.
- *A boundary defined by population characteristics*. The survey may have as its target all members of the public within the spatial and temporal boundaries of the study. Quite often though, there may be good reasons for including some members of the population and excluding others. In Survey 2, for example, people aged 15–24 years old were required to be represented in the sample because of the significance of this age group to HIV/AIDS health educational campaigns. Conversely, it was decided to screen out retired

people from the survey. Such inclusions or exclusions need very careful consideration as they have implications for sampling procedure. How can such an age-based target population be 'operationalised'? No list exists for residents of a given area by age group – how will contact be made with such a target population? This difficult problem will be considered after we have considered more straightforward sampling problems.

Probability sampling

The ideal source of information from which to sample any population is an up-to-date list of all the members of that population for the study area. Such a list, called a sampling frame, gives each member of the target population an equal and non-zero probability of being selected (a 'probability sample': see sampling terminology in Box 6.3). If the list is in no particular order, it is relatively easy to obtain a random sample of the required size from it. Be careful, though: there may be important systematic differences between one part of the list and another that may not be apparent at first glance. If in doubt it is always better to randomise the selection. When a sampling frame is arranged alphabetically, or by street, a standard sampling interval can be calculated so that the required sample is obtained but scattered systematically from the whole list. This method of drawing a sample is called systematic sampling (Box 6.3).

Assuming that the sample drawn from the list is reasonably large, there are good grounds for expecting the survey results to be representative of the target

BOX 6.3

Sampling terminology

A) Probability sample

Each member of the population has a known (and non-zero) chance of being selected into the sample.

Probability samples require a *sampling frame*. This is a list containing all members of the target population. The frame should be complete, up-to-date and without duplication of items. A Yellow Pages directory would be a good example of a sampling frame containing duplication: many businesses appear under more than one heading.

Simple random sample

All the population members are listed and numbered and the sample is drawn in one stage.

Systematic sample

A method of drawing a probability sample from a sampling frame. A sampling interval is calculated by dividing the total number of items on the list by the

Box 6.3 continued

sample size required. The start point should be randomly selected by numbering the beginning section of the list and selecting a number using the random number generator on a calculator. The sampling interval is then added to the number of the randomly selected member to identify sample number two. The process is repeated until the required sample has been drawn.

Stratification

The population is divided into homogeneous groups (strata) whose relative size is known. These strata must be mutually exclusive. A random sample can be taken in each stratum: either proportionately or disproportionately.

Proportionate sample

A uniform sampling fraction is applied to all the strata – in other words the proportion of the population sampled is the same for all strata.

Disproportionate sample

A variable sampling fraction is applied to strata of significantly different size, such that the smaller are over-sampled relative to the larger. This is a more efficient use of resources if statistical comparisons are to be made between the strata. If estimates are required for the population as a whole, the strata have to be weighted accordingly.

Multi-stage sample

The sample is drawn in more than one stage, usually after stratification by region and type of district. Three-stage drawing is quite common – first, constituencies; second, wards/polling districts; third, addresses from the electoral register.

Probability proportionate to size

Used in multi-stage drawing and associated with the use of a systematic interval. A range of numbers, equivalent to its population, is attached to each item on the list (e.g. each constituency, each polling district) before the draw is made. A number between one and the total population, divided by the number of sampling points, is drawn at random (or generated by computer). This indicates the starting point: the list of items is then systematically sampled, the probability of selection being proportionate to the size of each item.

B) Purposive/non-random sample

Selection of sample members is dependent on human judgement.

Quota sample

A method of stratified sampling in which selection of sample members within strata is non-random.

population as a whole. This ideal situation where the sample is drawn in a single step is called simple random sampling. It may result in a highly scattered selection of points to be sampled if the sampling frame covers a large geographical area. This may be one good reason for adopting a more sophisticated approach than simple random sampling.

There are many alternative ways of drawing a probability sample. The sample can be drawn so as to concentrate the respondents in discrete geographical areas: this is called a clustered sample. This technique is a useful means of cutting the costs of fieldwork if an interviewer-administered questionnaire is to be used. The population can be divided up by some important characteristic, such as type of industry or a key socioeconomic variable, such as age or income. For instance, a sample of 150 firms may be stratified by manufacturing and non-manufacturing sector, with the number drawn from each in proportion to the number of firms in each sector. If the research hypothesis requires comparisons to be made between firms in manufacturing and non-manufacturing sectors and one sector is ten times larger than the other, then the individual strata will have to be sampled disproportionately in order to achieve a sufficient sample of both in an economical way. These sorts of sampling decisions are governed by the priorities set by the research in terms of the key comparisons that the study needs to make.

Survey 1 can be used to illustrate the full mix of factors that can impinge on sampling strategy. As a reminder: the object of the survey was to interview householders in the vicinity of recycling sites and to analyse factors that might explain local variations in recycling activity. The main sampling questions were which recycling sites to choose and how many addresses to sample in the vicinity of each. The initial selection of the three recycling sites was a form of purposive (i.e. based mainly on judgement) cluster sampling. The sites were selected from a larger number of district council-run recycling sites, but there were very limited data by which to characterise the sites to form the basis of the selection.

It was decided to take the three largest throughput sites in the area, which happened also to be in three very different wards: two within the city (pre-war housing) and one on the edge of a sprawling post-war estate. As the site outside the city had a far larger potential catchment population than the two inner sites, it was decided to set a target of 200 interviews at this site and 100 each at the two smaller sites. The areas from which the samples were drawn corresponded to a notional catchment radius of approximately 1 km around the two inner sites and 2 km for the outer site. Major features, such as the outer ring-road, parks or playing fields, were used to define the exact boundaries of the sampling areas – these were marked up on 1:2500 scale maps.

A systematic random sample with a variable sampling fraction (proportionately more addresses were required at the two inner sites compared with the outer) was then drawn from the electoral register for streets within the selected areas. A contingency factor of about 30% was built into the sampling to allow for anticipated refusals and non-contacts. Even with a call-back régime of at least four visits before discarding a contact, one study area had a higher than anticipated refusal rate and further addresses had to be drawn.

Non-probability sampling

It has already been noted that sampling for Survey 2 was not straightforward. A sample was required which was stratified by age but the electoral register does not allow us to sample by age: it is only a list of names and addresses of those eligible to vote (and who have remembered to register). In such cases probability sampling cannot be pursued: instead a non-probability (or purposive) method has to be devised. It is worth observing that owing to changes in UK legislation in 2002, electoral registers now take a rather different form, with householders having the option to opt out of the 'edited' version that is available for sale to researchers and marketers. Although the registers may still be inspected at local authority premises, their assessment for research project use must now include new coverage and cost constraints.

The non-probability method most frequently used by researchers is quota sampling. This method involves sending interviewers out to find respondents of particular types so that the profile of the sample matches that of the target population within the area of study.

In the example of Survey 2 the sampling procedure involved two stages: the selection of the study areas from a possible target population from the 119 wards of the study area and the setting of quotas within each selected ward. The clustering routine that was adopted is no different to one that can be applied to any other sampling procedure (probability or non-probability). A total of 35 variables from the 1981 Census were obtained for each ward. A hierarchical clustering routine (available on the statistical package SPSS) was used to categorise the 119 wards into 7 main subgroups or clusters. Survey resources were sufficient to sample one ward per cluster.

Table 6.1 shows the quota samples set and achieved for 2 of the 7 wards that were sampled. Filling such a quota is more difficult than it sounds: some types of people are easier to locate than others. Interviewers had to continue knocking on doors within the study areas until the entire quota for the particular ward had been filled. Control of quota within each cell is difficult to achieve in practice, as all interviewers have to be in constant contact with a fieldwork controller, or with one another, so that over-shoot does not occur. Conversely, some quotas may remain unfilled, for example, the Pilling Ward, female, 45–64

Table 6.1 Quota samples: set and achieved for Survey 2, example of 2 of the 7 wards in which fieldwork was conducted

Ward name		*Age Categories*		
		15–24	*25–44*	*45–64*
Bispham	Male	6 (6)	8 (8)	11 (10)
	Female	5 (5)	8 (8)	12 (12)
Pilling	Male	5 (6)	10 (11)	9 (8)
	Female	6 (7)	11 (10)	9 (3)
Overall for 7 wards		84 (99)	132 (137)	134 (90)

Figures in brackets are the number of interviews achieved.

LIVERPOOL JOHN MOORES UNIVERSITY
LEARNING SERVICES

category. Interviewers are largely left to their own devices in locating the people that fit the quota, a process prone to interviewer bias. Stories abound from the market research industry of interviewers who claim to be able to match respondents to quota largely on the basis of their shoes (style and state of cleanliness) and by their curtains (Laura Ashley versus the rest)! Quota samples are therefore not such a soft option; they can be open to large interviewer biases, and they suffer from the overriding disadvantage that the method does not allow an estimation to be made of the size of the sampling error.

Choice of survey technique

The appropriate choice of survey technique is of great importance to the overall success of the research. Each means of gathering survey data has advantages and disadvantages, particularly in relation to the different types of error that have been discussed. The main methods compared here are observational studies, structured questionnaire surveys administered by interviewer, postal or 'self-completion' surveys and telephone interviews. Some observations will also be offered on the use of internet-based questionnaires, which may take a variety of hybrid forms, but to which important cautionary considerations apply. There are also circumstances in which a highly structured approach is not appropriate and a more flexible 'unstructured' interview is more useful. The use of such qualitative interviews, the subject of Chapter 7, will also be mentioned briefly here in the context of the other options discussed.

Observational studies

Some sorts of research problem may only require data that can be collected by observation and do not require interviews. One such example would be the observational survey at recycling sites that was a component of Survey 1, which collected data on what was brought to the recycling sites on specific days. Such data only require a simple grid in which observations are recorded.

The limitation of observational technique is that nothing is learnt about the underlying factors in the behaviour observed nor the attitudes, motives and explanations. However, they are simple to carry out and not very resource intensive.

Unstructured interviews

Sometimes an informal approach is more likely to produce the data of the type required: how did the members of the county council planning committee reach their decision over the new brickworks? What sorts of factors were taken into account? A series of unstructured interviews with elected councillors, officers, and interested parties for and against the development will provide more insight than a formal questionnaire. This was the option pursued in Survey 3 in order to find out about the conservation advice network operated by a diverse group of agencies.

The term 'unstructured' is a little misleading. Such interviews require a good deal of preparation and the key issues that are to be addressed require careful thought and a thorough knowledge of the subject matter from desktop research. The interviewer will need to ensure that the main focus of the interview does not shift away from the research topic: therefore polite interventions are required. This option is not amenable to formal statistical analysis, but the analysis of interview transcriptions can often give greater insights into certain types of research topics than use of quantitative techniques. A detailed discussion of such interviews is available in Chapter 7.

Unstructured interviews can also help in the development of a structured questionnaire. The attitude statements used in a quantitative questionnaire can be generated from 'brainstorming' interviews with members of the target population. So too can the precodes for any closed-format questions. The research hypothesis can be sharpened up by a 'toe in the water' series of unstructured interviews. However, use of such interviews as preliminary work to a quantitative survey should not take the place of a proper pilot survey.

What sort of questionnaire survey?

Much has been written about the relative merits of interviewer-administered, self-completion (postal or otherwise) or telephone interviews. Much newer is internet-based questionnaire delivery, although this may be of considerably less utility for the student research project than for commercial market research. E-mail offers a low-cost alternative to the delivery of questionnaires by conventional post, and most of the same design considerations apply to the actual sampling and questionnaire design. A further option is for the researcher to place the questionnaire on their own website and invite respondents to complete it online. This has the great advantage that responses are automatically coded in computer-readable form as the questionnaire is completed, but requires website programming skills and access privileges to the web server that are not usually available to students through their institution's websites. However, the enormous growth of unwanted junk e-mail (spam) and justifiable suspicion of viruses in unsolicited e-mail will significantly reduce response rates. Most forms of electronic delivery offer the researcher very limited control over the layout of the questionnaire as it appears to the respondent. Any internet-based option is also handicapped by the potential bias introduced by its availability only to those with internet access, which may make it totally inappropriate for many research topics. Speculative questionnaire delivery by e-mail is thus strongly discouraged, and you should seek advice within your own institution if you believe that you may still have good reason to pursue this option. All the more general principles discussed in this chapter will still apply, and the considerations applying to postal questionnaire delivery offer the closest analogy.

The match between research topic, resources (including time) and the use of a particular survey technique is one that should be made carefully. The choice has implications for all of the factors in survey design which have been discussed so far: the relative importance of different error sources, questionnaire content, question type, length and layout of questionnaire and sampling strategy. De Vaus (1991) provides a useful summary of the advantages and disadvantages of the different methods. This is shown in adapted form in Box 6.4.

The main choice facing a survey conducted within the constraints of a dissertation is that between postal or face-to-face interviews. Self-completion questionnaires are relatively cheap in comparison to interviewer-administered. Unless some sort of incentive for return is provided, or the topic is addressed to a highly motivated specialist target population, then response rates tend to be low (30–40% is quite typical). The self-completion questionnaire (including the internet-based options) requires extra care in its design and layout. All instructions must be clear and unambiguous as no interviewer will be there to guide the respondent through. There are greater constraints on the complexity and length of the questionnaire: if it is not short, simple and easy to complete the respondent is likely to bin it. A stamped addressed envelope or a freepost number can be used to facilitate return. A stamped addressed envelope would be wasteful if a low response rate is likely; with a freepost number the researcher only

BOX 6.4

Advantages and disadvantages of interviewer-administered, telephone and postal surveys

	Interviewer	*Telephone*	*Postal*
Response rates			
General samples	good	good	variable
Specialised samples	good	good	good
Representative samples			
Avoidance of non-response bias	good	good	poor
Control over who completes the questionnaire	good	satisfactory	poor
Gaining access to selected person	satisfactory	good	good
Locating the selected person	satisfactory	good	good
Effects on questionnaire design			
Ability to handle:			
Long questionnaires	good	satisfactory	satisfactory
Complex questions	good	poor	satisfactory
Boring questions	good	satisfactory	poor
Filter questions	good	good	poor
Question sequence control	good	good	poor
Open-format questions	good	good	poor
Quality of answers			
Ability to avoid distortion due to:			
Interviewer biases	poor	satisfactory	good
Influence of others on respondent	satisfactory	good	poor
Implementation			
Speed	poor	good	good
Cost	poor	satisfactory	good

Adapted from de Vaus (1991), in turn adapted from Dillman (1978)

pays for the number of questionnaires returned. Response rates can be improved by follow-up letters and phone calls.

In summary, the strengths of the postal questionnaire are the lower costs and the anonymity of sending and returning the questionnaire by post. Respondents also complete the interview at their own pace – they have more time to consider their responses and they are not influenced by the presence, personality and intonation of an interviewer (no interviewer bias). If the target population is scattered across a wide area, postal surveys permit a wider geographical coverage at a lower cost.

These advantages must be weighed carefully against the disadvantages compared with face-to-face interviewing. The self-completion format only permits simple and straightforward questions and complex filter instructions are likely to fail. Generally the longer the questionnaire and the more open-format questions it has, the greater the disadvantages of the self-completion option. Subtle question sequencing cannot be certain to work, as there is no control over the order in which the different sections of the questionnaire are read by the respondent. It is difficult to weed out insincere or patterned responses that would otherwise be detected by an interviewer. Furthermore, there is no control over who answers the questionnaire within the unit sampled. Potentially the most serious drawback is that it is difficult to check non-response biases. Since respondents are required to read the questionnaire and to write down their responses, this immediately biases the questionnaire against people who, for whatever reason, are poor at reading and writing (they may have poor eyesight, English may not be their first language and so on).

In many ways telephone interviewing may represent a good compromise between postal and interviewer-administered questionnaires. They are not that much more expensive than postal questionnaires, especially if the contact numbers are local rather than long-distance. The response rates tend to be relatively high. Call-backs are easy to make and the risks of the interviewer travelling to remote or unsavoury neighbourhoods are avoided. Of the three options, telephone interviews are the least susceptible to the responses being influenced by the presence of another person. As with face-to-face interviews, the question sequence can be controlled and filtered questions can be reliably administered.

There is no scope for using visual aids or written prompts in asking questions over the phone. Telephone questionnaires must therefore not contain attitude scales with a large number of response categories: respondents cannot be expected to remember what all of the points on the scale represent. If attitude scales are to be used, it is always a good idea to ask the respondent to write down the possible answers (e.g. 'agree', 'neither agree nor disagree', 'disagree') on a piece of paper. Further disadvantages are that, because of the increasing loss of privacy from speculative phone calls, particularly from sales teams, some people are ex-directory. Not everyone has a phone, although differences in ownership of a fixed telephone line are now small, with a range between 84% and 98% from the lowest to highest income groups (Walker *et al.*, 2001). A gender bias remains in that the entry in the directory is likely to be in the name of a male member of the household.

Execution of fieldwork

Conducting a pilot survey

Skimping on the pilot phase of a questionnaire survey is never a good idea as this is the only means of putting right any major defects in the questionnaire before its final printing. In the long run, if any sections of the questionnaire fail to be understood or produce ambiguous responses, vastly more time and effort will have been wasted than the time it would have taken to do an adequate pilot survey in the first place.

A pilot sample of at least twenty interviews is therefore essential. These interviews should be used to check a number of design aspects:

- *Question design and format.* Are they properly understood by respondents? Do the instructions work? Are the filtered questions properly specified? Is there enough space to fill in the responses to open-format questions? Are the precodes on closed-format questions working? Do they need adjusting? Were there any questions that would have benefited from the use of written prompts? – if so, show cards should be made.
- *Questionnaire length.* It is essential that you record how long the interviews take (face-to-face, telephone) and work out the implications for your fieldwork schedule. If there is a problem, think seriously about which questions to drop.
- *Questionnaire output.* The pilot questionnaires should be used to test out how the data are to be processed and analysed. Are the codes used on the questionnaire clear enough to be read off reliably? What about multiple responses: should respondents be permitted more than one answer? If so, how will the coding scheme cope with this?
- *Classification questions.* Have one last think about the classification data and its role in the analysis: do you really need income? Have you recorded respondent gender? Were there any difficulties with the classification questions during the pilot?
- *Serialisation and other information.* Each questionnaire should have a unique identification number, and telephone/face-to-face record sheets should have space to record the time, date and interviewer name.

Interviewing

Response error can be avoided or minimised by careful consideration of interviewing technique and, if the fieldwork is to involve a team of interviewers, an interviewer briefing and a clear set of instructions on the questionnaire will help to ensure standardisation of the way in which the questionnaire is administered. If you intend to do all of the interviewing yourself, it is still important to write yourself instructions so that you do not unwittingly vary your technique.

There are numerous practical problems which require careful thought before fieldwork starts. Personal safety of interviewers conducting personal interviews is of utmost importance; it is certainly not a good idea to work alone in areas which have a reputation for being troubled, especially if interviewing is being conducted after dark. Be safety-conscious and if it is necessary to work alone, tell someone where you are going, how long it should take and arrange to meet up at a specific time once you have finished. See Chapter 1 and your own institutional guidelines concerning health and safety considerations and the requirements for risk assessment.

Interview location will be an issue if you are not conducting a random sample of households but interviewing respondents in the street or 'on location' (for example, at a nature reserve as part of a survey of visitors). Choice of the exact interviewing location and time of day may introduce biases to the survey: these must be controlled by drawing up a balanced timetable of interviewing shifts and variations in interview location. If you intend to interview members of the public 'on the street' you should bear in mind that the refusal rate is likely to be higher than for a door-to-door survey and that there are likely to be more significant biases in the sorts of people willing to give interviews. The results of general street-based interviewing are therefore less likely to provide a representative sample of the intended target population.

Time of day is also important in door-to-door interviewing, particularly if the target population contains people who are likely to be out at work when you call. For instance, much of the interviewing for Survey 1 was done in the evening and at weekends. Even so, many of the sampled addresses did not result in an interview at the first time of calling. Interviewers were given a contact sheet containing the name and address to be sampled and a series of boxes in which the time of visit and outcome were recorded. Interviewers were instructed to make up to four visits to an address before rejecting it as a non-contact. This helped to ensure that the survey had a good chance of interviewing the sorts of people who were less often at home. Contact sheets that resulted in either a refusal to be interviewed or a non-contact were kept so that an analysis of contact rates and response rates could be made. The last column in Table 6.2 shows that the response rates for Survey 1 differed significantly between interviewing areas, largely reflecting differences in the proportion of economically active households between the areas.

When conducting door-to-door interviewing some respondents may be upset or unnerved by an unexpected visitor. This situation can be made worse by the personal characteristics of the interviewer. It is after all an invasion of their privacy and those living alone may feel particularly vulnerable. Similarly, street interviewing is highly sensitive to the demeanour of the interviewer when approaching someone to request an interview. It could be that some sections

Table 6.2 Survey 1: sampling and response analysis by site

Interviewing area	*Sample*	*Sampling fraction*	*Interviews target*	*Interviews completed*	*Response rate*
Site 1 (sprawling urban fringe)	255	3.5%	200	214	84%
Site 2 (outer city, pre-war estate)	133	10%	100	103	77%
Site 3 (inner city, 19th-century housing)	155	10%	100	100	65%
Total	543		400	417	77%

of the population would be intimidated by a male six-footer wearing army surplus clothing and dark glasses. Although an interview might be granted, it might well be rushed if the respondent feels uneasy and wants the whole thing over as quickly as possible. If more than one interviewer is to be used, they should be given the same instructions on approach to the interview, as well as to their own general appearance (a suit and tie is likely to get a different set of responses from a T-shirt and jeans).

A great aid to securing trust on contacting a potential respondent is to have some form of identification and a covering letter explaining what you are doing and why. If the survey is likely to be of interest to a local institution, see if you can win their confidence and get them to write you a covering letter. In any case you must let any relevant authorities know about your proposed fieldwork. If you are working on a health-related topic, for instance, it will be important for the local health trust to know about it, and equally important that you should know about any survey work that they might be carrying out.

The aftermath

Editing, coding and data input

The main editing and data processing tasks begin once the survey is complete. However, it is important to make a number of quality checks on completed questionnaires as fieldwork progresses (even if you are the sole interviewer). Particular problems that can be caught and corrected include the failure of interviewers to follow instructions on filtered questions and lack of legibility and detail in recording responses to open-format questions. Vigilance at this stage will reduce the chances of unpleasant surprises during the aftermath.

The task of editing and coding questionnaires centres on checking the logic of different sets of responses, looking for insincere or patterned responses (particularly a problem with postal questionnaires) and the creation of new codes. The latter arise from responses to open-format questions that require categorisation and possibly also from responses to closed-format questions that were not covered by the precodes on the questionnaire but which were recorded against an 'other answers' category. An example of the process of coding open-format questions is given in Box 6.5. This example is taken from a question in Survey 1.

First of all a crude listing of the main categories of answer is made and the number of responses recorded against each. Similar answers are then combined and those with low frequencies (in this case 4% or less) are put into an 'other answers' category. The less frequent responses are left in a 'catch-all' category, such as 'other answers', which is also given a code. As the coding of open questions proceeds, more categories are usually created, in which case all of the 'other answers' categories have to be sifted through again in order to pull out any that correspond with the newly created code.

The final set of categories and codes is shown in the second listing in Box 6.5. The coding process is a compromise between summarising the data as concisely as possible and minimising the loss of information that this process

BOX 6.5

Code frame development

Question: *'What do you believe to be the main advantages to you of having a recycling centre in your neighbourhood?'*

	Frequency	*%*
Preliminary listing		
1. 'It's somewhere convenient to recycle household waste'	131	31.4
2. 'Waste would be thrown away otherwise: it's good to reuse it'	67	16.1
3. 'It's good for the environment'	26	6.2
4. 'Encourages people to recycle'	9	2.2
5. 'Advantage to people with a large volume of waste to get rid of'	1	0.2
6. 'Prevents dustbin getting full so quickly'	29	7.0
7. 'It saves resources	6	1.4
8. 'None/ nothing/ nothing I can think of'	51	12.2
9. 'Safe way to dispose of glass'	8	1.9
10. 'Recycling is a good idea'	18	4.3
11. 'Prevents litter'	6	1.4
12. 'Don't get such a smell in the dustbin'	1	0.2
Final list of codes		
1. 'Convenient for recycling'	131	31.4
2. 'Waste would be thrown away otherwise: it's good to reuse it	67	16.1
3. 'None/ nothing/ nothing I can think of'	51	12.2
4. 'Recycling good/environmental comments'	50	12.0
5. 'Prevents dustbin getting full so quickly'	29	7.0
6. Other answers	25	6.0

entails. If categories of responses are carelessly combined, the resulting code descriptor (called a value label) may not do justice to the range of different responses in the original data, thus weakening any conclusions from the analysis.

The transcription of codes from the questionnaires into the computer package to be used for data analysis requires a methodical approach so that data entry errors are minimised. Some of the most popular PC-based packages (e.g. PC versions of SPSS and Minitab) are built around a spreadsheet into which data are entered directly. Whichever package or type of computer facility is available to you, it is important to draw up a scheme with all of the data items from your survey listed sequentially on it. Each data item requires a variable name and, for categorical data, a set of value labels for each code. Variable names and labels should be carefully chosen so as to convey as much as possible about the data that they describe. You will soon forget what is what if the names and labels chosen are obscure. Using the example of Survey 1, the first few variable names, variable labels and value labels are shown in Box 6.6 in relation to the first six cases (i.e. each row contains data from a single questionnaire). Notice that there are two main variable types: alpha data (i.e. containing letters, as in the values given to the variable called 'INT') and numeric data (all of the other variables). It is far easier to analyse data that do not contain a mixture of alpha and numeric variables.

BOX 6.6

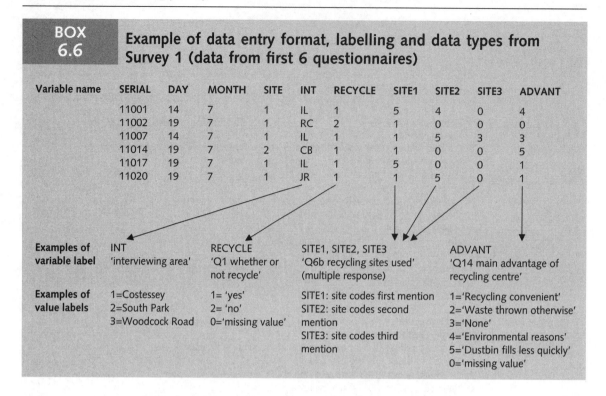

Example of data entry format, labelling and data types from Survey 1 (data from first 6 questionnaires)

Variable name	SERIAL	DAY	MONTH	SITE	INT	RECYCLE	SITE1	SITE2	SITE3	ADVANT
	11001	14	7	1	IL	1	5	4	0	4
	11002	19	7	1	RC	2	1	0	0	0
	11007	14	7	1	IL	1	1	5	3	3
	11014	19	7	2	CB	1	1	0	0	5
	11017	19	7	1	IL	1	5	0	0	1
	11020	19	7	1	JR	1	1	5	0	1

	INT	RECYCLE	SITE1, SITE2, SITE3	ADVANT
Examples of variable label	'interviewing area'	'Q1 whether or not recycle'	'Q6b recycling sites used' (multiple response)	'Q14 main advantage of recycling centre'
Examples of value labels	1=Costessey 2=South Park 3=Woodcock Road	1= 'yes' 2= 'no' 0='missing value'	SITE1: site codes first mention SITE2: site codes second mention SITE3: site codes third mention	1='Recycling convenient' 2='Waste thrown otherwise' 3='None' 4='Environmental reasons' 5='Dustbin fills less quickly' 0='missing value'

Two other features of coding are illustrated in Box 6.6: the treatment of 'residual' categories (missing values and 'don't knows') and questions for which more than one answer is permitted from each respondent (multiple response questions). It is very important to create specific codes for missing data and to distinguish these from respondents who gave 'don't know' as their answer. The choice of these codes should be consistent throughout to avoid confusion.

Multiple response questions require special treatment in the coding scheme. First of all it is necessary to establish the maximum number of responses to be coded. The example from Survey 1 relates to the use of more than one recycling site by respondents. Up to three sites were used, thus three separate variables were created (SITE1, SITE2 and SITE3). The code frames for each were kept consistent. The manipulation of multiple response data sets is particularly easy in SPSS whereas other statistical packages may not have this capability. It is therefore an important consideration if you are likely to have such data to check that the data package that you intend to adopt can handle it.

Specifying the data analysis

Rather than just analysing everything by everything else, you can save reams of paper (and your time) by writing out a data analysis scheme that specifies the data tables needed to answer your research hypothesis. Refer back to the original research plan: what were the main dependent and independent variables?

A convenient way to tackle this is to draw up three columns on sheets of paper. In the first column put the question number and variable name, in the second put details of the question and the codes associated with it and in the final column write in the main data items that you wish to analyse the question by. Important classification data, such as respondent age or housing type, is likely to be used frequently and can be written out once and then referred to by a reference letter (e.g. A = main classification variables). You should also consider which types of statistical test might be required in analysing the data and make a note of these on your data analysis scheme (see Chapter 12 for guidance on the analysis of categorical data).

What to include in the final report

Although writing up your research is the subject of Chapters 17 and 18, there are certain requirements when writing about questionnaire surveys which are worth mentioning here. Fieldwork locations and dates should be included in the methodology section as well as details of sampling. You should provide a justification of the survey technique used and be honest about any failings: few surveys go without hitches. Include a copy of the questionnaire, any interviewer briefing instructions and letter of introduction at the back of your dissertation. Be selective in the data tables which you include in the main body of the text: nothing is more boring than hundreds of tables when only a handful have anything important to say. You can always relegate the secondary tables to an appendix at the back of the document.

Above all, do not get frustrated if there are few statistically significant results to write about or if your favourite hypothesis is not supported. This is the nature of survey research and is an occupational hazard of attempting data analysis on relatively small sample sizes. Lastly, be sure to acknowledge any institutions which gave assistance and do not forget to send them a brief summary of your main findings.

BOX 6.7

Further reading

Black, T R 1993 *Evaluating social science research: an introduction*, London: Sage

Bryman, A and D Cramer 1994 *Quantitative data analysis for social scientists*, London: Routledge

McCrossan, L 1991 *A handbook for interviewers – a manual of social survey practice and procedures on structured interviewing*, London: HMSO

Smith, H W 1975 *Strategies of social research – the methodological imagination*, London: Prentice Hall International

LIVERPOOL
JOHN MOORES UNIVERSITY
AVRIL ROBARTS LRC
TITHEBARN STREET
LIVERPOOL L2 2ER
TEL. 0151 231 4022

7 Tell me about . . . : using interviews as a research methodology

Gill Valentine

Synopsis

This chapter outlines some advantages of using conversational-style interviews as a research methodology. It provides advice on who to talk to, how to recruit research participants and where to hold your interviews. Particular consideration is given to the actual process of asking questions, the ethics and politics of interviewing, and ways to avoid common problems experienced by interviewers. The chapter also highlights some of the differences between conducting social interviews, interviewing élites and interviewing in different cultural contexts.

Why talk to people?

Try answering the questionnaire set out in Box 7.1. Like me, did you find that the questions did not let you give the answers that you wanted to? For example, in answer to the question 'How many items of chocolate do you eat?' I did not want to choose a number. I wanted to ask 'Well, what do you mean by chocolate?' Does that include chocolate cake or chocolate milkshakes or just bars of chocolate? And I wanted to say I do not regularly eat a fixed quantity. How much I consume depends on whether it is summer or winter, whether I have got easy access to a shop, whether I feel happy or depressed, fat or thin. The questionnaire does not allow for me to explain my experiences, whereas an in-depth taped interview would allow me to express all the complexities and contradictions of my chocolate eating habits. This tendency of questionnaire surveys to ask a rigid set of simple questions which 'force' or push the respondents' answers into particular categories which they may not have thought of unprompted or may not want to use, is just one of the reasons why researchers often choose to use interviews either as a supplement or as an alternative to a questionnaire survey.

Questionnaires are also usually standardised; they are not tailored to individuals' circumstances. Rather they rely on a 'now go to' format to try and suit individuals' responses (Eyles, 1988). This standardisation means that they can easily be replicated. Questionnaires are usually analysed using various statistical

LIVERPOOL JOHN MOORES UNIVERSITY AVRIL ROBARTS LRC TITHEBARN STREET LIVERPOOL L2 2ER TEL. 0151 231 4022

BOX 7.1

Chocolate consumption

Q1. *How many items of chocolate do you eat each week? Ring an answer below.*
A. 1–3 3–5 5–7 7–9 9–11 11–13 over 13

Q2. *Where do you buy most of your chocolate? Ring an answer below.*
A. supermarket newsagent vending machine other (please state)

Q3. *How much chocolate do you eat relative to other forms of confectionery? Using the scale below, identify the response which best reflects your opinion.*
A. 1. a lot more 2. a little more 3. the same 4. a little less 5. a lot less

techniques which generate numerical data, although they may also include open-ended short written comments. The aim of using a questionnaire is often to survey a representative sample of the population so that you can make generalisations from your responses. However, in some cases the explanatory power of questionnaires can be limited.

Interviews, in contrast to questionnaires, are generally unstructured or semi-structured. In other words, they take a conversational, fluid form, each interview varying according to the interests, experiences and views of the interviewees. They are a dialogue rather than an interrogation. Eyles (1988) describes an interview as 'a conversation with a purpose'. The advantage of this approach is that it is sensitive and people-oriented, allowing interviewees to construct their own accounts of their experiences by describing and explaining their lives in their own words. This sort of conversation offers the chance for the researcher and interviewee to have a far more wide-ranging discussion than a questionnaire would allow. In the course of the interview, researchers have the chance to go back over the same ground, asking the same questions in different ways in order to explore issues thoroughly; and interviewees can explain the complexities and contradictions of their experiences and can describe the mundane details of their everyday lives (Bryman, 1988). One of the additional strengths of this approach is that it allows respondents to raise issues that the interviewer may not have anticipated (Silverman, 1993). The material generated in this way is rich, detailed and multi-layered (Burgess, 1984), producing 'a deeper picture' than a questionnaire survey (Silverman, 1993: 15). It is analysed using a textual approach, relying on words and meanings, rather than statistics.

Unlike a questionnaire, the aim of an interview is *not* to be representative (a common but mistaken criticism of this technique) but to understand how individual people experience and make sense of their own lives. The emphasis is on considering the meanings people attribute to their lives and the processes which operate in particular social contexts. The fluid and individual nature of conversational-style interviews means that they can never be replicated, only corroborated by similar studies or complementary techniques.

Positivists often criticise in-depth interviews, claiming that interviewers bias the respondents' answers or that interviewers are not or cannot be objective or detached. Those who take a humanist or post-structuralist approach to research argue, however, that there is no such thing as objectivity in social science research. Rather they argue that all research work is explicitly or implicitly informed by the experiences, aims and interpretations of the researcher who designed the questionnaire or the interview schedule and that researchers should treat participants in their research as people, not objects to be exploited or mined for information (England, 1994). As Stanley and Wise (1993: 157), two sociologists who have been at the vanguard of feminist critiques of objectivity argue: 'Whether we like it or not, researchers remain human beings complete with all the usual assembly of feelings, failings and moods. And all of those things influence how we feel and understand what is going on. . . . Our consciousness is always the medium through which the research occurs; there is no method or technique of doing research other than through the medium of the researcher.' This is a view echoed by O'Connell Davidson and Layder (1994: 125) who explain: 'Interviewers are not losing their "objectivity", becoming partial or imposing a particular world view on the respondent, rather they are using the interview as an opportunity to explore the subjective values, beliefs and thoughts of the individual respondent.'

The final thing to remember when trying to decide whether to use in-depth interviews as a research technique for a project is that they do not have to be used in isolation but that they can be employed as just one part of a multi-method approach to a research question (see Sporton, 1999 on 'mixed methods'). Often researchers draw on many different perspectives or sources in the course of their work. This is known as triangulation. The term comes from surveying where it describes using different bearings to give the correct position. In the same way researchers can use multiple methods or different sources to try and maximise their understanding of a research question.

Who to talk to

Researchers using questionnaires often use random sampling techniques to select a representative sample of the population under investigation to be surveyed (see Chapter 6). The aim, however, in recruiting informants for interview is not to choose a representative sample, rather to select an illustrative one. Choosing who to interview is therefore often a theoretically motivated decision. For example, if you were studying fear of crime, you may anticipate that gender may be an important influence on people's perceptions of danger, because crime statistics suggest that men are the group most likely to be the victims of interpersonal violence whereas women are the group subject to most sexual assaults. You may also recognise that 'men' and 'women' are not homogeneous social groups but that the identities 'man' or 'woman' are cut across by other identities such as age, class, ethnicity and so on which may also have a bearing on perceptions of fear. Therefore, you may choose to interview men and women of different ages and from different ethnic backgrounds in an attempt to explore which processes are important in shaping their perceptions of crime. As geographers, we are also concerned with the difference that place makes and so you may choose interviewees who live in both rural and urban areas and who live in different types of housing (e.g. council flats, cul-de-sacs on private estates and so on) so that you can explore the role of place in your interviews.

When you are thinking about who you want to interview it is important to reflect on who you are and how your own identity will shape the interactions that you have with others. This is what academics describe as recognising your *positionality* and being *reflexive*. Kim England defines reflexivity as 'self-critical sympathetic introspection and the self conscious *analytical* scrutiny of the self as a researcher' (England, 1994: 82). In particular, it is important to recognise the different power relationships that exist between yourself and your informants (see for example Mohammad, 2001; Morris-Roberts, 2001). As Schoenberger explains, 'questions of gender, class, race, nationality, politics, history, and experience shape our research and our interpretations of the world, however much we are supposed to deny it. The task, then, is not to do away with these things, but to know them and learn from them' (Schoenberger, 1992: 218). For example, it is very difficult for a man to conduct interviews on women's fear because the very act of a strange man turning up to talk to a woman can itself be perceived as threatening. Women are also likely to feel uncomfortable and hesitant about talking to a man about experiences of fear or violence. It is important therefore to be sensitive to the power-laden nature of interviewing encounters. Rather than risk upsetting or intimidating women in this way, it would be more appropriate for a man interested in researching fear to explore why men are the gender least fearful of crime despite the fact they are the group who experience most violence. Likewise, Horton (2001) has described his unease as a man carrying out research with children against a backdrop of popular concerns about paedophiles and children's safety.

Sharing the same background or a similar identity to your informant can have a positive effect, facilitating the development of a rapport between interviewer and interviewee and thus producing a rich, detailed conversation based on empathy and mutual respect and understanding. As Dyck (1993: 54), a mother who carried out a project on motherhood, explains 'the sharing of a common frame of reference and familiarity with culturally specific terms and nuances around the practice of mothering facilitated the wording of interview questions and allowed me to respond to comments in a way that promoted the flow of conversation'. Similarly, you may find it easier to build a rapport with your research participants and conduct interviews if your project is linked to your own interests or you are interviewing people who you have something in common with. However, it is also important to be aware that as a researcher you may still fail to 'connect' with an interviewee despite the fact that you share aspects of your identity and interests in common; and that it is possible to develop a positive relationship and understanding across social differences (Rose, 1997). Indeed, our prior assumptions about our sameness to, or difference from, an informant can often be challenged in the course of the interview, because the way that we as researchers and our informants perform our identities and read those of others is something that is negotiated in the relational moment of the interview (these issues are discussed with examples in Kobayashi, 2001; Valentine, 2002). Allied to this growing recognition of the way that research positions are fashioned through interactions is an awareness of the need to acknowledge the role of emotions in shaping, and being shaped by, research encounters (Widdowfield, 2000; Laurier and Parr, 2000). Laurier and Parr (2000: 99) claim that 'acknowledging emotions and emotional exchanges orientates us *differently* within our research interview'. They demonstrate this by drawing on Parr's experience of interviewing a schizophrenic man with a twitching leg to illustrate how different corporealities can lead to 'othering' within the interview context. Likewise, Widdowfield (2000) also draws on her own fieldwork to show how emotional judgements can cloud our interpretations of research encounters.

Postmodernism has contributed towards making research on marginalised groups fashionable. On the one hand, this recognition of diversity and difference has led researchers to try to allow 'other voices' to speak in their work. On the other hand, some writers argue that this sort of research is no more than a form of academic voyeurism in which marginal groups, for example lesbians and gay men, are treated as 'exotic others'. Liz Bondi (1990: 163) sums up these anxieties when she writes: 'the postmodern venture is a new kind of gender tourism, whereby male theorists are able to take package trips into the world of femininity, in which they "get a bit of the other" in the knowledge that they have return tickets to the safe, familiar, above all empowering, terrain of masculinity.' Reflecting on these sorts of concerns led Kim England (1994) to abandon her attempt as a heterosexual woman to carry out research on lesbians in Toronto. Writing about this experience in the *Professional Geographer*, she points out the dangers of appropriating the voices of 'others' and representing yourself as an expert in their lives. However, able-bodied researchers working in the field of disability have argued that it is possible to establish genuinely collaborative research partnerships with disabled people by involving disabled people in the design and conduct of the research, and in interpretation and dissemination of the results (Chouinard, 1997; Gleeson, 2000).

If you are embarking on research in the developing world it is particularly important to be aware of your privileged position in terms of wealth, education and so on, in relation to those you will be working with and to recognise that your research is embedded in the context of colonialism (though similar issues come up in the context of other cultures too; see Smith, 2003). As Cindi Katz (1992: 496) neatly articulates, the research relationship is 'a peculiar relationship – unequally initiated, situationally lop-sided, spatially dislocated, temporally isolated, extrinsic in purpose – it oozes with power'. Anthropologists have been particularly reflective on power relationships in the field and the ethics of conducting fieldwork in developing countries. This is also increasingly being debated by geographers (see for example Robson and Willis, 1994; Sidaway, 1992; Katz, 1992; Townsend, 1995; Women and Geography Study Group, 1997; Skelton, 2001).

Often when academics talk about power relations the assumption is that it is the interviewer who is in the dominant position. However, if you are interviewing élites and business people, it is often they who have the upper hand, by controlling access to knowledge, information and informants. Indeed, they often want to have some influence on the research process, refusing to allow interviews to be tape-recorded or demanding the right to vet interview transcripts and influencing the way that research findings are presented (see for example McDowell, 1992; Herz and Luber, 1995; England, 2002).

Finally, when you are doing household research it is important to reflect on whether different household members should be interviewed, and if so whether these interviews should be conducted together or separately. Traditionally, research on the 'family' has tended to be based on interviews with women because the home, domestic work and 'family' have been regarded as women's domain. Yet a number of studies suggest that there is often widespread disparity between household members' (both other adults' and children's) accounts of a range of topics. As such it is now recognised that in studies of domestic situations it is important to interview all the household members rather than rely on one person's testimony. Interviewing families together can provide more spontaneous, richer and validated accounts than those with individuals alone because different household members can corroborate each other's stories or challenge inaccuracies in each other's memories. As a result joint interviews can give you a clearer picture of the contested nature of household life. However, there is also a danger that collective interviews can silence individuals, particularly children, who may find it difficult or be unwilling to contradict other household members. One household member can also take an authoritarian role speaking for the others (especially children) or reinterpreting the questions for them (these issues are discussed in more detail in Valentine, 1999a).

Recruiting interviewees

There are a number of different ways to approach potential informants.

Via a questionnaire

A common practice is to carry out a simple questionnaire survey to gather basic factual information which includes a request at the end asking respondents who are willing to take part in a follow-up interview to give their address and telephone number. From those who provide this information it is possible to choose interviewees on the basis of the answers they give to the questionnaire. The advantage of this method is that it is a quick, easy and painless way of recruiting people and you know something about the informants and their views on the topic before you talk to them, which can help you shape your interview cues. The disadvantage of this approach is that you may find it is largely only one group of people who offer to be interviewed, for example middle-class professionals, when you may want to talk to people from a wider range of backgrounds.

Cold calling

The electoral register lists everyone over 18 living in each house (but see the more detailed discussion of the register in Chapter 6). Implicitly, therefore, it provides basic information about the social characteristics of every household because, for example, from names it is possible to guess whether occupants are

BOX 7.2

Electoral register

10 Springfield Gardens	Jones, Brenda
	Jones, Douglas
11 Springfield Gardens	Walker, Edna
12 Springfield Gardens	Page, Margaret
	Page, Russell
	Page, Christopher
	Page, Amy

married, living alone and male or female. Due to changing fashions in names it is also possible to make an educated guess about whether the occupants are young, middle-aged or elderly. For example, from the list in Box 7.2 it would be reasonable to assume that Brenda and Douglas Jones are a middle-aged married couple, whilst Edna Walker is most probably an elderly woman living alone. Consequently, if you decided that for the purpose of your research you needed to interview married couples and elderly people in a rural village you could obtain the electoral register for the village from a public library. By studying the names and number of occupants in each house you could find addresses for those who would appear to be suitable interviewees and then knock on their door and ask if they would be prepared to be interviewed at a later date; or even go one step further and use directory inquiries to trace their telephone number and ring them. This is what salespeople term 'cold calling'. It is a very problematic recruiting strategy. Firstly, it does not always guarantee results because in areas of high population turnover the electoral register can be out of date and educated guesses about the social characteristics of occupants may be inaccurate. Secondly, cold calling is very intrusive and so interviewers often get a high refusal rate. Thirdly and most importantly, the elderly and women living alone can feel very intimidated if a stranger arrives on their doorstep or rings up unannounced, especially if this occurs after dark, the optimum time for 'catching' people at home if you are cold calling.

Gatekeepers

Gatekeepers are 'those individuals in an organization that have the power to grant or withhold access to people or situations for the purposes of research' (Burgess, 1984: 48). It is important therefore when you contact 'gatekeepers' in organisations and institutions that you make it clear exactly what sort of information you want and who you would like to talk to (preferably in writing). Gaining access to the right people through gatekeepers is very important because who you talk to will affect what information and perspective you get. Beware, however, particularly when you are relying on gatekeepers for an introduction to members of a social group (for example the chair of the local Women's Institute), that they do not try to direct you to a narrow selection of the members (probably their friends) and discourage you from talking to others. In this situation you may want to make a discreet effort to talk to other people in order to make sure the gatekeeper is not trying to steer you away from them in order to prevent you hearing a dissenting voice and is therefore distorting your understanding of the issue you are researching. However, this has to be done very sensitively and only in appropriate circumstances because the potential repercussions of deceiving gatekeepers can be very serious. If you are considering this course of action talk it over first with your project supervisor or course tutor.

Snowballing

This term describes using one contact to help you recruit another contact, who in turn can put you in touch with someone else. The initial contact may be a friend, relative, neighbour, or someone from a social group or formal organisation. As the term implies, through this method, recruiting gains momentum, or 'snowballs' as the researcher builds up layers of contacts. For example Donovan (1988), a white researcher, describes how she used black friends to help her get an introduction to Afro-Caribbean groups. She was then able to snowball further using one informant to introduce her to another. The strength of this technique is that it helps researchers to overcome one of the main obstacles to recruiting interviewees, gaining their trust. It also allows the researcher to seek out more easily interviewees with particular experiences or backgrounds. It is important, however, to make sure that you use multiple initial contact points when you start snowballing so that you do not recruit all your informants from a very narrow circle of like-minded people. Interviewees recruited through snowballing will often try and persuade you to tell them what their friends said when you talked to them. It is essential, however, that you maintain the confidentiality of each interview.

Some groups are particularly difficult to gain access to. Anxieties about children's safety, changes in the way schools are managed, and the introduction of the national curriculum, which means school activities are more tightly timetabled, have all contributed towards making it more difficult for researchers to gain access to schools (for a more detailed discussion of the ethical issues involved in recruiting and working with children and young people see Valentine, 1999b and a special issue of *Ethics, Place, Environment*, 2000). If you want to work with children you need to allow plenty of time in your research schedule for organising this as you will have to write to the head teacher, who may in turn need to seek permission from the Educational Authority, parent-governors or parents before they will allow you into a classroom. It is also difficult to get elderly people to participate in research projects because of their anxieties about their personal safety. Rather than approach individuals at home it is more effective and more ethical to contact those running senior citizens' clubs, or those who provide sheltered accommodation. They in turn may then be able to introduce you to potential informants. Likewise, contacting those who coordinate organisations and support groups for lesbians, gay men and bisexuals, and negotiating access with members of the group through them, is a more sensitive and ethical way to try and recruit sexual dissidents for a research project than blundering into gay pubs and clubs.

The main obstacle to recruiting informants from organisations is often knowing who is the most appropriate person to talk to. Requests for interviews (by letter or phone) are more effective if they are targeted at a named individual rather than sent to 'whom it may concern' at a company. One way to find out who it would be most appropriate to speak to is to research the organisation as thoroughly as possible beforehand, perhaps by obtaining company or institutional publicity material or by looking it up in directories in the public library. Alternatively, telephone the public relations department (if the organisation has one) or a secretary (although these gatekeepers can be more obstructive than helpful) to ask who you should direct your enquiries to. Whenever you are approaching people for interview it is important to set out clearly in writing the aims of your research, who you would like to talk to, the issues you wish to discuss, an estimation of how long the interview is likely to take and when and where you would ideally like to conduct it; and to include a letter of introduction from your tutor. If you do not get a reply (which is often the case when contacting organisations) follow up your letter with a phone call or a personal visit.

Where to hold interviews

Where an interview is held can make a difference (Denzin, 1970). In most cases if you are talking to business people or officials from institutions and organisations you will have no choice but to interview them in their own offices. There are often more options when interviewing members of the public. Try to avoid conducting taped conversations in busy, noisy social spaces like pubs, clubs, or leisure centres unless it is possible to find a quiet private office or room where you can talk to one person in depth without being disturbed or distracted. Inviting people to the University is a possible alternative way of finding a quiet space. The drawback of this location is that the formal setting often contributes to producing a more stilted, formal, interview. Talking to people on their own 'territory', i.e. in their own home, can facilitate a more relaxed conversation. It also offers you the possibility to learn more about the person from seeing them in their own environment.

For your own safety never arrange interviews with people you do not feel comfortable with or agree to meet strangers in places where you feel vulnerable. For example, O'Connell Davidson (1991) reports her own unsettling experiences of carrying out research on the water industry which involved meeting unknown men in isolated places in the countryside. Skeggs (1994) also points out that for women, sexual harassment can be 'a built in part of the research process' whether the research is carried out in interviewees' homes or in offices. Regardless of whether you are male or female or interviewing in 'public' or 'private' places, it is wise therefore to always take certain safety precautions. Firstly, always take some form of identification with you and a letter from your tutor explaining what you are doing. Secondly, always tell someone where you have gone and what time you expect to return. Thirdly, make sure that you are aware of and abide by any safety guidelines that your own department or institution may have.

Asking questions

There are no hard and fast rules about interviewing. Each interviewee is an individual and therefore each interview will be different. 'The interview is a social encounter, and how the respondent answers the questions will depend to some degree upon what the respondent and interviewer think and feel about each other' (O'Connell Davidson and Layder, 1994: 118). Self-presentation, in other words managing your appearance and behaviour, is important (Smith, 1988) in helping you as a researcher to get people to warm to you and establish a rapport. McDowell (1992), for example, points out that when interviewing business people it is important to dress in a professional way. Similarly, smart dress can reassure elderly respondents who may be cautious about letting strangers into their home. In contrast, if you are interviewing squatters or people involved in environmental protest groups, for example, it is more appropriate to dress in a very casual way.

The key to conducting a successful interview is to brief yourself on the topic carefully. Interviews do not rely on a rigid set of questions worded beforehand because you want to understand the issues in the interviewees' own terms; however, it is important to work out a list of themes that you want to cover in each interview. This may include a few factual questions which you want to ask everyone. There are several ways of preparing an interview schedule, two are outlined in Box 7.3. If you are confident about interviewing, one approach is just to make a list of themes which you will develop as the interview progresses. But if you are more anxious about losing your train of thought or drying up during the interview, it may be advisable to work out some key questions (hangers) on which you may then hang a series of follow-up ideas. These may include descriptive questions which ask for information on activities/experiences; structural questions which focus on how and when events occurred; and thoughtful questions which explore meanings, feelings and opinions. This will help you to remember all the different angles you want to explore with your informants, although obviously you do not have to stick to this format in the interview or to ask the questions in the order which you prepared them.

Most people's experiences of being interviewed will have been a questionnaire in the street. Often they do not expect to be engaged in conversation: they expect to give very short answers to set questions. It takes time therefore to 'warm up' an interview and to develop a rapport with the participant. Try to start your interview with general descriptive or factual questions which signal to the interviewee that you would like them to talk freely and not give short 'yes no' style answers. Griffiths (1991), for example, found that when interviewing teenagers for a study of friendships her interviews were more successful when she began by asking factual questions like 'so how did you get to know each other?' rather than abstract questions like 'why are you friends?' Asking respondents to provide a 24-hour retrospective diary of their day is also an effective way of pinpointing the structure of a respondent's everyday lifeworld and gleaning information which you can use to inform questions later in the interview. As you develop a rapport with the interviewee it then becomes possible to ask

BOX 7.3

Interview schedules

Schedule 1

A detailed list of key questions (not necessarily asked in the order listed). Possible follow-up questions are listed in parentheses.

- Tell me about your household's usual food shopping routine.
- Who usually does the food shopping in the household? (Why? Why doesn't *x* or *y* do the shopping?)
- Why do you shop there or in those places? (When did you first start shopping there?)
- Do you ever vary your routine? (How and why?)
- Tell me about what you normally buy on your regular shopping trips? (What sort of treats do you buy?)
- Tell me about how you go around the supermarket and decide what to buy? (Do you ever make lists?)
- How does your shopping differ if other people are with you? (Who and how?)
- What about on special occasions? (Like Christmas? Birthdays?)
- Tell me about how you feel about food shopping. (Have you ever suffered from trolley rage?)
- How would you describe yourself as a shopper?
- Do you ever use the special facilities (e.g. crèche) or the coffee shop? (Why?/why not?)

Schedule 2

A rough list of key themes

- Shopping routine
- Who shops/Where/Why?
- Nature of shopping
- Process of shopping
- Shopping routine/how it varies
- Special occasions
- Experience of shopping
- Facilities used

more abstract, sensitive or difficult questions. As the interview draws to a close, try and 'warm down' the interview by asking more relaxed, or light-hearted questions so that it ends on a positive note.

Usually each interview is different and will have its own pattern. It is important to follow the conversational flow of your interviewee. If they start talking about your last theme first, follow their train of thought, otherwise the spontaneity of the interview is lost. This often means that you will have to complete feats of mental gymnastics in order to remember which of your themes you have covered and which you need to bring up later. It is important to get the balance right between keeping the interview focused and letting it flow and take its own course.

Try not to phrase questions which impose an answer (Burgess, 1982) on your interviewee. 'Tell me about . . .' is an effective way of encouraging interviewees to talk about an issue in their own words. Graham (1984) argues that allowing interviewees to tell their own stories elicits more information and detail than a string of questions and answers, and does not fracture the interviewee's experiences as much as an inquisitorial approach. Use silence, 'uhuh', or repeating back to your interviewees their own statements as questions to encourage them to elaborate on what they are saying. Unfinished questions or mumbled questions which encourage interviewees to 'step in' to help out the incompetent researcher are all interviewing techniques used by journalists. Also monitor your own gestures, facial expressions and general body language as these can also be conveying particular meanings to interviewees, encouraging them to talk, or accidentally signalling to them that you are bored (Burgess, 1984).

Some researchers (Hammersley and Atkinson, 1989) also advocate asking questions which deliberately lead interviewees in the opposite direction to that which you would expect. Provoking controversy can be particularly effective if you are interviewing couples or groups. For example, when I was carrying out research about parents' perception of children's safety I kept getting very bland answers when I asked mothers and fathers to tell me about their roles in looking after the children, despite the fact that it had or would become apparent in the interview that there were strong gender differences in parenting. When I switched to assuming a difference by asking 'so tell me which of you is the strictest?', this question would open up a floodgate of rich detailed anecdotes which often exposed disagreements between parents on their relative responsibilities. However, it is an interview strategy which should be used sparingly.

Feminist researchers have stressed the importance of interacting and sharing information with participants rather than treating them as subordinates from whom you are extracting information (Oakley, 1981). As O'Connell Davidson and Layder (1994: 121) argue: 'The interviewee is not a research "subject" to be controlled and systematically investigated by a "scientist", but a reasoning, conscious human being to be engaged with.' The principle of reciprocity, sharing your experiences and exchanging ideas and information with participants, is now widely accepted as good research practice. Burgess (1984), for example, recalls how when he was doing research in a school, people he talked to frequently asked him about his biography, and his views on schools and teaching in the course of interviews. Oakley (1981) became so involved with women she interviewed about pregnancy, birth and motherhood that four years after the study was completed she was still in touch with a third of the women she interviewed. Be careful, however, that you do not express your own views and experiences in such a way that your interviewees feel unable to declare contrary opinions.

Interviewing business people poses a different set of problems. Rather than befriending corporate interviewees it is often more appropriate to demonstrate your professionalism and knowledge of what you are talking about. Participants in this type of research are often themselves experienced at interviewing and being interviewed and consequently know how to subvert interviews, control them or deny interviewer access to key information. 'In some situations interviewers can and do exploit the negative stereotype that certain people hold of them to their advantage. A young woman interviewing relatively powerful male respondents, senior managers for example, can live up to their expectations by "acting dumb" which can encourage them to disclose more information to her than they would to an older male interviewer, whom they might assume would know how to use the information against them' (O'Connell Davidson and Layder, 1994: 146).

Asking the questions is only half the skill of interviewing: the other half is learning to listen and respond to your participants. One of the advantages of this sort of interviewing is that themes often emerge in interviews which were not anticipated by the interviewer. You therefore have to be alert and quick-witted to pick up on these ideas and follow them through. As Glesne and Peskin (1992: 76, quoted in O'Connell Davidson and Layder, 1994: 122) explain: 'At no time do you stop listening, because without the data your listening furnishes, you cannot make any of the decisions inherent in interviewing . . . Has your question been answered, and is it time to move on? If so, move on to what question? Should you probe now or later? What form should your probe take?' The spontaneity and unpredictability of the interview exchange precludes planning your probes ahead of time: you must, accordingly, think and talk on your feet. 'Given that the main aim of interviewing in ethnographic research is to allow people to reveal their own versions of events in their own words, it is important to ask follow up questions in such a way as to encourage and critically question the stories told' (Cook and Crang, 1995: 36). The most frustrating side of interviewing is reading through your transcripts and realising that your informant brought up new and interesting ideas but because you were not listening carefully or critically enough you did not pick up on them and probe further but rather stepped in with a question that changed the topic under discussion. It is worth constantly remembering the golden catchphrase of journalists – 'who, what, when, where, how and why?' – as you listen to your participants.

One last thing about asking questions – interviewing can be a stimulating process. But remember, however much you may be enjoying yourself and obtaining exciting material, you cannot expect to talk to people for more than one to two hours at most without outstaying your welcome.

Potential problems

It is important to try and anticipate potentially difficult issues and situations that may arise in your interviews. For example, talking to people about crime, racism, or lesbian, gay and bisexual identities may prompt interviewees to recall distressing experiences of violence and discrimination. Always tell interviewees that they do not have to answer any questions which they do not want to and remind them of this during the course of the interview. Be sensitive to their responses, and never pressurise people to talk about anything which is obviously making them uncomfortable or distressed. If you are planning to carry out sensitive research it is advisable to find out about local counselling and support services, so that if necessary you can pass this information on to your informants.

Although undergraduate dissertations are rarely published in a public forum, it is still important to respect the wishes of interviewees who want their identities to be kept anonymous in your study by giving them pseudonyms and disguising key locations. Never repeat confidential comments and informants' personal stories to other interviewees or use their experiences to entertain friends in the pub!

Sometimes in the course of an interview participants may express racist, homophobic or other offensive views. This poses an awkward dilemma for researchers. On the one hand, writers such as Griffiths (1991) and Tronya and Hatcher (1992) argue that researchers should not ingratiate themselves with participants but rather must challenge offensive comments because to remain silent is to reproduce and legitimise the interviewees' prejudices through collusion. Some feminist academics even go so far as to argue that feminist interviewing should be a consciousness-raising process and therefore 'educating' participants is an important goal of research. On the other hand, Skeggs (1994) acknowledges that challenging an interviewee in this way is unlikely to change their views and will merely destroy the rapport between the researcher and the interviewee and make some types of research impossible. Indeed, some academics argue that we will never understand racism if we do not try to examine how it operates in practice by studying institutions and individuals who hold these views. For example, Michael Keith (1992) made only selective self-disclosure in the course of his research on the police and he did not challenge officers' comments which he considered to be racist and offensive. Although this was personally difficult, it enabled him to collect a body of material exposing the institutional racism of the police force. Likewise, recounting his experience of researching people who were organising to oppose human service clients and facilities being located in their community, Robert Wilton (2000) argued that his own commitment to social justice for people with disabilities, and his political reading that the opponents of service facilities were drawing on and reproducing negative stereotypes of disability, justified his decision not to be open with his interviewees about the purpose of his research.

Taping

There are a lot of advantages to using tape recorders rather than trying to make notes. Recording equipment allows the researcher to concentrate on the interview without the pressure of struggling to get the interviewee's words down on paper, and allows the interviewee to engage in a proper conversation with the researcher without trying to pause and talk slowly so that the notetaker can keep up. A tape also produces a more accurate and detailed record of the conversation (including capturing all the nuances of sarcasm, humour and so on) than notes, and of course it has the added advantage that the researcher can listen to the interview again and again and therefore pick up on ideas and inferences which they may have missed or which did not seem important when the conversation first took place.

Not everyone likes to be taped. Some people may be just shy, some have cultural objections to having their voice recorded; and others, particularly élites, may not want their comments on the record. It is important therefore to be sensitive to the interviewees' wishes and always to carry a notebook in case taping is refused. Indeed, business people may be more forthcoming if their comments are not being taped.

One of the perverse laws of interviewing is that in any research project at least one tape will turn out blank because of a technical mishap and it will always be the tape with the most effective and longest interview on it. There are a few things, however, that you can try and do to prevent this disaster. The first is to use a tape recorder with an external microphone because internal microphones are often not very effective in picking up voices any distance from the machine. Always familiarise yourself with the tape recorder before using it. Test the record levels at the start of each interview by taping a few sentences and replaying them to make sure the machine is working properly before you commence the 'proper' interview because vital controls like the pause button or the record level dial can easily accidentally become altered whilst the recorder is in your bag. Carry plenty of spare batteries and tapes. If possible (and your tape recorder has a mains adapter) plug your recorder into the mains as this avoids the risk of batteries running out in the middle of your conversation.

When setting up your tape recorder think about the positioning of the machine carefully. Place it close to and in front of the interviewee, not to the side, as the informant may turn away from the machine and so their voice will be muffled. Beware of radios and televisions in the background. These often sound unobtrusive at the time of interviewing but can drown out voices on the tape. Gas fires, fish tanks, animals and children can all produce disruptive background noise which can hamper effective tape recording. In these circumstances, do not be embarrassed to ask if the noise can be subdued or if you can move the interview to another room because if the tape is inaudible you have wasted not only your time but also the informant's. Finally, after each interview always make notes of your thoughts on how the interview went and the key points of the conversation. Then, if the perverse law of interviewing strikes, you should at least have some record of your discussion.

Interviewing in different cultural contexts[1]

Interviewing in different cultural contexts, particularly in less developed countries, requires a heightened sensitivity to the complex power relations which exist between researchers and interviewees, and to local codes of behaviour (see

[1] I am very grateful to Tracey Skelton for sharing her experiences of interviewing in the Caribbean with me.

Robson and Willis, 1994; Townsend, 1995; and Skelton, 2001 for a detailed discussion of these issues).

The cultural and economic power of first-world countries casts a shadow over relationships between North American and European researchers working with interviewees in third-world countries. As Howard (1994: 20) explains: 'There is a tendency for white angolophones to be perceived as powerful and even superior. Thus, in some ways for a British or US researcher . . . the relationship between the researcher and the researched is a continuation of the relationship between coloniser and colonised.' She goes on to explain that one consequence of this unequal power relation is that informants may often feel beholden to cooperate with researchers. It is vital therefore to make it clear to interviewees that they do not have to answer everything which they are asked and that they can terminate an interview at any time. Howard (1994: 21) also points out that 'the very presence of the researcher, by virtue of the respondents' perception of his or her being a powerful person, can generate a whole host of expectations on the part of the respondents'. This she argues can lead interviewees to tell the researcher what they believe she or he wants to hear. Interviewees may also assume that the researcher is affluent and expect to receive money, gifts or photographs from them.

Gender, age and marital status are all aspects of a researcher's identity that can limit access to informants or situations. Howard (1994) points out, however, that being an outsider means that researchers can often be positioned outside local gender 'norms'. It is also important to recognise local political agendas and not, through your choice of interviewees, to be seen to be taking sides or belonging to a particular group. It is also important to maintain the confidentiality of each informant as gossip can be a powerful weapon when wielded in particular social contexts.

Remember that the way people are expected to dress and behave also varies from place to place. As Lloyd Evans *et al.* (1994: 17) point out: 'In some societies, shorts are strictly for the beach and jeans for after work. In the Caribbean, for example, smart dress is often the norm in universities and government offices. Meeting a local official in casual dress is rarely the best way to obtain assistance, as in many societies this shows a lack of respect.'

Misunderstandings are common when interviewing in different cultural contexts because of slight differences in meanings. For example, in the UK/USA, interviewers often say 'uhuh', meaning 'carry on' or 'tell me more'. A similar sound in the cultural context of the Caribbean means 'can you say that again?' Such slight differences can obviously be very significant in the context of an interview. Some languages also have different grammatical forms for polite and informal use. If you are interviewing in the local language it is important to ensure that you are aware of these linguistic codes of behaviour and use the appropriate form in your interviews (Lloyd Evans *et al.*, 1994). In order to pick up on these subtleties it is important to try and spend plenty of time settling in and talking to people in general social situations before formally conducting any interviews. In some cultures, people do not like their voices to be recorded. Pilot interviews with local people you befriend are one practical and effective way of sensitising yourself to these sorts of issues, and facilitating your ability to negotiate ethical relationships in the field.

Working with an interpreter can result in even more complicated linguistic and cultural misunderstandings (see for example: Smith 1996, 2003; Twyman *et al.*, 1999). The presence of an interpreter can influence the behaviour and response of both the interviewer and informant. In particular, informants can be very wary about the interpreter breaking their confidences. For this reason it is best where possible to let the informant choose or nominate an interpreter. Changes in message can also occur as a result of errors when one spoken language is interpreted into another. For example, the interpreter may use a wrong word or pitch the register incorrectly. (There is a further inevitable loss of meaning when a visual language such as British or American Sign Language is interpreted into a linear spoken language and the written word. These issues are discussed in Valentine and Skelton, 2003.)

What next?

After you have completed each interview, try and transcribe the tape as soon as possible. Firstly, if your tape recorder is not working, the interview will still be fresh in your mind and you can salvage some of it in note form. Secondly, it will stop you ploughing on and recording a whole series of blank interviews. Thirdly, if you do not transcribe each tape as you go along you will complete your interviews and be confronted with a huge backlog of tapes that will take weeks to transcribe before you can start your analysis.

It is also worth keeping a research diary about the context of your interviews, how you felt about them and any inspirations that you have about things to ask in later interviews or about ways of analysing your material. When the interviewing is over, this can remind you of things that seemed important at the time of interviewing which you may have forgotten once you start analysing material.

If possible try to write a thank-you letter to your participants and provide each one with a short (one page) summary of your findings.

How to practise your interview skills

A good way to practise your interviewing technique is to get into pairs. Agree a topic for an interview. One of you should then research it and prepare some interview cues. The other should invent a character for themselves. This character should have had a key experience that the interviewer does not know about but which if they are a skilled interviewer they will uncover in the course of the discussion. When the interviewee is questioned they should answer the questions they are asked but should not volunteer the 'key experience' unless the interviewer probes deeper and asks an appropriate question to elicit this information. The interviewee may also test the interviewer's skills in other ways, for example by wandering off the point, not addressing the question or by appearing nervous or distressed. Tape your mock interview. Play it back and analyse your strengths and weaknesses as an interviewer.

Acknowledgement

I wish to acknowledge the support of a Philip Leverhulme Prize Fellowship which enabled me to work on this chapter.

BOX 7.4

Further reading

There are many books on qualitative methods. The texts I have found most useful are: Kitchin, R and N Tate 2000 *Conducting research into human geography*, Harlow: Prentice Hall; Limb, M and C Dwyer (eds) 2001 *Qualitative methodologies for geographers*, London: Arnold; and Clifford, N and G Valentine (eds) 2003 *Key methods in geography*, London: Sage.

The feminist literature is a particularly good source for material on the ethics and politics of interviewing, notably: Stanley, L and S Wise 1993 *Breaking out again: feminist ontology and epistemology*, London: Routledge; and Moss, P (ed.) 2002 *Feminist geography in practice: research and methods*, Oxford: Blackwell.

More attention is now being paid to the problems of interviewing élites. Useful references are: Herz, R and J Luber 1995 *Studying elites: using qualitative methods*, London: Sage; and England, K 2002 Interviewing elites: cautionary tales about researching women managers in Canada's banking industry. In P Moss (ed.) *Feminist geography in practice: research and methods*, Oxford: Blackwell.

Relatively little has been written within geography about interviewing in different cultural contexts. A useful publication is a monograph by E Robson and K Willis (eds) (1994) published by the Developing Areas Research Group (DARG) of the Royal Geographical Society with the Institute of British Geographers. This considers everything from the ethics and politics of research in the Third World to the practicalities of mosquito nets, money, insurance and health and safety. Although it was put together for postgraduate researchers, most of the issues are also pertinent for undergraduate studies in developing areas and it is written in a very accessible style. Other good sources include: Skelton, T 2001 Cross-cultural research: issues of power, positionality and 'race'. In M Limb and C Dwyer (eds) *Qualitative methodologies for geographers*, London: Arnold; Smith, F 2003 Working in different cultures. In N Clifford and G Valentine (eds) *Key methods in geography*, London: Sage. The journal *Ethics, Place and Environment* is a good place to start for articles debating the ethics of human geography research involving techniques such as interviewing. Special issues of this journal have focused on disability, geography and ethics (2000, 3 (1)) and the particular ethical dilemmas involved in working with children and young people (2001, 4 (2)).

8 Focus groups

David Conradson

Synopsis

In recent years, focus groups have enjoyed growing popularity as a research method in human geography. Across a diverse range of settings, they have been employed to explore the complex understandings and interactions that people have with their everyday environments. Subjects investigated in this manner have included recreational forest use, childcare provision, cyberspace, fair trade and urban leisure practices amongst many others. In recognition of this work, this chapter outlines the distinctive features of focus groups as a research method and then considers some of the practical issues associated with their use in human geography. Particular attention is given to the process of assembling a group (considerations of size, homogeneity vs mixed composition, recruitment incentives) and to managing group discussion effectively (using focusing exercises, the role of the moderator, negotiating power relations, audio recording strategies). Examples from undergraduate student research projects are used to illustrate some of the different ways in which these issues can be handled. The chapter concludes with a consideration of the qualitative materials typically generated by focus groups and the analytical strategies appropriate to exploring them.

Introduction

Given their relatively widespread use in school and university geography, many of you will be familiar with questionnaires and interviews as research methods (discussed in Chapters 6 and 7 of this book). But focus groups are often somewhat less familiar. While they have been much employed by Western governments in recent years, and indeed have enjoyed a measure of popularity in the social sciences, they are nevertheless not the most well known or intuitively understood of research methods. In methodological textbooks they also often receive less coverage than questionnaires, interviews, participant observation and textual analysis (see, for example, Mason, 1996; Seale, 1998; Silverman,

2000; Denzin and Lincoln, 1998). In view of this situation, this chapter provides an introduction to focus groups and their use in human geographical research. With reference to a series of student projects, it discusses some of the conceptual and practical considerations associated with this method. The aim is to provide an introductory framework that will allow you to employ focus groups in your own research projects. The chapter concludes by outlining a set of resources where questions and enquiries regarding points of detail can be pursued.

What, then, is a focus group? As R Powell *et al.* (1996: 499) note, the basic format is 'a group of individuals selected and assembled by researchers to discuss and comment on, from personal experience, the topic that is the subject of research'. If a research project concerns the shifting understandings of home amongst university students, for example, then this issue will likely constitute the broad focus of the group discussion. In terms of group size, many commentators suggest somewhere between four and ten participants as an ideal, with a researcher (or two) to facilitate and moderate the conversation (Bloor *et al.*, 2001). After an introduction the focus group typically proceeds as an issue-based discussion, with participants contributing in the ways and at the times they wish to. The conversation is usually audio recorded – although in some professional settings video-recording may also be undertaken – and this recording is then transcribed as a written text. The text may be supplemented by moderators' notes regarding group dynamic, mood, gesture and so forth (matters which a transcript can only partially convey). The result is a rich qualitative record of a focused group conversation.

As a research method, focus groups are useful in two main regards. They provide the social scientist with a way of gaining insight into (a) the spectrum of views that individuals hold regarding a particular issue, and (b) the nature of their interaction and dialogue over that issue. A focus group might thus be used to understand how business commuters perceive public transport, both in terms of their individual views and with regard to the arguments between these perspectives. How, for example, do the perceptions of those who commute by train compare with those who make the same journey by car? As Cook and Crang (1995: 56) write, 'a [focus] group is not just a way of collecting individual statements, but rather a means to set up a negotiation of meanings through intra- and inter-personal debates'. While one may periodically have the opportunity to observe this kind of interaction in a 'natural setting' such as a workplace, perhaps through ethnographic research, focus groups are about intentionally fashioning the conditions within which such debate and discussion may emerge. In conceptual terms, focus groups thus lie somewhere between individual interviews (one research participant in a relatively structured setting) and participant observation (many participants in a relatively unstructured or 'natural' setting).

The origins of focus groups

To understand the diverse ways in which focus groups are currently employed by businesses, the public sector and universities, it is helpful to consider their historical development. The origins of focus groups are generally held to lie in work undertaken at Columbia University in the 1940s. At this time Paul Lazarsfeld and Robert Merton were conducting market research for a range of businesses that explored, amongst other things, listeners' views of particular radio soap operas (Bloor *et al.*, 2001). They began with an experimental procedure that brought twelve people together in a studio and then required them to employ yes/no buttons to register their responses to simple questions about this radio material. Over time, however, it became apparent that this methodology was insufficiently capable of apprehending the complexity of the respondents' views Bloor *et al.* (2001: 2):

> *Dissatisfied with an approach which simply quantified positive and negative responses, Merton set about developing an interviewing procedure for the groups, which would help researchers to describe the subjective reactions of group members to the programmes they heard.*

In a series of subsequent studies, an alternative procedure for conducting these group-based interviews was therefore developed. In place of a method that recorded frequency counts as indicators of a social group's preferences, greater attention was paid to the unstructured and qualitative dimensions of individuals' views as articulated in their own words. The essential details of this approach were published in a now classic paper by Merton and Kendall (1946) entitled 'The focused interview' in the *American Journal of Sociology* (see also Merton *et al.*, 1990). Despite the significance retrospectively attributed to this publication, however, its academic influence at the time was relatively short-lived. The 'focused interview' faded into relative obscurity within American social science in the 1950s; the interactive techniques it encompassed were taken onboard in the development of the *individual* qualitative interview, but group interviews received less sustained attention (Bloor *et al.*, 2001).

Outside of the university, however, the commercial potential of the focused interview technique attracted growing attention. In the 1960s a number of companies began to employ focus groups as part of their market research strategy. The work of Thomas Greenbaum was especially significant in this commercial diffusion and development (see Greenbaum, 1998), although his company is now only one amidst a sizeable international industry. Entire floors of Manhattan office blocks have been developed as dedicated focus group suites, complete with moderators and recording facilities that can be hired for specific projects. Alongside their use in business, Western public-sector organisations have also made increasing use of focus groups for gauging public opinion on various matters. In Britain, for example, the New Labour government has used focus groups to probe societal views of matters such as military action, health spending and education policy. The aim in all these cases has been to develop a better understanding of the multiple and sometimes conflicting perspectives held by the public on particular issues.

It was not until the 1980s that focus groups began to re-emerge as a prominent method in the social sciences. In disciplines such as marketing, where the external interface with business had always been strong, the adoption of focus groups as a methodological practice was relatively unsurprising. For human geography and other social sciences, however, their uptake had a more complex provenance. In addition to the influence of the commercial sphere, group psychotherapy was also a significant stimulus. Such therapy typically involves 'a series of meetings [in which] group members explore their own personalities and identities, providing support for each other and the sharing of experiences helps them to come to terms with themselves' (Cook and Crang, 1995: 56). The work of Jacquie Burgess has drawn upon this psychotherapeutic practice to explore the complex nature of environmental perception and practice (Burgess, 1996; Burgess *et al.*, 1988a, 1988b). As Bedford and Burgess (2001: 121) write 'we regularly use focus groups as an efficient and interesting way of gaining insight into the ways in which people construct environmental and social issues; share their knowledge, experiences and prejudices; and argue their different points of view'. Other examples of geographical research that have employed focus groups include Jarrett (1993), Jackson and Holbrook (1995), Longhurst (1996) and Goss and Leinbach (1996).

The potential uses of focus groups

When designing a research project one needs to consider carefully which methods are best suited to generating the information needed. To evaluate focus groups in this regard we need to have a clear sense of what they enable us to do, in terms of their possible functions within a wider research strategy, as well as an appreciation of their limitations. As Box 8.1 suggests, focus groups are useful for gaining insight into the multiple understandings of participants regarding a particular social or environmental issue. They allow us to go beyond quantitative measures of support or opposition regarding, say, a proposed new motorway, and to begin to look at *why* such views are held. They can also provide insight into the debates and arguments that exist between these different views and, if used effectively, enable us to consider how such understandings differ by social groupings (on the basis of age, gender, profession and so forth). It is for these reasons that Morgan (1988a) argues that focus groups are particularly well suited to exploring topics where complex patterns of behaviour and motivation are evident, or where diverse views are held.

BOX 8.1

Uses of focus groups in research

As a core research method to:

- gain insight into participants' understandings and views of an issue
- gain insight into how these views relate to each other (debates and arguments)
- look at how such views differ by social grouping
- probe the gap between what people say and do

Also useful for:

- preliminary brainstorming: generating potential lines of enquiry
- developing survey or interview questions
- disseminating research findings and receiving feedback

Focus groups also provide possibilities for exploring the gap between what people say and what they do. An important element of methodological discussion regarding survey research, for instance, concerns the problematic nature of the relationship between reported attitudes and actual behaviour (see Chapter 6 of this volume; De Vaus, 2001). When asked to give an opinion of waste recycling, for example, many western people would acknowledge its significant environmental merits. The degree to which recycling is actually practised, however, is not always strongly correlated to this declared approval. Focus group discussion is a useful context for exploring this divergence. If a person believes recycling to be a good idea, why does he or she actually recycle so little? In an interview, where the primary relational dynamic is between interviewer and interviewee, an individual may be reluctant to discuss this discrepancy. In a focus group setting, however, where the primary interaction is between participants rather than with the moderator, the dynamic is arguably more conducive to an open discussion on why this might be the case. Individuals are able to contribute points of view without them necessarily having quite the same degree of personal association. The focus group environment also allows the researcher to return to comments and to cross-reference between participants in a more interactive manner than that afforded by a typical questionnaire or individual interview.

There are a number of other ways in which focus groups might be used in a research project. Firstly, they can be employed as a method of preliminary brainstorming, so as to generate potential avenues of enquiry for a new project. One might ask participants to identify what they consider to be the most significant influences upon the quality of contemporary family life for example. Their responses will likely provide a useful and time-efficient summary of the key concerns around this issue, and these could then be incorporated into the research design. Secondly, and in a similar fashion, focus groups can be used to generate survey or interview questions, such that their relevance to the population concerned is established beforehand. Finally, focus groups can be a useful way of disseminating research findings to user communities. This might involve making a draft report open for comment and discussion, with the research narrative then adjusted, subject to the feedback given. In this way some degree of authorial collaboration may be achieved such that the research becomes a more collective endeavour. In a policy setting, such feedback is often valuable in terms of shaping the development of future practice.

When thinking about focus groups, it is also important to be clear about their limitations. In contrast to a series of interviews, for instance, a focus group does not allow each individual's perspective to come through equally. In seeking to facilitate 'natural' forms of conversation, some participants will inevitably have more to say than others on the topic at hand. Some may contribute relatively little to the discussion at all. For these reasons, focus groups should not generally be considered as a quick solution to the time and resource constraints associated with undertaking a series of individual interviews (the 'speak to them all at once' strategy). In addition to the likelihood of uneven coverage between individuals, the group dynamic may well lead to an under-reporting of those views and opinions that individuals perceive to be controversial or significantly different from those of the others present. Such effects are perhaps particularly likely to emerge with sensitive topics such as (say) gender roles, sexuality, race and ethnicity.

What focus groups offer, then, is a way of observing individual views as they emerge within a social context. This enables the researcher to develop an understanding of the debates which occur around and between individual attitudes and positions.

Setting up a focus group

Once you have decided that focus groups are an appropriate method for your particular project, the central issues are determining a focus, recruiting participants, and facilitating discussion between them. Establishing a focus is likely to be relatively straightforward, as this should flow out of the broader research aims of the project. For a project on National Park user conflicts, for example, the focus groups might concentrate on identifying the perspectives of different user collectives. The key matters are then the recruitment of individuals to discuss the issue; the formulation of question schedules or other forms of discussion prompts; the nature of the moderator's role during conversation; and how to analyse the qualitative materials generated by the focus group. With reference to a series of student projects, the following pages consider these issues in turn.

A general design principle in setting up focus groups is one of intra-group homogeneity (in terms such as gender, race, age, class background or occupational sector) and between group comparison. In a student project on the geographies of home, for instance, gender was a key variable in terms of an interest in how male and female students experienced the move to university (see Box 8.2). The research design involved conducting four focus groups: two solely with women, two solely with men. This internal homogeneity provided an environment where each gender could speak of their experiences relatively freely, whilst the wider design also permitted the students to explore the differences between groups. As Morgan (1988b: 61) argues, 'the sharing of ideas and experiences is at the heart of focus groups, and this requires a climate of mutual respect. At a minimum, the composition of each focus group should minimize suspicion and open disagreement.' Seeking broad social compatibility in this manner is not, of course, about finding people who think exactly the same: that would be counterproductive. Rather the emphasis is upon bringing together people who have enough in common to allow the development of a productive conversational dynamic.

As noted earlier, focus groups typically have between four and ten participants, along with one or two researchers to facilitate and moderate the conversation. This is an ideal configuration; however, there may be instances where one is required to work with fewer or more participants. In practice the group size one ends up with is likely to emerge from the intersection of research design preferences (such as the nature of the social groups that are sought) and the practical constraints of who is available and willing to participate. It is generally a good idea to 'over-recruit' for focus groups, as not everyone who agrees to participate

BOX 8.2

Student geographies of home

This group project focused upon the shifting geographies of home amongst university students. In moving to a new town and beginning a degree, many students understandably experience some initial sense of dislocation. A lot seems new and different. Often as not, the uncomfortable aspects of this change are tempered by the excitement of fresh opportunities and meeting new people. During their studies, however, the place that many students consider home, and their feelings towards it, often change. When returning to their families for the summer, for instance, this once familiar place may no longer seem so comfortable. It may even feel staid or somehow restrictive. In these situations, conceptions of 'home' and 'away' are likely to undergo significant reworking.

In researching the way in which students' conceptions of home undergo change as part of their move to university, the project gave particular attention to gender differences. Two male and two female focus groups were accordingly assembled, with 6–8 participants in each group. This set-up allowed the researchers to explore the ways in which male and female students negotiated this transition, and to consider whether there were notable differences between them in this regard. Intra-group homogeneity in terms of gender, the key research variable, was maximised. At the same time, the composition between groups allowed the differences between genders to be explored.

may be able to attend. Beyond the influence of unforeseen events that prevent involvement, focus group participants may at times drop out on the assumption that others will be attending anyway, and that their individual presence or absence is of little overall significance. This can lead to unexpected group attrition. It is thus common practice to over-recruit by two to three individuals (be sure your venue is able to accommodate the full group, however, in the event that everyone does turn up).

Within the recruitment process, an important issue to consider is whether participants will be drawn from a pre-existing social group – such as a workplace, sports club, faith-based organisation or neighbourhood – or whether they will be a set of relative strangers. These two options are known as 'natural' and 'assembled' focus groups respectively. Circumstances may well force one of these models upon you, but in either case it is important to consider the implications for group dynamics and the nature of the conversation. Within natural groups, individuals are likely to be familiar not only with the issue but also with each other. Such acquaintance will, at one level, usually facilitate conversation. Equally, however, familiarity can be a barrier to disclosure, in that existing social groupings are likely to come complete with their embedded hierarchies, dynamics of dominance and submission, as well as anxieties regarding the expression of unusual or unpopular views. Here the focus group is very much connected to participants' wider lives, in that what is said may have ongoing ramifications for both the sociability of the group and the position of individuals within it.

On some occasions, therefore, a researcher may have to work with a natural focus group in order to facilitate discussion that goes beyond the reproduction of established or consensus narratives. In contrast, an assembled focus group is one where the individuals share an engagement with the issue at hand but have not interacted in any significant capacity outside the group. Here the moderator is likely to have to work in a different way to facilitate conversation, as the individuals will effectively be strangers to one another. There is clear potential for unfamiliarity to inhibit participation in these circumstances, as individuals may be concerned about group perceptions of them or their contributions. Allowing time for the development of intra-group trust will therefore be important, and prompting and drawing out may also be required. These are all elements of the moderator's role, and are explored in more detail in the following section.

In commercial or funded academic research it is common practice to provide financial incentives for people to participate in focus groups. Monetary compensation is not normally feasible for student projects, however, and so one has to consider other ways to encourage participation. A number of points are relevant here. The first is that for certain issues some people will genuinely value the opportunity to meet and discuss their views and experiences with others, irrespective of payment. This was the case for a student project that explored social attitudes towards mobile phone use in public places (Box 8.3). As a subject

BOX 8.3

Social attitudes towards the use of mobile phones in public places

The interruption caused by a loud ring tone on a train, or the distraction of someone loudly discussing their food purchases with a housemate whilst in the supermarket are new social phenomena that have arisen through the spread of mobile phone technologies. As mobiles have taken telephone conversations beyond the home and office, so concern and feeling over the encroachment of public space has grown. Slowly a body of unwritten etiquette regarding the use of mobile phones in public places is emerging, but this of course is neither centrally determined nor uniformly practised. Different people have widely differing perspectives, and therein lies one source of the social tension around mobile phones.

This group project was interested to explore how views on mobile phone use in public places differed by generation. The students undertook screening questionnaires with people on the university campus, asking them if they would be willing to be involved in a focus group discussion on the issue. This generated respondents from the researchers' own generation. The parental generation was accessed through networking outwards from adults they knew (parents included), whilst the grandparental generation was accessed through their own grandparents. Two focus groups for each generation were held, and some interesting differences in perspective were revealed between them. An important point that became clear to the students afterwards was that it may have been advantageous to try to structure the groups around those who possessed a mobile phone themselves and those who did not, as this appeared to be a key determinant of personal views that to some extent transcended generational strata.

that aroused strong feelings, many of the respondents were both articulate and eager to discuss their views. For them the opportunity to talk through the topic appeared to be sufficient incentive itself for participation. Secondly, even on a relatively constrained budget, it is usually possible to provide refreshments of some kind that add to the perceived attractiveness of the focus group event. Tea and biscuits, for example, may prove attractive to some, whilst if one was meeting in a bar, then the offer to buy a round of drinks might also be persuasive. Finally, an additional incentive for some people will be the opportunity to see an overview or summary of the research findings. They may be interested to know, for example, as were the transnational employees in another student project (Box 8.4) about the experiences of others in similar situations. In this case, it will be important to obtain contact details (postal or e-mail) of those who are interested so that you can fulfil any commitment made in this regard.

Once the group composition and incentive details have been determined, there are a number of ways of contacting potential participants. Whether this is done by letter, telephone, e-mail or word of mouth, it will be important to summarise the research, to provide individuals with information regarding the pre-arranged time and place of the focus group, and to outline anonymity considerations (regarding the use of pseudonyms in any publication of transcript material, for example). If you have chosen to recruit from an existing social group, it will be useful to consider who might provide you with access to individuals within this group. In the case of a walking club, for instance, you might wish to write to the organisational secretary, outlining your project and

BOX 8.4

Relocation experiences of transnational employees

Many large corporations now operate across national borders, splitting production and management functions for example, or simply seeking to have a presence within a series of key market regions. For the employees of these global companies, transnational relocation is at times an opportunity to be seized and in other instances a requirement to come to terms with.

This project concentrated on a British hi-tech company with operations in France and Germany. The interest was in how British workers experienced the employee relocation programme – which typically involved being away for a period of 2–3 years – and in particular how they dealt with the associated processes of familial adjustment. Rather than conducting questionnaires or interviews, it was decided to undertake focus groups with employees of the company. An initial point of contact was made through a family friend in each location, and other participants were enrolled by snowballing outwards from this person. Approval was gained from company management to undertake the focus group in each case, although the groups themselves were in the end conducted outside of work hours. Interesting findings emerged regarding feelings about local integration (and the significance of language to this), dealing with children's schooling and maintaining relations with people and places back 'home' in the UK.

the sort of research involvement you would hope to negotiate. Equally your own involvement or that of a friend may provide useful personal contacts on a more informal level. In a workplace setting it will generally be appropriate to approach the relevant manager, explaining the project and asking whether he or she might be willing to provide you with some form of access to staff (or to distribute letters for you). Working through such 'gatekeepers' will in many cases facilitate participant recruitment, but it is also critically important from the point of view of obtaining informed consent from those who have the power to give it. There is perhaps nothing more likely to draw a project to an abrupt conclusion than offending someone in a managerial position. If such individuals are surprised or feel that something significant, perhaps even potentially disruptive, has been taking place without their knowledge (such as talking to employees about their work experiences), then they are unlikely to look favourably upon the continuation of the research.

If you are seeking to assemble a group from a larger social collective, recruitment strategies will necessarily differ. In the absence of a clear point of social contact, advertising in a local newspaper or special interest publication that is read by the community of (say) students, sports people or local residents that you are interested in may be an effective way of making contact. One might also consider web advertising, as some internet bulletin boards permit project-related advertisements. Recruitment on this basis would clearly exclude those who do not have web access, but one could make a judgement about the extent of this exclusion for the social group concerned, and if necessary seek to offset the bias by employing other recruitment strategies. Once contact has been made with a small set of individuals, the group might also be enlarged by 'snowballing' along their existing social networks so as to enrol additional contacts (see the chapter on interviews for further details on 'snowballing').

With respect to how many groups one should run, there are two main considerations. The first relates to the range of subgroups you are interested in. In a project on recreational use of a national park, for example, one will want to recruit individuals from major user groups such as: walkers, mountain-bikers, horse riders and climbers. Each of these user collectives will need to be consulted, generating a need for at least four focus groups. The second issue is how many groups should be conducted for each type of user. Here the general consensus is that the researcher should aim for a number that makes 'theoretical saturation' possible, in terms of reaching a point where one has a sense of having heard all the main points of view on the issue, with decreasing instances of significantly different or novel arguments. Two or three meetings per subgroup may be sufficient to obtain this, but if time or resources are limited then a single meeting with each subgroup will still be of some value in its own right.

Running a focus group

Once you have recruited a number of willing individuals and arranged a suitable venue – somewhere with enough chairs and conditions appropriate to audio recording – you will be ready to run the focus group. The participants and moderators will normally be seated in a circle, perhaps around a table, in such a way as to facilitate interaction. Environments such as cafés, workplaces, school halls, home living rooms, and student bars are thus all possible venues in which to conduct focus group research. For some of these sites, permission will clearly be required prior to using them for focus group purposes. Perhaps the most important thing in preparing the venue is to ensure there is nothing obvious about it that is likely to significantly inhibit conversation. It is thus advisable to ensure the room is sufficiently well lit, heated and ventilated before beginning.

It is also worth taking time to consider whether the ambient or background noise conditions will be conducive to audio recording. This is important, as the recording will form your primary documentation of the conversation; without it, much of the detail of what is said is likely to be lost. You should not therefore hesitate to have a practice run at the start of the focus group, in terms of recording 30 seconds of conversation, and then playing it back to ensure that (a) the tape recorder is working correctly and (b) that it is adequately picking up what is being said. As is noted in Chapter 6, transcribing audio material is a time-consuming business, and whatever you can do to enhance the recording quality at the point of production will be time well spent. If a particular individual is speaking very quietly during the focus group, it may also be appropriate to ask them to speak a little louder for this reason.

Once everyone is seated, a good way to begin is by introducing yourself and, in a manner that seems appropriate, asking participants to introduce themselves to the group. To help initiate a focused conversation, researchers then typically employ a number of different strategies known as 'focusing exercises'. One approach involves presenting participants with statements that express a particular point of view about the topic. 'Genuine asylum seekers should be allowed unrestricted access to western European countries' might be used as stimulant for thought and conversation amidst a group considering international migrants in Europe, for example. 'Food companies systematically obscure the poor conditions in which animals are kept from the consumer' and 'A quality public health system is only possible if individuals are willing, in financial terms, to lay aside private gain for collective good' are other examples. Such statements clearly need to relate to the topic at hand, and you may wish to provide a copy for each participant to read, reflect upon and even make a few notes regarding their thoughts on it. A similar but more visual approach would be to use images as a way of initiating thought and discussion within the group. In the case of a student whose project concerned acceptable levels of public use (and by implication crowding) in the scenic natural environment of a British National Park, photographic montages were produced which showed the same landscape scene with differing numbers of individuals walking upon a path. Participants were then prompted to discuss the level of usage they considered appropriate or acceptable. A final strategy that some focus group researchers have used is to distribute preliminary questionnaires to participants, with these serving the dual purpose of obtaining participant profile data – valuable for subsequent analysis of what was said – as well as requiring individuals to begin thinking through a particular issue by virtue of having to answer questions about it. The answers to these questionnaires may also allow the researcher to observe perspectives or views that, while not explicitly raised within the conversation, were nonetheless part of the group's spectrum of opinion.

Group conversation should then be allowed to develop around the general topic or, in more managed cases, follow a series of specific questions that the moderator presents to the group (see Krueger, 1998a). In both cases the moderator may need to perform a number of different roles in order to maintain a productive and well-focused discussion. First, and perhaps most obviously, if the conversation is drying up because the group dynamic has shifted, then some form of *prompting* by the moderator may be necessary to facilitate further interaction. This could take the form of asking a question, or picking up on an earlier comment made by one of the participants. *Guiding* might then be understood as providing direction or clarity if the group appears uncertain regarding how to proceed. In some cases, *intervening* may be appropriate if the conversation appears to be getting overly heated or aggressive. The moderator may decide to gently interrupt with a view to dissipating some of the tension in the group, so as to create a short interval in which people can find a different approach to the subject. In all these forms of engagement it is important to recall that the moderator's primary role is not one of interviewing, but rather of facilitating the flow of conversation amongst group members. As noted earlier, focus groups are about observing relatively natural forms of conversation, and it is generally thought that the moderator's involvement should centre around (a) ensuring the conversation does not significantly diverge from the predetermined focus and (b) intervening when there is an unproductive conversational dynamic emerging between participants, such as when the volume or intensity of one contributor is notably inhibiting others from speaking freely (Krueger, 1998b; Morgan, 1988c).

In terms of length, most commentators suggest that focus groups should last for somewhere between 60 and 120 minutes. There is a balance to be struck between providing enough time for a group to warm up and begin talking productively about the issue and continuing for so long that attention begins to wane or individuals have to leave for other commitments (which will significantly alter the group dynamic). It is good to have a prior idea of how long you wish to proceed, and to give the respondents a broad indication of this before starting (whilst also leaving open the possibility for an extended conversation if particularly interesting issues or points emerge). The precise duration will in large part depend, however, on the group dynamic and flow of conversation. It may be the case that a group effectively covers the schedule of issues one had wanted to explore in less time than expected. Here, there is generally little value in pursuing things further, especially if the moderator senses that a point of 'saturation' has been reached, where the same points are beginning to recur, or the group is evidently flagging in energy. In such a situation it will likely be appropriate to draw the focus group to a close. In other instances, the group may have more to say, but in the interests of timekeeping you may wish to stop at, say, the 60-minute point. At whatever time you decide to finish, be sure to thank participants for their involvement, and to provide them with a way of contacting you should they have further queries or thoughts about the topic. If appropriate, you may also want to request their contact details so that you can send or e-mail them a copy of the summary research findings at the appropriate stage.

Transcribing and analysing focus group material

The qualitative material generated by a focus group is primarily, but not exclusively, textual and usually takes the form of an audio transcript. The quality of this transcript depends significantly upon the quality of recording, which in turn is related to the successful operation of the recording device (such that those speaking can be heard most of the time). An important consideration here is that to a much greater extent than in a two-person interview, a focus-group conversation is likely to involve people talking over each other. This can create difficulties with the transcribing process, and in any group there will usually be elements of spoken exchanges that are too quick or quiet to be fully transcribed. Small omissions of this kind need not be a major cause for concern, but it is useful to be aware of the symbols conventionally used to depict various forms of conversational overlap and interruption (Box 8.5). In this way it becomes

BOX 8.5

Transcribing focus group conversation

As Silverman (2001) notes, there are a number of symbols commonly used to depict pauses, overlap, tone and emphasis in spoken speech. These can be used in transcriptions, which are typically set out something like a play script, to provide a richer account of a focus group conversation. The following example illustrates how some of them are commonly used:

John:	But I thought you were going to the cinema tonight with Rachel
Kate:	(0.3) Yeah I was [but
Megan:	[What's on at the movies at the moment anyway?
Paul:	(0.7) Umm <u>nothing</u> interesting that I can see. There [was that
Megan:	[But what about that new sci-fi film? Isn't that out on Friday?
Kate:	Yeah (0.8) .hhh But Rach <u>DOES NOT DO</u> the whole sci-fi movie thing. All a bit surreal in her view. We'll probably end up going for a drink instead.

These and other symbols are explained by Silverman (2001: 303) as follows:

[	Left brackets indicate the point at which a current speaker's talk is overlapped by the following speaker's talk.
(0.7)	Denotes a period of silence in seconds.
(.)	A tiny gap, less than one-tenth of a second
<u>nothing</u>	Underlining indicates emphasis in terms of intonation or stress
DOES NOT	Capitals indicate increased volume relative to the surrounding talk
.hhh	A row of h's indicates an outbreath; when prefixed by a full-stop (.), it indicates an in-breath.

Source: Silverman (2001: 303), with some adaptation of phrasing

BOX 8.6

The significance of tone

The significance of conversational tone can be seen in the following excerpt from Krueger (1998c: 33). A textual record alone would observe that an individual spoke the words, 'This was good. But depending on tone, these words could mean many different things':

Comment	*Translation*
This was GOOD!	(It was good)
This was GOOD?	(It was supposed to be good but wasn't)
THIS was good!	(This one was good, but others were not)
This WAS good.	(It used to be good, but not anymore)

possible to show when a person interrupts the speech of someone else, as well as to represent pauses and rising and falling intonation (Box 8.6; see Silverman 2001). This allows the researcher to capture a little more of the context and emotional dynamic present in the group conversation.

There is much discussion in human geography at present regarding the methodological limitations of focusing upon talk and text as proxies for the rich complexity of social life (see Latham and Conradson, 2003; Rose and Thrift, 2000). As part of this non-representational work, new methodologies are being developed that seek to give greater attention to the embodied, unspoken and habitual dimensions of everyday life. Within the limits of audio transcription it is thus useful to think about ways of supplementing the textual record of what is spoken – the sequences of words which form particular sentences – with details of emotional tonality, forms of expression and notes on general group dynamic. Some of these dimensions will be partially registered on the audio tape, whilst others might usefully be noted by hand during the meeting. One thing that will likely prove useful for subsequent analysis is a sketch of the group configuration, indicating participants' names and their position in the room. This may be of help in trying to differentiate between individual speakers on the recording at a later date. When gathering this extra information the presence of two moderators is invaluable, as one can focus on facilitating the conversation whilst the other attempts to record some of the broader dimensions of group interaction. Some of the qualitative analysis software packages available at present, such as NVivo, usefully allow the researcher to append things such as digital photos, video clips and observations regarding mood to the standard textual record of the conversation (see Richards, 1999).

This brings us to the matter of analysis. In many areas there are significant similarities between the analysis of focus group and interview materials. First, the researcher is centrally interested in the themes that are evident within the discussion of the research issue. How, for example, did the shift in conceptions of home differ between male and female university students (Box 8.2)? How did different generation groups view the use of mobile phones in public places (Box

8.3)? What sort of difficulties did transnational employees face as a result of relocation (Box 8.4)? One will want to search for instances where individuals or the group touch upon particular experiences and concerns in these regards. It will also be important to be attentive to the unexpected issues and ideas that emerge within the conversation, whether these fit with the *a priori* research framework interests or not. Given that a large part of qualitative research is directed towards understanding a person or group's subjective world, such unforeseen responses may be of particular significance. The thematic analysis can be conducted in the same line-by-line coding fashion that is now commonly used for qualitative text, and which draws upon notions of grounded theory (see Chapter 13; Glaser and Strauss, 1967).

The analysis of focus group material also differs from that associated with interviews in that we are additionally interested in the *interactions* between different positions and the *arguments* marshalled in these exchanges. So alongside the coding of individual themes, it will also be important to consider whether there are certain forms of conversational exchange that recur throughout the focus group, and to look at how particular positions are related. Are the individuals in agreement? Do their views conflict? In what ways? Is there any significant connection between types of people and the nature of their points of view? These are all matters worth exploring in regard to the conversational data that focus groups produce.

The overall aim of the analysis should be an understanding, as noted earlier, of the spectrum of different views a group of individuals has on an issue (Krueger, 1998c). For the National Park example mentioned earlier, we would thus seek a greater awareness of the different concerns of walkers, mountain-bikers, horse riders and climbers as significant users of this environment. We would also want to see how these groups perceive each other's activities and the nature of the tensions and arguments between their perspectives. With this kind of knowledge, we may be better able to develop policy that takes into account the concerns of each group.

Concluding remarks

Focus groups are a method that has seen growing popularity in human geography in recent years (Crang, 2002). In this chapter the aim has been to outline the basic nature of focus groups and to work through some of the practicalities associated with their use. Within a student research project, the major issues that warrant consideration have to do with recruitment, group composition (and how this relates to a project's analytical concerns with profile variables such as, say, gender or profession) and the role of the moderator. We have also touched upon some of the issues associated with analysing focus group transcripts. This discussion should have provided you with the basic starting points needed to employ focus groups in your own research. In conclusion, it is worth reiterating that focus groups are able to generate rich conversational data that provides insights into complex social issues. This material is relatively unstructured, however, is not statistical and is not readily generalisable to a wider population. It is thus important to consider whether focus groups are the most appropriate method for addressing your research aims. In some cases they can be productively used in combination with other more extensive methods such as questionnaires. There is no *a priori* need to do so, but it is worth bearing in mind how the strengths and limitations of focus groups might be balanced by those of other research methods.

BOX 8.7

Further reading

This chapter should provide sufficient information for you to make a start on using focus groups in your own research project. For those who wish to explore the conceptual background or practicalities of administering focus groups in more depth, the works below should prove valuable.

Barbour, R and J Kitzinger (eds) 1999 *Developing focus group research*, London: Sage

Bloor, M, M Frankland, M Thomas and K Robson 2001 *Focus groups in social research*, London: Sage

Krueger, R 1998d *Focus groups: a practical guide for applied research*, 2nd edn, London: Sage

Morgan, D 1988c *Focus groups as qualitative research*, London: Sage

Stewart, D and P Shamdasani 1990 *Focus groups: theory and practice*, London: Sage

9 'Participatory' approaches and diagramming techniques

Mike Kesby, Sara Kindon and Rachel Pain

Synopsis

In this chapter we invite you to consider taking a participatory approach to some or all aspects of your research project and provide you with a practical introductory guide to help you achieve this. Drawing on our own experiences in both 'developed' and 'developing world' contexts, we discuss group participatory diagramming techniques as one way to undertake participatory research. We suggest that the advantages of adopting these techniques and approaches are that they will enable you to work with hard-to-reach groups in ways that can both generate rich information and affect positive change in participants' lives.

Introduction

In this chapter we introduce you to the philosophy and techniques associated with a growing group of research approaches, loosely referred to as 'participatory', which are just beginning to make an impact in our discipline (e.g. Rocheleau *et al.*, 1995; Kindon, 1998; Kesby, 2000; Kitchin, 2001; Pain and Francis, 2003). They emerge at a time when a broader 'relevance' debate is resurfacing within critical human geography (see the special issue of *Area* edited by Kitchin and Hubbard, 1999, for a useful introduction) and offer one means to try and make geographical research more relevant to the lives of ordinary people. Our intention here is to encourage you to think about the possibility of 'doing research differently' not simply to add another useful technique to geographers' ever-expanding 'toolkit'. Ideally, participatory approaches are about working *with* rather than *on* people; about generating data and working in ways that increase participants' ability to bring about positive change in their own lives. Like other forms of 'action research' they are about changing, not simply describing or analysing, social realities (Pratt, 2000). Thus we encourage you to recognise a distinction between alternative research *techniques* such as participatory diagramming, and a participatory *approach* to research; one does not necessarily imply the other. Our intention is to introduce you to both, and to encourage you to think about how you might apply them in your own research project.

This said, we recognise that a fully participatory approach presents major challenges, particularly for undergraduates who must work within strict time-frames and limited resources. Rather than being puritanical about the need to 'do participation deeply' or not at all, we recognise that the road to 'doing research differently' has to begin somewhere. Therefore, we encourage you to use diagramming techniques to generate data, even though the level of 'participation' in your project may be relatively low and its action-oriented outputs limited. Likewise, while we will show you how you can make every aspect of your research project participatory, we expect that you will only feel able to use some of these ideas, at least initially. As we will explain, the key is to be honest with yourself and your participants about what your work might realistically achieve. While participatory approaches and diagramming techniques may not offer panaceas for all the ethical, political and practical issues of field investigation, we believe they do present you with many exciting possibilities for doing and thinking about research differently.

After reviewing some of the main ethical and epistemological issues, we present the basic elements of participatory diagramming techniques and a participatory methodology, including issues of access, sampling, design, sequencing of techniques, analysis and dissemination of information. We illustrate our arguments by offering some examples and reflections from our own experiences of practising participatory research in Indonesia, Britain and Zimbabwe. We then move on to offer some practical advice about how you might build participation into some or all aspects of your research project (whether or not you choose to utilise diagramming techniques to generate data). Finally, we discuss the advantages of using diagramming techniques and participatory approaches in terms of the richness, quality, rigour and validity of the data produced.

'Bothering with philosophy'

As Elspeth Graham argues in Chapter 2, it is important for you to grapple with the philosophical issues and underlying assumptions that forge and give purpose to the methods you use. Without such a starting point, 'participatory' diagramming techniques might seem familiar enough; the only innovation being that participants, not you the geographer, produce the maps, transects, charts and matrices that describe the research results. Without an understanding of the philosophy of participation, you might think that respondents' collective production and discussion of diagrammatic data is just an interesting variation on wholly discursive focus group interviewing (see Chapter 8 and the special issue of *Area* 28(2), 1996). However, such an interpretation overlooks the 'scare quotes' we carefully placed around the term 'participatory'. These imply that techniques alone will not make your project genuinely *participatory*; for that, techniques must be guided by an epistemology that conceives the process and purpose of investigation, and your role as the researcher, rather differently from most conventional research.[1]

We might characterise conventional research as a process that begins with an externally developed research design, proceeds with the extraction of data from 'the field' and their transportation to distant research institutes for lengthy processing by experts, and ends in the presentation of results at scholarly conferences and in academic journals and, occasionally, in reports to policy-makers. The knowledge produced in this way is seen as valid because it has been generated rigorously by specialist researchers and vetted by their peers and is useful to other experts and decision-makers. However, because this approach centralises control in the hands of the researcher, it tends (regardless of which particular techniques are used) to distance respondents from the process of knowledge production, and minimise the benefit researchers can gain from local people's insights and understandings. Moreover, while it may be guided by the ethic 'do no harm', it usually produces few positive impacts for those who are researched (see Katz, 1994; Hagey, 1997; Kesby, 2000).

An alternative participatory research epistemology has emerged over several decades among some researchers studying or based in the 'developing world' who have become dissatisfied with the intellectual and practical limitations of conventional research. Within the family of techniques and approaches they have developed (grouped under the umbrella term Participatory Action Research or PAR: see Whyte, 1991; Chambers, 1997; Selener, 1997), the term 'participant' rather than 'informant' or 'respondent' is significant. This signals a shift in understanding about *who* should instigate, conduct, analyse, present, act on and benefit from a research project. As in other qualitative and ethnographic approaches, the knowledges and experiences of participants are recognised as valuable and they are encouraged to give voice to them. However, PAR takes this acknowledgement of participants' capacities a step further by recognising that they also have the ability to take an active role in controlling and directing the research itself. Given the opportunity of a more reciprocal relationship with researchers, ordinary people can work together to identify pertinent research questions, design and utilise techniques and generate and analyse data about their own lives, not merely answer questions posed by experts (for example see Kindon and Latham, 2002).

In this mode of knowledge production, researchers take on new facilitative roles (i.e. making it possible for people to participate in this way) as well as the usual inquisitorial role. The researcher's job is to create a physical and social environment in which self-research can take place, by developing and helping to monitor systems and ground rules that enable all participants to express their opinions (we show what we mean by this later in the chapter). Researchers also perform more familiar roles by challenging participants' perceptions, asking them to explain or justify their responses, and recording the data generated and

[1] A simple definition of terms would be: (a) *method* = a technique for gathering evidence, (b) *methodology* = a theory of how research should proceed through its various stages and phases and (c) *epistemology* = a theory of knowledge defining its source, limits and criteria for evaluation, and the strategies used to justify these claims (see Harding, 1986).

the discussion and processes through which it was produced. These interventions are intended as a catalyst to the thinking and learning of participants as well as that of researchers'. This is because PAR's central objective is to promote the kind of positive outcomes that other qualitative approaches regard as fortuitous side effects. These might include improved communication skills among participants, a sharing of experiences, a process of collective learning and the challenging of harmful myths and behavioural norms (see Goss and Leinbach, 1996; Longhurst, 1996; for a PAR example see Kindon, 1998). So the validity of a PAR project is gauged on the quality of the data generated *and* by the extent to which the process of research itself develops the skills, knowledge and capacities of participants to use the results *themselves* to tackle problems that *they* have identified (Maguire, 1987; Cornwall and Jewkes, 1995; Chambers, 1997). Many variants on PAR have been developed which have different emphases, but all share this central tenet.

In the following section we want to give you a basic introduction to participatory diagramming. This is *one* of the alternative research *techniques* developed within PAR. Other techniques have involved researchers and their partners experimenting with the use of creative and performing arts techniques such as drama, song, photography and video. These can also be extremely effective in giving groups the opportunity to develop and transmit their ideas in ways that are appropriate and meaningful to them (for an example of participatory video see Kindon, 2003). These techniques are worthy of their own chapter, but given the timescale, resources and experience usually available to undergraduates, we believe it is sensible for us to concentrate on participatory diagramming as a suggested first step into 'alternative' methods and participatory approaches.

Doing participatory diagramming: the techniques and a methodology in outline

The basics of visual diagramming techniques

Participatory diagramming describes a set of techniques increasingly used in participatory research. While diagramming techniques have similarities with focus group interviewing in that they are group activities, the differences are significant. First, in a diagramming session the researcher is not the focus of attention, at least initially, because after raising a question and suggesting a visual means of addressing it, he or she takes a 'back seat' while participants generate the diagram. Secondly, because the task of diagramming is tactile as well as verbal, it facilitates the contributions of less eloquent individuals in the group. Because the diagram becomes the focus of everyone's attention, quieter or less confident participants can contribute without having to do so in a 'face-to-face' verbal exchange (a difficult skill, as you will know from tutorials). There is a school of thought that strongly believes that diagrams should always be constructed on the ground, to put everyone 'on the same level' and further reduce the likelihood of intimidating eye contact. We suggest you use your initiative: this approach is unlikely to be popular with everyone, e.g. older or disabled people may find it difficult. Finally, diagrams can incorporate words or, to ensure that literacy is

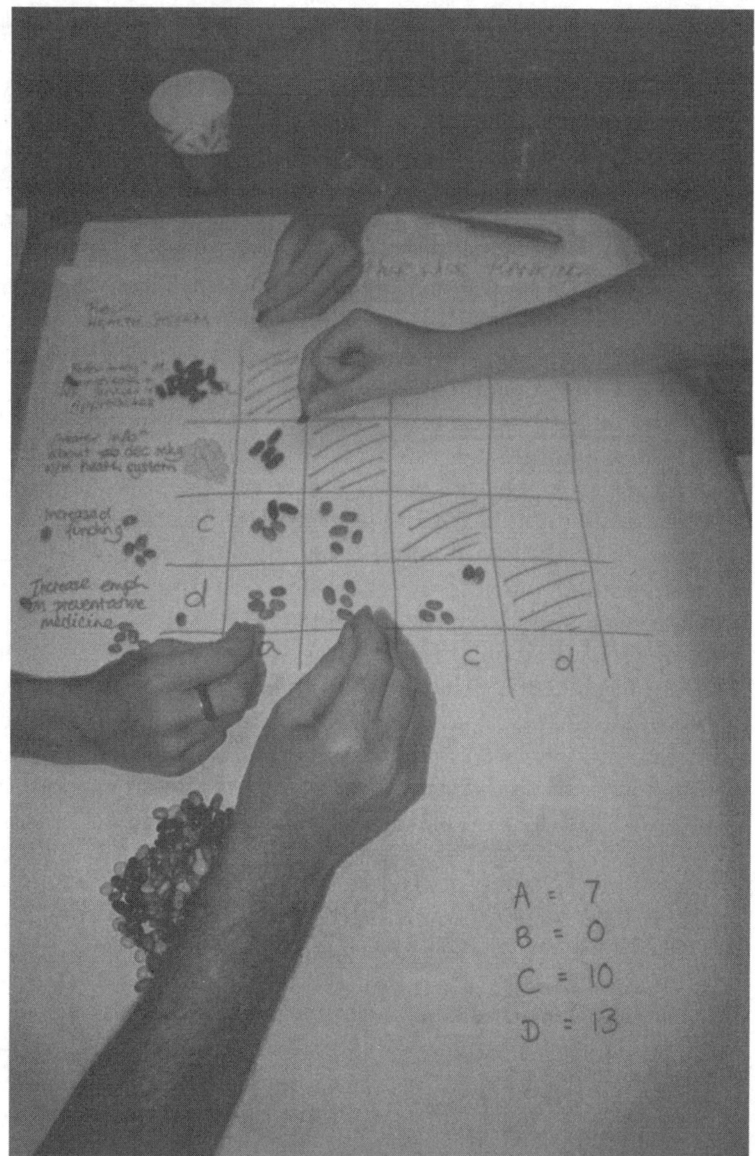

Figure 9.1 Pair-wise ranking: women compare, and give a relative score to their health needs
Source: Kindon: Bali,1991–92

not a barrier to some people's participation, can be entirely constructed using symbols or objects that represent key ideas and phenomena. They can be made using locally available materials (e.g. seeds, bottle tops, pebbles, sticks, straw, household objects or representations drawn in the soil) or stationery provided by researchers, e.g. large pieces of paper, marker pens, Post-it notes and coloured sticky dots (see Figure 9.1). Much depends on the environment, the resources and preparation time available: the key point is that participants should be comfortable with the materials chosen.

There is an abundance of diagramming techniques to choose from, ranging from sketches, cartoons, transects and maps, to pie charts, flow diagrams and matrixes (see Mikkelson, 1995 for some examples). All are highly visual, and most are highly flexible – the use of loose materials gives participants fewer inhibitions about changing and adapting the diagram as they go along. Some good manuals and study packs are available from the Institute of Development Studies at the University of Sussex (IDS, 2000) but these are only guides and need not be followed rigidly. One of the great attractions of diagramming as a technique is that it enables you to develop specific innovations as you go along which suit your research questions and the participants. We will give some examples of techniques later in the chapter, but the general point is that diagrams can be used to plot and describe a great variety of socio-economic, material and discursive phenomena. They can be used to elicit and represent historical or contemporary data, as well as predicted future patterns or aspirations (the resources produced by IDS and IIED give more details).

A basic methodology

Five points about the methodology in Box 9.1 are worth elaborating. First, the initial diagram should not be viewed as an end product; rather you should see it: (a) as a mechanism through which to gather further qualitative interview data, and (b) as a means to stimulate further discussion and self-analysis among the participants. The visual, flexible and tactile nature of diagramming helps, as it gets sensitive issues 'out in the open' while at the same time presenting them as unfinished and open to debate. Making time to 'interview the diagram' is crucial. Notes you have taken while participants worked (point 5, Box 9.1) may help as a guide to identifying differences of opinion or issues that you would like to explore in more detail.

Secondly, plan your *methodology* so you have mapped out in advance: (a) the particular diagramming session (Box 9.1), but also (b) its place within the project as a whole (which may consist of a number of diagramming sessions, and perhaps other methods too). You should give careful thought to the sequencing of questions and diagrams. The usual practice is to start with fairly broad questions and issues, giving people the opportunity to raise new research questions independently, and then to go on to ask them to focus on key topics. In addition, think beforehand about what kind of information a particular diagram will generate, and try to ensure that the topic, data and discussion from one session lead naturally into the next. For example, you might move from (a) brainstorming what the group's problems are, to (b) focusing on those associated with the research project, to (c) thinking in more detail about what the causes and impacts of these are, to (d) prioritising the most important of these, to (e) suggesting or evaluating solutions to these priority concerns. Your methodology might have a third level if you choose to use participatory diagramming as part of a 'mixed method' approach (perhaps in combination with more conventional methods like secondary literature reviews, semi-structured interviews or participant observation – see Chapters 4, 7 and 10). In this case you will need to think about the sequencing and role of the participatory diagramming stage within your overall research design (as well as the philosophical assumptions

BOX 9.1

A methodological guide to working with diagramming techniques

1. Research design: Ideally done in consultation with research partners and participants (see Box 9.2) so that you generate a series of research questions and diagramming tools together. Both of these should be written down in a clear and simple form as your aide-mémoire to fieldwork. While questions and techniques are always open to negotiation with participants during sessions, diagramming will go much more smoothly if you have prepared well beforehand and explain your intentions clearly to participants.
2. Ask a question, and explain the diagrammatic format through which you would like participants to address that question. Provide or suggest the materials they might like to use.
3. Encourage participants to discuss their possible responses with each other.
4. Look for a good moment to encourage them to begin diagramming and working visually if they are not already doing so.
5. While participants are doing this, observe and record group dynamics, and be ready to facilitate if difficulties arise.
6. Once the diagram is 'completed', photograph and/or sketch it.
7. Now 'interview the diagram' (see below): ask for explanations of its elements and how they fit together. Ask about the conflicts and consensus that arose during its creation, and compare it to other diagrams produced in the study or to other sources of data. Record these discussions.
8. Encourage participants to respond to your questions by analysing their own responses and fine-tuning their diagram.
9. Once discussion is exhausted, encourage participants to make their own record of their final interpretation by transferring their data to a large poster for reference and use in the future. Make your own copy of this final diagram.
10. Move on to the next question and its diagram.
11. Decide with participants what to do with the data generated (for example, hold further sessions to explore particular points, elect delegates to present the data to other groups or influential decision-makers and/or act on the data locally).

that underlie each of your multiple techniques when deciding which data is most valuable to your project – see Graham, 1999).

A third and related point refers to the dual purpose of your methodology. Ideally, in a participatory project, diagramming techniques should aim to achieve two goals simultaneously: (a) obtain the best and fullest impression possible of the subject you are investigating, and (b) facilitate participants' own learning, self-reflection and analysis. Diagramming has the advantage of allowing participants (not just researchers) to immediately see the provisional results of research, to analyse and discuss them and ultimately, it is hoped, to act upon the findings. So both in terms of your own interpretation and in terms of participants' self-analysis, individual diagrams should not be seen as standing alone.

A fourth point is that if alternative techniques and participatory approaches are to run smoothly, considerable forward planning is required (see point 1 in Box 9.1). This should begin, of course, with a thorough literature review (see

Chapter 4) that gives you the contextual sensitivity and background knowledge to interpret and critically challenge the 'local knowledges' of participants. Do not think that the 'flexible' and 'relaxed' approach recommended by advocates of participation means that you can turn up at a group meeting without a clear idea of what you will be doing. Even when well organised, sessions will always be quite lengthy and demanding, particularly if you are undertaking most research roles yourself (facilitator, interviewer, data recorder and 'anti-saboteur' [i.e. dealing with disruptive individuals and events which can sometimes arise]). It often helps to have a friend or colleague to help facilitate, especially with large groups. If you are going to bring groups together for a 'plenary', you need to prepare strategies for dealing with conflicts and dominant characters and groups.

Finally, just to reassure you that although what we have described above may sound complex and daunting, if you prepare thoroughly, diagramming is often very rewarding and enjoyable for both you and your participants. You can cover a wide range of issues relatively quickly and get on to difficult subjects more readily than if you use more established methods. Participants are likely to be far more involved, and the techniques can accommodate a wide range of local environments and levels of skill (including literacy). While sessions can be short, participants frequently volunteer to continue for lengthier periods (as long as you have remembered to provide them with refreshments!).

To illustrate what we have been saying so far, we will now present some examples from our own experience (since the approaches are quite new to geography there are few undergraduate examples we can draw on). The first example is taken from Sara's Master's research, the second and third draw on Rachel's and Mike's work as full-time academic researchers. Their context, topics, research design, techniques and action outputs may seem more demanding than you would expect to encounter within your own project; however, they will give you ideas and guidance which you can adapt to fit your own context. You may wish to read them with an eye to what we say later, particularly in Box 9.2.

Using a 'participatory' research methodology in 'developing' and 'developed world' contexts

Case Study One

Gender and sustainable development planning in Bali, Indonesia (Sara 1991–1992, with Putu Hermawati)

Aims: This project aimed to find out about the roles, relationships and needs of women and men in rural Bali, Indonesia, so that these could be integrated into development planning and policy at the provincial level. In an effort to make planning more appropriate and sustainable, the provincial government was keen to understand the realities of people's lives and particularly to integrate women's needs into their plans. The project was carried out with representative groups of men and women in two villages in different types of farming environments, and with a range of government officials at different levels (Kindon, 1995).

Case Study One continued

Research design: I worked and lived with a Balinese co-researcher, Putu Hermawati, in both villages where we selected two to three diverse temple communities for inclusion. Within them, women and men representing the range of ages, marital status and occupational characteristics of their communities were invited to participate in a series of gender-segregated workshops. A range and sequence of participatory techniques were used in these workshops to generate information and analyses about gender and development. Information from these workshops was discussed at a village-level meeting and priorities for action were identified. Representatives of each village then accompanied Putu and Sara to present their findings and analyses to regional government officials and request support to meet their gendered development needs. The process of selection, workshops and meetings took about three months for each village. This time period also included other aspects of contextual research (see Kindon, 1998 for more detail). In addition, we held interviews with key government officials and a workshop at provincial level where all information generated was discussed and participatory training in gender analysis for government workers and researchers was facilitated.

Access: We worked through local male leaders to contact villagers. The support of these headmen was critical to the project's success, but in some cases also inhibited the involvement of certain people, particularly women. Other participants were sought through personal visits, or through impromptu research sessions which occurred while we participated in daily activities. These additional informal activities ensured that the perspectives of more marginal community members were acknowledged. In total, a sequence of six workshops was held with each group and over 300 people from village to provincial government level took part in the research.

Techniques: Techniques used within the workshops complemented the wider contextual research we conducted while living in the village communities and enabled us to tease out issues and relationships in more detail. In terms of diagramming, three techniques were used. (a) *Cartoon picture cards*: participants used pre-drawn cartoon picture cards of typical village and household scenes to stimulate the telling of their own stories about family life, gender relationships and environmental issues in their communities, and to stimulate discussion about development needs that they would like to meet. (b) *Needs assessment and priority selection*: I made drawings of the development needs identified under the guidance of participants who then selected their three priority needs by placing a coloured sticker next to the relevant drawings (Figure 9.2). Women's and men's needs were compared in a subsequent joint meeting. (c) *Mapping*: participants drew a map of their community and placed coloured stickers in appropriate locations which identified the natural (green), cultural (red) and institutional (blue) resources that could help them meet some of their needs. Discussion was encouraged during and after each technique, and Putu and Sara recorded the key points for future reference.

Analysis: Men and women were brought together in the last meeting in each temple community and in the village-level meetings. This enabled them to communicate their findings and analyses to each other. The village, regional government meetings and the provincial-level meeting all helped to refine, cross-check and validate

Case Study One continued

Figure 9.2 Needs assessment and priority selection: men indicate the development priorities
Source: Kindon: Bali, 1991–92

data generated and involve a more diverse range of people in analysis and discussion. Putu and I then wrote recommendations based on these discussions for a sustainable development policy for Bali, which was presented to the Provincial Planning Agency.

Reflections on the fieldwork: Issues of access, rapport, cross-cultural communication and intra-communal diversity are common within participatory research in 'developing world' contexts. In Bali, it was important to hold meetings at convenient times for both women and men, which meant we first had to find out the daily time activities and movements of people within each community. Most meetings were held in the late afternoon or early evening in the local community balai (hall). Participants enjoyed the techniques and found them easy to relate to, particularly because they did not require them to read or write. Alongside the workshops, it was critically important to live with the people with whom we were working, sharing their food and helping out when needed. This developed rapport and provided informal opportunities to involve a wider range of people. Being able to speak Bahasa Indonesia was essential to my understanding and communication but, as most participants also used Bahasa Bali, Putu's involvement was vital to the groups' smooth facilitation and the analysis. In some groups, it became clear that caste differences inhibited the full participation of some lower-caste men and women, and in hindsight, it might have been beneficial to have subdivided the women and men into smaller, less diverse peer groups.

Case Study One continued

Outcomes: When Putu and I visited the two villages in 1998, a number of participants were keen to show us how the development needs identified through the workshops had been met by the regional government or themselves. In the upland village, the government had provided a sorely needed local health clinic to their northern-based community (a need identified by women) and paved a considerable stretch of the village's main road (identified by the men). The government also instituted training in water system maintenance, which had been identified as an issue particularly by women. For the first time in Bali, women were involved in training so they could repair the systems independently of men. From their own initiative, participants had set up a revolving credit scheme and generated enough money to pave a minor road and enable the easier transportation of agricultural produce to market. This had increased the incomes of more remote families and strengthened connections within the community as a whole. While these outcomes were encouraging, most of the needs met were practical and did not challenge the more fundamental gender inequalities also identified within the research. The more strategic issues identified within the workshops, particularly by women, remained unchanged.

Case Study Two

Nothing to do and nowhere to go – young people's needs in County Durham (Rachel 2002, with Sally Gill)

Aims: This project aimed to find out what young people thought of the provision of services and facilities for their age group in some of the more deprived villages of County Durham in north-east England. It was carried out for a local youth project that wanted evidence about the need for new drop-in points to provide a range of advice (e.g. on sexual health, counselling, training and employment) in relatively remote areas. In several villages, youth clubs had been closed down. Some community facilities were available, but they tended to be controlled by older people who did not always welcome young people.

Research design: Ten young people were recruited as peer researchers, who already had links with the youth project and were keen to get involved in the research. They had a good knowledge of the areas and knew the best places to access other young people. Peer researchers were provided with a half-day training session in research methods, materials and refreshments. Each was given a certificate of achievement from the University.

Access: Some young people were contacted via pre-arranged sessions in schools and colleges, and others were accessed on the streets by the peer researchers. This helped ensure that the views of so-called 'hard to reach' young people were included. In total, 178 young people aged 11–25 took part in the research.

Techniques: Because some sessions were informal and tended to be short (especially the street work), techniques were kept to a minimum and aimed to elicit as much data as possible in a short time. Discussion was encouraged during and after the

Case Study Two continued

diagramming, and researchers recorded the key points. (a) *H-form*: participants drew a large 'H' on flipchart paper. They marked on the horizontal line a score between zero and ten to show how they rated the things for young people to do in their area. They then marked down one side of the 'H' diagram the good points about the area for young people, and on the opposite side the bad points. Underneath the horizontal line, they marked on the improvements that they thought could be made to make things better for young people in their area. Finally they were asked to write on the back of the form their age, gender and the area they lived in. (b) *Mapping*: participants mapped their village. They were asked to mark on their map the places that young people go at the moment, and then to stick Post-it notes on their map where they thought new facilities were needed (Figure 9.3).

Analysis: A final meeting was held for all peer researchers and other interested young people to bring together the main points in the findings. Here young people had input into the final recommendations. Their views and evaluation of the significance of the results (for example, which villages had the greatest problems) were given priority.

Reflections on the fieldwork: There are always questions about how representative of the wider group participants are, and how far they represent the range of viewpoints rather than prioritising their own – in this project, for example, the

1. 'Byker Grove' type place–open all day during holidays and after school
2. Need lights
3. Need a shelter
4. Pond with benches
5. Need lights
6. Need lights
7. Not allowed in, not even the toilets

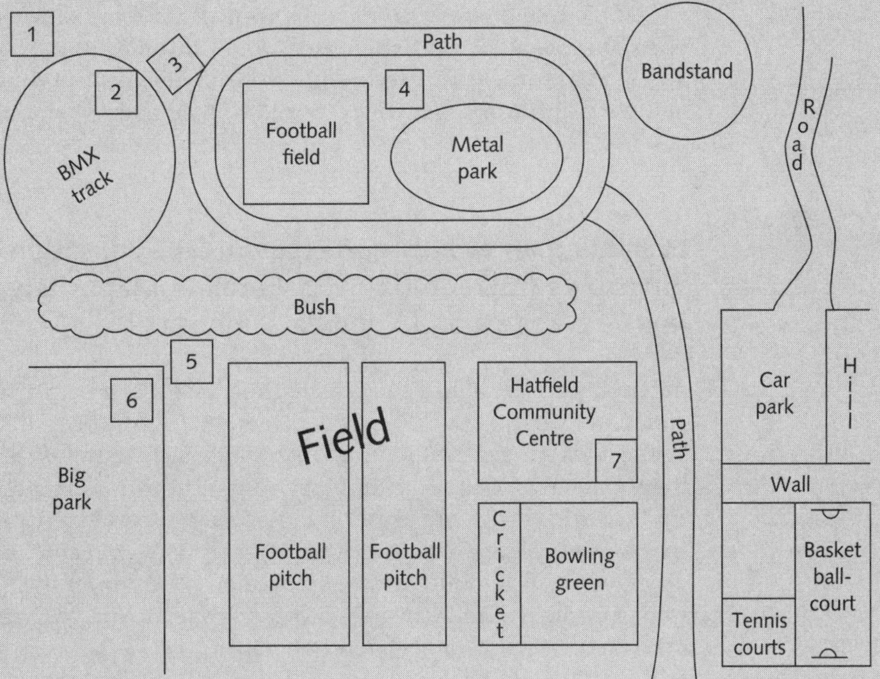

Figure 9.3 Mapping: a young woman's map of her local park with Post-it notes identifying changes she would like to see
Source: Pain: County Durham, 2002

Case Study Two continued

peer researchers were mostly young women. However, these concerns are greater still when 'professional' researchers undertake the research alone, as we are often even less 'like' a study's participants in terms of age, social background, level of education and area of residence. As peer researchers were taking part voluntarily, there were problems organising meetings, and not everyone turned up every time. Inevitably, with a large number of researchers on one project, there is variation and flexibility in how the research is carried out and in the depth of material gathered. Much of this, though, is to be expected when undertaking diagramming outdoors in the wind and rain on the Durham moors and, to us, the benefits of accessing such a wide group of young people outweighed the costs. Whatever the context, participatory diagramming almost always involves adapting to circumstances and to the particular group taking part.

Outcomes: The youth project was able to use the final report to seek funding for drop-in points and outreach work in the villages which the young people identified as having the greatest need. The fact that the local Council and police beat-officers gave the project their support from the start helped in achieving this. Young people will be consulted further about the exact nature of the changes. However, like much participatory research which is primarily led by organisations rather than the marginalised groups they represent, the project did not involve the deepest level of participation: young people did not take forward findings and action themselves. While as academic researchers our involvement was relatively brief, when we left the field it was with some confidence that some benefits would come to the young people who had participated, based on their views and ideas. Knowledge and skills in participatory techniques were also left behind which the young people and the youth workers involved would be able to employ again in the future.

Case Study Three

Evaluating an NGO's participatory HIV education project in rural Zimbabwe (Mike 2001, with Caroline Maposhere, Irene Moyo, Irene Tavengwa and Themba Mhlanga)

Aims: This project aimed to evaluate the impacts of a participatory HIV education and training project run by a local non-governmental organisation (NGO) the year before. I was interested to know what the impacts of the programme had been on the social–sexual behaviour of participants and to compare these with those who had not attended. More broadly, I was interested to explore the sustainability of the 'empowerment' facilitated through involvement in participatory projects. While diagramming techniques were employed and the project was designed in partnership with the NGO, the main goal was to evaluate the programme: the reflection, participation and action of research subjects were only secondary objectives on this occasion.

Access: I gained access first through the international NGO that developed the training programme, and then via the local NGO they had sponsored to deliver

Case Study Three continued

the training package. Both organisations became research partners and together we formed a research team. Political instability meant access on the ground via the local NGO's trusted volunteer field-officers was essential and they helped us convene a community meeting at which we explained the evaluation and sought informed consent from participants. However, we were careful to distinguish ourselves from the local NGO, when talking about the outcomes of the research and when handing out refreshments and compensation payments so as not to create future expectations that our partners could not meet.

Research design: Over a three-week period we attempted to work with as many people who had participated in the original NGO programme as we could find and to compare their responses with those of non-participants from the same area (including among others, elders, clinic staff and commercial-sex workers) and a control group from another area. We used a mixture of methods, including questionnaires (phase 1), participatory group diagramming (phase 2), and semi-structured interviews (phase 3). The three phases of research and the sequencing of diagrams in the second phase were cumulative and rolling, so that researchers and participants took data and ideas generated in one session and developed them in the next. This enabled triangulation[1] of data and enhanced the rigour of the results, but also facilitated self-reflection among participants.

Techniques: (a) *Ballot style questionnaires*: a series of questions (on HIV-related knowledge, attitudes and practice) were read out to a peer group and individuals privately marked a symbol on their own ballot paper representing a 'yes', 'no' or 'don't know' response. This generated descriptive statistical data very rapidly and enabled the research team to give immediate general feedback to groups in which some individuals displayed poor levels of knowledge about HIV. (b) *Flow diagrams:* using cards, words and pictures, string and bottle tops, participants were able to construct flow diagrams that illustrated, grouped together and interrelated the contexts in which sex commonly took place as well as to indicate their perceived relative frequency. When 'interviewing the diagrams' we were able to ascertain that participants believed that coercive sex and multiple partnering had decreased among participants of the programme as a result of the training, while mutually negotiated sex and monogamy had increased. (c) *Tree diagram:* In this diagram responses were structured in the shape of a tree. The 'trunk' was a card indicating 'what strategies do you use to avoid risk?', while the string and card 'roots' and 'branches' represented (respectively) the strategies utilised and their likely positive and negative outcomes. Each 'root' and 'branch' was given a score out of ten using bottle tops to indicate the frequency of the phenomenon. (d) *Stepping-stones diagram*: participants brainstormed about their hopes for future sexual health and indicated these on cards on the floor. Next, the 'banks of a river' were constructed on the floor using string, dividing the participants on the side representing 'where we stand today' from their 'goals for future sexual health', which lay on the other

[1] Triangulation is a term borrowed from surveying and refers to the cross-referencing of one piece of evidence with another in order to better determine what the actual position is.

Case Study Three continued

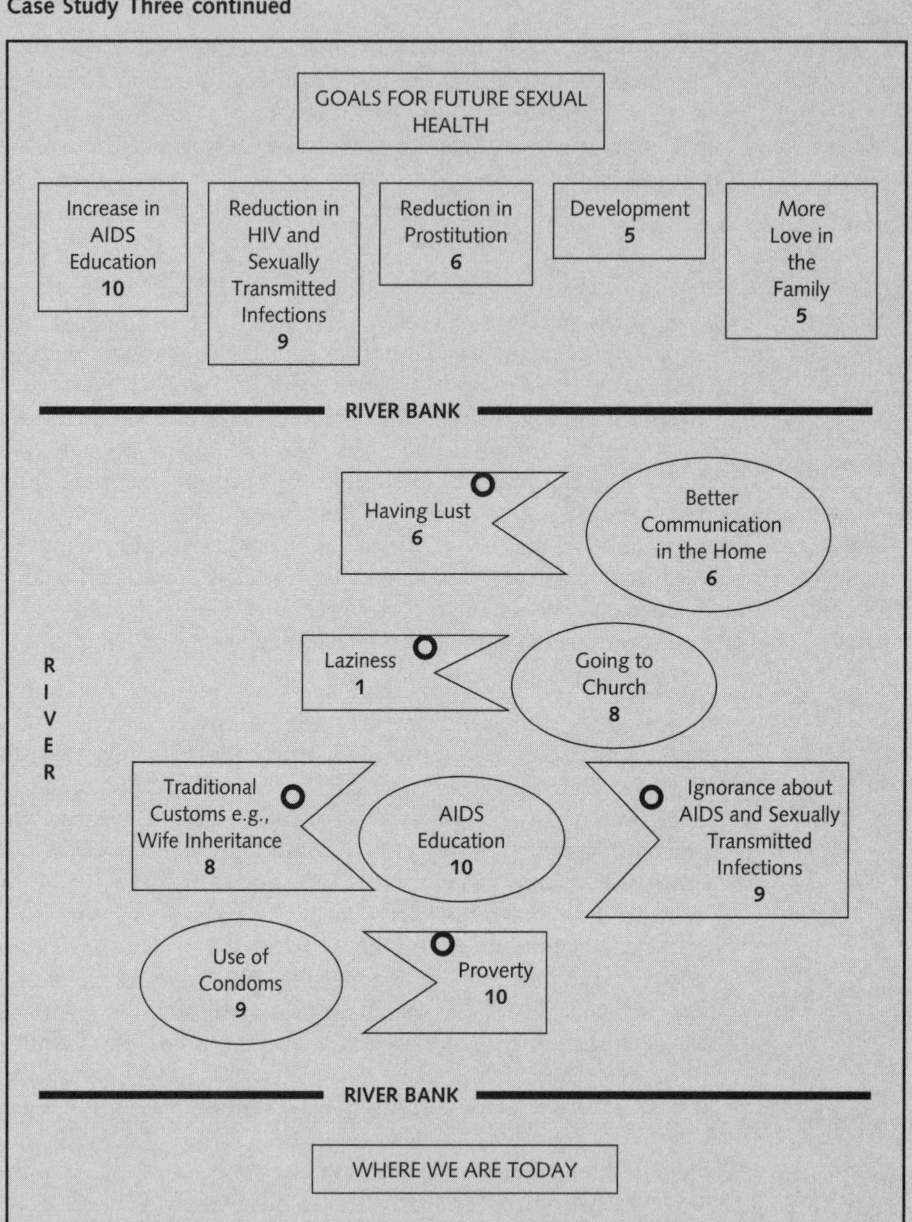

Figure 9.4 Stepping-stones diagram: formalised representation of a diagram produced by Zimbabwean women (numbers represent the relative importance given to each factor on a nominal one to ten scale)
Source: Kesby: Zimbabwe, 2001

'bank'. Participants used large cards representing things that would help them attain their goal and laid them in the river as 'stepping-stones'. Cards shaped like crocodile heads were used to indicate the factors that would get in the way of them reaching their goals. All cards were scored to indicate their importance (Figure 9.4).

Case Study Three continued

Analysis: Participants reflected on and analysed the diagrams and sessions in their peer groups as research progressed; however, no final community meeting was held to discuss the findings. The research team acted as advocates for participants by reporting their suggestions and requests to the local and international NGOs.

Reflections on fieldwork: Four observations are worth highlighting: first, diagramming techniques can be used to approach very sensitive topics relatively easily, speedily and in some considerable depth. This has much to do with the 'back seat' position researchers adopted after introducing each session, the tactile and engaging nature of the exercises and the fact that difficult issues become easier to discuss once they are (visibly) 'out in the open' on the diagram. Secondly, diagramming sessions were very social and enjoyable occasions for all concerned. Afternoons spent diagramming on the floor of the community hall and under a big tree were often punctuated by laughter and once by singing and drumming. Thirdly, it was important to adapt techniques to fit the local context. The symbol on our ballot questionnaire representing 'no' (an 'X') had to be compared to that used by teachers at school ('yes' was a 'tick') so that participants did not confuse it with the 'X' used to indicate 'yes' on political ballots and, more importantly, that local party officials did not think we were engaging in voter education, which was banned. Finally, using diagramming to evaluate a participatory training project made us reflect on the sustainability of participatory projects and the need to try and build in mechanisms that enable participants to continue the work of an initiative once the official programme has ended and the researchers/facilitators have withdrawn.

Outcomes: I reported the findings to the two NGOs. These have been used to improve the participatory HIV education programme that will now be run in other communities. The evaluation also galvanised those who participated in the initial programme to remobilise themselves and to appeal to the local NGO for support to undertake further follow-up activities in their own area.

Practical suggestions for 'doing' participation 'deeply'

Having focused on *techniques*, we return in this section to the question of *approach* by offering you some practical suggestions about how you can integrate participation into one or more stages of your own research project. A useful first step is simply to go out and talk to relevant organisations, community groups or individuals at an early stage in your research design, to see if collaboration is possible. If it is not, you are still likely to learn something useful from the meeting. If the response is enthusiastic, you can begin to involve participants at a number of stages in your research: (a) in setting priorities and generating research questions, (b) giving input into research design, (c) carrying out the research with you (perhaps as peer researchers), (d) analysing and making sense of information generated with you, and most importantly (e) using the findings to make changes themselves, or to campaign for others to make changes on their behalf. In Box 9.2, we summarise some ideas about how you might involve

BOX 9.2

'Deep' Participatory Action Research: how to involve participants at every stage of a project

Participation in project design and question development

1. Undertake web-based searches of organisations involved in the topic you are interested in. They may have posted useful agendas for urgently needed research (see Chapter 4).
2. Seek an early association with an NGO, community organisation or statutory body (e.g. local government department) at the same time as you begin to formulate your topic; try and solicit from them ideas about any burning questions that need addressing.
3. Arrange early pilot focus group discussions (see Chapter 8) with potential participants. This gives them the opportunity to raise research questions *they* see as relevant and gives you a chance to see whether *your* questions are perceived as pertinent.
4. Work together with partner organisations and participants to establish questions, research tools and methodological processes that are appropriate for the context in which you are working. Pay attention to cultural issues as well as temporal and spatial opportunities and constraints.
5. Negotiate a 'memorandum of understanding' with organisations and/or participants which details research priorities and desired outcomes, participants' and researchers' rights and responsibilities and which clarifies use/ownership rights over the information generated before beginning data collection.

Participation in data collection

1. Encourage participants to become actively involved in data generation, discussing and debating with each other, not simply answering questions posed by you.
2. If possible and appropriate, train some participants in facilitation and investigation techniques so that they become 'peer researchers' able to assist you or to work independently with their contemporaries on topics that might not be so accessible to outsider researchers.

(See also Box 9.1)

Participation in analysis

1. Make time to 'interview' diagrams and data generated by specific exercises. Both you and your participants will gain further insights if you spend time analysing and fine-tuning initial results.
2. Encourage participants to organise and categorise the information and to undertake higher order forms of analysis: e.g. encourage them to make connections between the different diagrams generated and to address unresolved or difficult issues that divide them.
3. Encourage participants to keep their own copies of the material generated (e.g. give them access to flip charts, Polaroid photos etc.) and to think about what they themselves would like to do next with the information they have generated.
4. Where a large number of participants have been involved in the research, organise a later event at which provisional findings are presented and provide

Box 9.2 continued

opportunities for participants (and others) to question, challenge and add to the findings.

5. Where participants have identified other groups or individuals as a problem (particularly where these other people are not well represented in existing research), try and find out their views and incorporate them into the analysis.

(See also Box 9.1)

Participation in writing up

1. Return drafts of any information generated (diagrams, interview transcripts etc.) to participants for 'member checking' and further comment.
2. Include participants' names on reports and articles (especially peer researchers) and acknowledge their contributions.
3. Organise and facilitate participatory writing workshops with participants to produce joint documents and reports (such activities may need to parallel rather than include your dissertation, which when submitted for assessment may require a signed statement that it is 'all your own work').

Participation in dissemination and actioning change

1. Return to partner organisations and/or participants' 'communities' with results and discuss ways of disseminating them in non-academic formats.
2. Make joint presentations for the general public and/or make representations to key decision-makers.
3. With their approval, make such presentations on participants' behalf (advocacy rather than participation).
4. Encourage participants and partner organisations who wish to write up the findings in other formats: e.g., a community newsletter, press releases, web pages etc.

Participation through reciprocity

1. Look for opportunities to pass on skills and training to partners and participants: e.g. in computing, data collection, research design, writing for different audiences and/or presentation of results.
2. Make your skills available to them: e.g. in longer projects write regular briefings about progress, write short accessible summaries for use in publicity material, help with fund-raising activities, grant-writing applications or web-page design.
3. Offer to undertake voluntary work for your partner organisation (this of course can double as a participant observation phase of your research: see Chapter 10).
4. Offer peer researchers some formal recognition of their effort and skills acquisition (especially younger people): e.g. a signed certificate or record of achievement, or provide a job reference.
5. If you are able to, provide refreshments and childcare at meetings, and pay participants' and peer researchers' expenses. If you promise money in any form make sure reimbursement is prompt and transparent. In larger better-resourced projects, think about covering the project-related management costs of partner organisations.

participants at any or every stage of your research project. These set out an *approach* within which you might decide to use a particular *methodology* (such as the one outlined in Box 9.1).

We would like to stress that even if you do not use diagramming techniques but choose instead to adopt other more conventional qualitative and even quantitative methods, you can still utilise many of the strategies set out in Box 9.2 – a participatory approach is not defined by particular techniques. For example, partners and peer researchers could help you design and analyse formal questionnaires (see Power and Hunter, 2001) and you could involve local participants in 'ground truthing' highly technical GIS data (Harris *et al.*, 1995).

Furthermore, we advise you to view Box 9.2 as an ideal 'gold standard' to aim towards, rather than as a checklist for things you *must* achieve. While adopting all the strategies would certainly enable you to move towards a very 'deep' form of participation that involved participants actively in all aspect of the research process (as opposed, say, to the 'shallow' use of diagramming just to collect data – see Cornwall and Jewkes, 1995; Hagey, 1997), only a few projects have achieved such depth in practice. Reread our case studies and you will see that we did not achieve every element of our own 'gold standard'. Nevertheless, this did not make our projects 'sub-standard' and they prove that valuable and action-orientated research can be achieved through *partial* participation.

So, even if you feel that you do not have the skills, resources or time to attempt some aspects of participation, much less to achieve all of them at once, do not be put off trying simply because you cannot 'do participation deeply'. Any single element of Box 9.2 will bring considerable benefits to your project, and some to your participants. Just remember the 'scare quotes' when you make claims about having 'adopted a [partially] participatory approach' in your project. Explain which elements were participatory (e.g. data collection) and which were more conventional and determined by you as the researcher (e.g. research design).

Finally, do not let the idea of a 'gold standard' make you think that deep participation is necessarily the *only* ethical, useful or 'relevant' research paradigm (see Cooke and Kothari, 2001) or that it offers a panacea for all the ethical problems of doing research. New problems about representing the voices of 'others' and intervening in their lives arise in participatory projects. They tend to produce data on collective, publicly held opinions, rather than private individual beliefs and without careful facilitation can lead to some elements of a community dominating proceedings while others do not even engage (see Guijt and Shah, 1998). Meanwhile, empowerment facilitated within a participatory arena can be difficult to sustain in people's everyday lives, especially once a project has ended (Kesby, 2005). With these facts in mind, we suggest you view Box 9.2 as a stimulus to your imagination and as a challenge to your thinking about ways in which participation might realistically and usefully be integrated into aspects of your own research project.

We would like to make four further comments to assist your thinking about undertaking participatory research. First, you will need to make some decisions about the *scale* at which you wish to operate (e.g. street, school, 'community', health authority, district etc.). The nature of the topic and the suggestions of potential partners and participants will help you identify the scale at which research and action would be most effective, but as geographers you should be especially careful not to impose an inappropriate scale on the phenomena under study. When working with a small geographically concentrated community (e.g. residents of a single village or members of an organisation or community group) it may seem obvious who to work with, but there may be important causal processes that work at scales greater than the individuals within a single community. For example, an HIV awareness project might work with a single African village community, but the influence of sexual networks that extend to urban areas and even across national boundaries should not be ignored. At the very least, forces working at larger scales will need to be recognised when participants in a given community attempt to develop strategies for actioning future change.

Allied to this is a second issue related to the *level* at which you attempt to enter the power/decision-making structures of society: a complex question that may be fraught with contradictions. Working at a 'grass roots' level helps ordinary people to get their voices heard, while working with a governmental or non-governmental organisation might have a more direct input into policy-making. A big organisation, on the other hand, may pay lip-service to supporting your 'participatory project' but fail to heed or prioritise the views and needs of the grass roots or be prepared to make its own organisational structure more inclusive and participatory. Conversely, some policy-makers are deeply committed to participatory models of change and genuinely concerned to understand their clients' needs and views, while many ordinary people do not want to participate 'deeply' and may prefer brief, researcher-led encounters (Kitchin, 2001). The nature of the organisation may also affect the types of participation and reciprocity you attempt. We cannot provide formulaic solutions; you will have to use your own judgement about the nature of the topic and the motivations of potential partners. Participation is a two-way street: you can play your part but rely on partners and participants to make the action elements of a project really work.

Who you work with raises a third issue related to the presence, or absence, of *pre-existing structures* through which to work and *gain access* to participants. Statutory and non-governmental organisations (e.g. local councils and charities) as well as self-identified communities (e.g. a neighbourhood action group) are likely to have well-established structures that you can hook into and utilise. Such structures will enable you to pre-arrange access even where your target group is large and dispersed. For example a study with children may be organised through participating schools, while work with homeless people can be arranged by setting up your research materials in a drop-in lunch centre. Be sensitive to the possibility that these venues may restrict the range of participants you can recruit. For instance, they may exclude children who do not go to school, the 'hidden homeless' or those in temporary accommodation. In many cases, well established structures and pre-existing venues and points of contact will not exist, particularly when working with 'hard to reach'/ outsider-identified 'communities', such as teenage drug users or homosexual men and women, and you will have to formulate your own. You could advertise for participants via a medium that the target group has access to (a magazine or by distributing flyers), access people directly on the streets (e.g. approach wheelchair users in the city centre) and/or 'snowball' other participants via an initial contact. While participatory research *can* take place in street environments if this is where people are accessible and comfortable, 'deep participation' may be difficult unless these locations become 'venues' for repeated contacts. Alternatively, people recruited in the street or via advertisements can be invited to venues arranged by researchers (e.g. university seminar room or local community centre) *if* they are happy to work in such environments.

In research that involves working with groups, a fourth set of issues arises. Some believe that focus groups should be composed of strangers who will be less inhibited about giving their opinions, while others suggest that not only is it more convenient to work with 'naturally occurring' groups of people who know each other but that this enables researchers to triangulate informants' responses (see Chapter 8; also Holbrook, 1996; Kong, 1998). Additionally, work with natural groups can also help facilitate action and the sustainability of a participatory project's impacts once researchers have withdrawn from the community, because participants already have a reason to continue to meet and communicate with each other. Nevertheless, it is often wise, at least initially, to break up natural groups into 'peer groups' upon advice from local partners about appropriate divisions (e.g. groups divided by age, gender, ethnicity, etc.). Such groups can help to reduce individuals' inhibitions, and enable them to share experiences and develop ideas independently of those with different or competing agendas.

The scientific validity of 'participatory' approaches and techniques

We have already highlighted the ethical reasons for pursuing a participatory approach and would also like to stress its 'scientific validity'. While there is insufficient space here to engage with all the arguments, we present some that readily meet recent demands for a quality audit on qualitative methods (see Baxter and Eyles, 1997).

First, if you want to obtain accurate in-depth data that reveal the perspectives and experiences of participants, especially on sensitive subjects or from difficult-to-reach 'marginal groups', it makes sense to use a participatory approach and associated methods. Involving participants in project design can legitimise the researcher's presence, and position the participants as stakeholders with a reason to help the project succeed. Involving them in the design of research tools can ensure that methods and questions are appropriate, meaningful and relevant to those who will use and address them (Kindon and Latham, 2002). Moreover, involving them in data collection (rather than simply asking them questions) means that they are more likely to feel able to address complex and difficult issues (Kesby *et al.*, 2003).

Secondly, if participants are involved in the data analysis, the method is likely to produce more 'credible accounts' that participants themselves recognise as accurate. Triangulation and 'member checking' (respondents certifying the validity of data) are built in; participants do not simply provide data which researchers interpret, they build on and develop earlier diagrams, interrogate the data produced and even help define analytical categories (Kesby, 2000).

Thirdly, because *process* (the level, quality and nature of participant contribution) is such a central concern and is meticulously recorded, participatory research is well placed to provide a full and open audit of how and under what conditions ideas and themes arose. This helps to avoid some of the vagaries that have traditionally (although not always fairly) been associated with qualitative research.

Fourthly, because the final written report incorporates multiple accounts, including the participants' own analyses of the data, it provides a narrative 'between' the perspectives of participants and those of the researcher(s). This is a form of positioned and reflexive objectivity increasingly favoured by feminists (see Katz, 1994; Townsend, 1995). Finally, the epistemological foundation that underlies all qualitative research is that researchers need to build trusting and reciprocal relationships with 'respondents' if they hope to gain really deep insights into their life-worlds. A participatory approach with its associated consultation, involvements and reciprocity will help you do this and so, ultimately, will help you produce a valuable research project.

Conclusion

In this chapter we have covered a wide range of issues around participatory research. We hope you will use it both as a 'how to' guide to alternative diagramming methods and as a stimulus to thinking about the possibilities of 'doing research differently' in terms of approach. While we have laid out something akin to a 'gold standard' for integrating participation into all aspects of a research project our own case studies illustrate that most projects fall short of this goal. As academics, our desire to pursue participatory research must be juggled with our responsibilities for teaching and the pressure to publish (Pain and Francis, 2003). You, meanwhile, will have to reconcile your aspiration to make your dissertation relevant to the lives of your research subjects with other timetable and assessment demands, and your commitments outside university life. Nevertheless we encourage you to explore how participation might be built into, and indeed improve, some aspects of your project, whether or not you utilise diagramming methods. In summary, we encourage you to explore the techniques and methodologies described in this chapter while simultaneously engaging constructively and critically with the philosophical debates that underpin them. If you do, we promise that your dissertation project will be a more challenging and rewarding experience for both you and the people you work with.

BOX 9.3

Further reading

The Institute of Development Studies at the University of Sussex is the largest and best-known organisation using these approaches in Britain. Several publications are available at their website at http://www.ids.ac.uk/ids/

The International Institute for Environment and Development (IIED) is also a non-governmental organisation based in London. It publishes widely on the use of participatory techniques, particularly through *PLA Notes*. More information is available at http://www.iied.org and http://www.planotes.org

10 Participant observation

Ian Cook

Synopsis

This chapter is intended to outline the method of 'participant observation' and how it can be constructively and creatively applied in human geographical dissertation research. It is a method which involves living and/or working within particular communities in order to understand how they work 'from the inside', and a number of practical and more academic considerations have to be taken into consideration to do this kind of research well. The chapter thus deals with issues of *access* to communities within which research can take place, of the *roles* or identities which a researcher may have to consider adopting in the process, and of *constructing information* which is appropriate to this kind of research encounter and which can later be worked upon in writing up the dissertation. Given that participant observation research often changes as it proceeds (initial expectations can often be found to be wanting and important 'discoveries' can lead researchers off in previously unexpected directions), the intention here is not to be too prescriptive in outlining what this method is about. So much depends on the context. Here, then, general points are illustrated through extracts from, and summaries of, three undergraduate dissertations in which this method has been used quite differently. The reader is shown how these general issues can become translated into concrete projects and is thus invited to think about how such translations could be made into a project of her/his own.

Introduction

Historically, ethnographic research has developed out of a concern to understand the world-views and ways of life of actual people from the 'inside', in the contexts of their everyday, lived experiences. The method of participant observation is the means by which researchers have often done this. As its name suggests, it involves researchers moving between *participating in* a community – by deliberately immersing themselves into its everyday rhythms and routines, developing relationships with people who can show and tell them what is 'going on' there, and writing accounts of how these relationships developed and what was learned from them – and *observing* a community: by sitting back and watching activities which unfold in front of their eyes, recording their impressions of these activities in field notes, tallies, drawings, photographs and other forms of material evidence (see Fyfe, 1992; Hunt, 1989; Maranhao, 1986; Tedlock, 1991; Wax, 1983).

In its basic form it can be described as a three-stage process in which the researcher somehow, first, gains access to a particular community, second, lives and/or works among the people under study in order to take on their world-views and ways of life and, third, travels back to the academy to make sense of this through writing up an account of that community's 'culture'. This chapter addresses the why, how, when and where you might want to adopt this approach in dissertation research. In order to do this, the following pages are divided into sections which discuss why it is important to think about:

- your (in)ability to *access* the kinds of research communities which you might consider studying,
- the way that you will inevitably have to take on a certain kind of *role* in the community to which you do gain access in order to get your research done and
- the kinds of data which you can construct and use through this method.

However, these sections will not provide you with a rigid list of 'dos' and 'don'ts' as so much depends on so many circumstances, some of which will be within your control and others which will not. So, to provide some impressions of participant observation research as a creative, evolving, and contingent process, a series of boxes have been provided which include key aspects of the dissertations written by three geography students – Catherine Hawkins, Zoe Howse and someone whom we shall simply know as 'Bob' – who used participant observation as their main research method. These boxes, starting with Box 10.1, can be

BOX 10.1

Three dissertations: subject matter

To give a flavour of the range of research that it is possible to do using participant observation methods, Catherine's dissertation – entitled 'A city for men? A study into the access of facilities available to women with young children' – was submitted in 1995 and involved her assessment of the town centre of Hull as an (in)accessible environment for women with children in pushchairs; Zoe's dissertation – entitled 'The Reading Festival and experiences of research' – was submitted in the same year and was a study of whether and how resistant-youth cultures centred around music could develop in and around this music festival; and Bob's dissertation juxtaposed the public and private faces of the McDonald's fast food chain, and was untitled at the time that this chapter was first written because Bob was in the process of completing his work for submission in 1996. Catherine's and Zoe's works have been quoted word for word. Bob's, however, has been somewhat altered because, first, I decided to tidy up the rough version which he gave me and, secondly, his research was into a topic where the 'covert' nature of his work (see later for a discussion of 'covert' versus 'overt' research) meant that reference to real names and places had to be changed, modified and/or deleted to preserve his anonymity.

read not only with the surrounding text to put some flesh on its bones, point by point, but also separately in order to build up more detailed impressions of these three dissertations and how your own may be similar to and/or different from them.

a) Access

Factors which you will need to take into account when choosing a feasible and interesting dissertation topic have already been discussed in Chapter 3. But, if we believe that 'geography is everywhere' (Cosgrove, 1989), the nature of participant observation means that a worthwhile project can be done wherever you may be and whatever you may be doing during the period set aside for your research (the ways in which Catherine, Zoe and Bob settled on their projects is shown in Box 10.2). In coming to conclusions about what to study and how to study it, for instance, many students choose the labour processes which become part of their own lives through part-time and holiday jobs like waiting in restaurants or picking vegetables (see P. Crang, 1994 and Thomas, 1985 for particularly good workplace ethnographies involving participant observation). As in Bob's case, one of the great advantages of doing research in your place of work is that, with the financial difficulties so often experienced by students, the need to earn cash during vacations will not necessarily get in the way of your dissertation. Yet other students have followed their interests, concerns and/or involvements in communities based in and around other kinds of locality such as Catherine's choice of the public spaces of a city centre or Zoe's choice of the semi-private spaces of a festival site. Indeed, there are countless other spaces in which you could consider doing participant observation research for your dissertation such as swimming pools, night clubs, golf courses, dog-training classes, rehabilitation centres, football terraces and/or the place(s) where you grew up. All of these have provided the basis for other students I know who have used participant observation in their dissertation research.

In making such choices, it is important to bear in mind that – contrary to the most well-known images of participant observation research – you do not have to gain access to an isolated community in some faraway part of the world and then live there for a year or more to do your work 'properly'. As the dissertations outlined in Box 10.2 suggest, this kind of research can legitimately involve anything from several months of sustained work during a summer holiday like Bob's, to a short but intense study of a 'community' which comes together for a matter of days like Zoe's, to a journey taken once through a particular urban landscape like Catherine's. Any or all of these, and plenty of others besides, have their own geographies which can be well worthy of study and, as discussed in the final section of this chapter, the 'data' constructed under these differing circumstances can be plenty for a dissertation.

As well as the 'what' of access as described above, many students find that *wanting* to study a particular community does not easily translate into *being able* to study it because this access has to be negotiated through various 'gatekeepers' who can control this. Some communities are much more guarded than others

BOX 10.2 Three dissertations: settling on a topic which is both practically possible and academically relevant

As with dissertations using any other method (and, as you will see below, often *including* other methods), you will have to explain in your introduction why you did what you did and why you did it the way that you did it. To give a flavour of how *practical* considerations of access and *academic* considerations of interest and relevance were weighed up by Catherine, Zoe and Bob in their choices of topics and methods, this is how they described them in their dissertations:

Catherine:

'The ideas for this research project really came from three different angles. The first originated while childminding a little girl of three years old called Victoria near to my home in Hull. On my suggestion of taking Victoria to the playground, which is about a ten-minute walk from us, her mother asked her if she wanted her pushchair (big mistake). Of course with this idea in her head she would not go without it . . . During this short journey alone I was faced with a number of problems: high pavement kerbs when crossing over streets, no crossings over streets – which meant racing across the road with Victoria and the buggy – works on the pavement – which meant having to walk on the roads – and roads with no pavements at all, obviously leaving us no option but to walk with the traffic. This was the only route that could have been taken and . . . I soon realised how mothers in this situation would and do feel. The second angle from which the idea grew was from Ian's "Living with disabilities" course that I took in my second year. From this course I realised that different groups in society existed with difficulties concerning urban mobility. By looking at the wider politics of access in the urban environment, I was prompted to carry out a research project similar to what had been discussed in the disabilities course. The final angle from where the idea came . . . was again directly related to the project in that after flicking through the channels on the television one evening I found a programme which was discussing the problems that mothers were facing when going out with children in pushchairs and prams etc. I remember the image of a young mother with a child in a buggy at the top of an extremely steep escalator . . . It was all these ideas mingled together that made me base the piece of work on this topic.'

Zoe:

'For the past five years I have liked the majority of bands who have played at the [Reading] festival, but even though I live only half an hour away from Reading, I have never before attended the festival. This was due to lack of money (tickets are £50) and a disinterest in the concept of festivals.

. . . I first chose the Reading Festival as my area of study because I was fascinated by music and youth subcultures. I had wanted to look at these aspects of the festival in relation to the rest of society – i.e. Police, Security and local people of Reading – to see how people reacted to youth and deviant behaviour.

When I chose this area of study, I was only in contact with a couple of people who shared my interests in music and unfortunately they weren't even friends,

Box 10.2 continued

just people I corresponded with. I had just found out about Kingmaker fanzine (Kingmaker being a band, and fanzine being a home produced magazine which is sold at gigs, festivals, or through the ads section of music papers). Although I wasn't initially interested in such a publication, throughout last year, I became involved. Two sisters . . . edit the fanzine, and back last February, I had started to plan how I could use them for my research. They had both been to Reading Festival for the past 3 years, so I was keen to keep in touch with them, so as to interview them for my research.

Last March, for the fanzine, I interviewed Laurence Hardy, the singer with the band Kingmaker. At the same time as asking him questions specifically for the fanzine, I also used the opportunity to find out his opinions about the Reading Festival, which Kingmaker had played at for the past 3 years.'

Bob:

'The problem of what to choose for your dissertation study can not be as much of a problem as you may first think. "What am I going to base my dissertation on?" was the question going on and on in my mind to which the answer seemed very elusive. One of the main essentials was that the dissertation would not only be interesting to carry out but it would enthral the reader making him/her want to read on further, wishing it would never end. With these two qualities in mind, the dissertation also had to contain a very real account on the geographical understanding(s) of the subject at hand. With this in mind, the answer to the question of what to study was blatantly obvious: McDonald's (the restaurant chain) would be the basis for the dissertation study. There are three reasons for this:

(1) A brief history: McDonald's is said to be perhaps the world's biggest single provider of food, with an estimated twenty-eight million people entering its 14,000 restaurants to consume food and drink every day. With its $24 billion annual profit, it is the most recognised brand name in the world after Coca-Cola . . .
(2) McDonald's to date: the smooth running of the McDonald's machine seemed set to continue asserting its domination world wide well into the twenty-first century, silencing any opposition or critics who could possibly halt the multinational giant with legal action . . . But, the recent "McLibel" court case has been more than they bargained for . . . as their use of the courts as a "silencing tool" did not deter these people's opposition to their power.
(3) Inside the multinational: having worked at McDonald's as a crew member for two and a half years, I was familiar with the basic operational aspects involved in running the store . . . Being in this position, I realised that I could produce an insider's account of the operation of this machine . . . I was able to put myself in this position in the 1995 summer vacation, making this a real possibility for a dissertation study . . . exposing the different layers of the "McDonald's myth" as promoted in the corporation's advertising campaigns and as highlighted in the McLibel trial.'

against the presence of certain kinds of people. For instance, so long as you have the money to enter a bowling alley, you are likely to be able to enter the community that comes together there on a competition day if this is what you want to study. Yet, you may find that you need more than money to enter, for instance, more élite spaces such as masonic lodges and gentlemen's clubs, or places of punishment or rehabilitation such as prisons and hospitals. However, the reasons given for why a community and/or situation which you want to research is or is not open to you may reveal vital clues to its character. So, it is important to note that *this* is where the 'fieldwork' starts. In the process of gaining access to a research community, you may have to endure days or weeks of doubt and frustration, before becoming quite suddenly overjoyed when things somehow start to work out, often better than could have ever have been planned. But in terms of the time needed to get to this stage, this can be very unpredictable. It may take a couple of weeks to arrange a first formal meeting with one potential 'gatekeeper' who may then refer you to another, and so on. So, if this second meeting takes as long to arrange, you could have spent a month on meeting just two people and, even then, you may not be sure to have gained access to the community with/in which you want to do your research.

In research such as that done by Catherine and Bob – where pre-existing avenues of access were built upon – this problem of negotiating 'gatekeepers' can rarely be much of a problem, except in terms of deciding what to tell those who you are working with about yourself and your role as a researcher (see later). However, if this community is one which you have had little contact with before, these can be very real problems which have to be dealt with. Take, for example, the excerpts from Zoe's dissertation provided in Box 10.3 where she

BOX 10.3

Zoe's research: getting into the Reading Festival

Unlike Catherine and Bob, Zoe did not enjoy either the easy access to public space in a city centre nor an already-gained access to a workspace to undertake her research. Instead, she needed access to a heavily policed, expensive to enter space of entertainment and encountered, like many others who do this kind of work, a considerable number of obstacles in doing this. In her dissertation, she described what she had to go through to get into the site in the days leading up to the start of the festival, before eventually getting hold of the tickets she had been promised by a band which she had interviewed for a fanzine. The extracts quoted below show, day by day, how practical problems of access to a site or a community can provide valuable insight to its characteristics. For Zoe, as for many other researchers having difficulty negotiating 'gatekeepers', participant observation began *during*, not after, this access period and showed the importance of the policing of this space as well as, more or less by accident, of the background preparations necessary for the festival to take place.

Tuesday 23 August:

'I forced myself to go into Reading today and take a look and talk to some of the people working on the preparations for the festival . . . Nervously I approached the

Box 10.3 continued

entrance to the festival site, and looked for security guards to annoy. I spied two men in blue anoraks with "Security" inscribed across their backs . . . After a gabbled explanation of being a student and studying the festival, I smiled sweetly and asked if I could have a look around the festival site. They seemed to think that this would not be a problem, but had to check with the person in charge of security. I was accompanied onto the festival site by one of the security guards who I found out was called Kevin, and whilst he was trying to find the right person, I tried to strike up a conversation. He was from Edinburgh, and had been working at the Summer Events organised by the Mean Fiddler organisation. After the festivals finished, . . . he was to return to Scotland where he worked as part of security at clubs and venues. He had been at the festival site here since August 9th when the preparation had first begun.

I found it difficult talking to him because he was actually asking me so many questions about my project that I didn't have a chance to ask him much about the festival preparations, it was difficult to ask these without seeming too demanding, or a cocky, patronising intellectual! So instead, I just allowed the conversation to take its own course . . .'

Wednesday 24 August:

'. . . Even though I was familiar with the site and security guards from yesterday, I . . . felt a sense of hostility as I approached the festival site. There were about three times as many security officers by the backstage entrance, a proper security cabin had been put in place, and everyone was rushing about and speaking into their walkie-talkies, and looking full of self-importance . . . I wasn't in the mood for dealing with new people, so I wandered into the backstage area, and hoped that nobody would question my activities . . .'

Thursday 24 August:

'. . . On arrival at the festival site, to my horror I discovered that the Guest List portakabin didn't actually exist, guest list tickets weren't going to be issued [until] mid-day on Friday – when the first bands of the festival came on stage. I could quite easily live without going to this damn festival . . .

Instead, trying to stay calm, I tried to find one of the security guards whom I had been talking to earlier in the week, and see if they would let me onto the site. Kevin came to my assistance, and on recognising me as "the girlie doing the project" he escorted me to the production office and left me to explain my situation to Jane, the production assistant. The production office was a portakabin packed full of people, all who had problems equally as trivial as mine . . . I found it interesting just being stood in the control centre for the festival. Whilst I was there, there were a multitude of activities going on around me . . . Eventually, I managed to find out that no one at all was being allowed into the arena today, and there was nothing she could do about the guest lists today . . . [But] I vaguely knew from where I was in the backstage area [that] I could go down a lane and come out not in the arena, but around the camp site. I tried this, and luckily it brought me out near to the entrance to the arena, exactly where I wanted to be. I had managed to get onto the festival site and without a ticket – quite an achievement!'

discussed the network of 'gatekeepers' that she had to negotiate in order to get into the Reading Festival site.

Under similar circumstances, it can often be a good idea to approach the community that you wish to study via, so to speak, the 'keepers' of several of its different 'gates' so as to avoid any dependency on just one person who may not return your calls, who may not turn up to meetings, or who may prove uncooperative or uninformative when you do get through to them. Other options, depending as usual on the circumstances, would include simply turning up at a place and trying to wing your access on the spot (as Zoe did, when pushed). But you should make sure not to leave things to the last minute when negotiating access so that such drastic methods do not have to be relied upon.

Unlike more controllable research methods, with participant observation you should not expect things to proceed according to any pre-planned schedule. So, perhaps the best advice here is that you should try to make the most of contacts you already have and to prepare to be flexible with any others. Gaining access to communities for participant observation research can be a messy and confusing business so, if you find yourself to be messed up and confused by this, these are not necessarily signs that your research is going badly (see Rowles, 1978). You will have to be persistent and resourceful to get where you want to go. If all else fails, you can always shift your sights and concentrate on what you learned about *not* being allowed to do research in a particular place or with a particular community, to do a project on gatekeeping and the policing of access into certain spaces and communities.

b) Roles

In terms of gaining access to particular sites and communities, it is not only *who* you contact during the gatekeeping and subsequent stages of your research that is important, but also the way that you present yourself when doing this: i.e. what 'role(s)' you take on in the course of your work. In preparation for the first meeting with your first 'gatekeeper', then, you need to have thought about what you should tell her/him about the role that you want to adopt in their community. After a number of initial enquiries, you will have had the chance to hone your 'purposes' in order to properly word any formal or informal application for access. But, in contrast to what you might expect, this does not necessarily have to be couched in terms of a set of barefaced lies! In the early access stages of more 'overt' research (see later), you may simply want to enquire what constitutes a community's everyday activities. Later, once this has been established, a vague idea of what your research might eventually be about or a watered-down version of your research question(s) will often suffice. In terms of what these might be, there is often a huge difference between what you might tell your advisor or fellow students what your research is about (e.g. the precise methods, substantive areas and theoretical standpoints you want to take) and what you tell various 'gatekeepers' to 'the field' (something which, as Zoe pointed out above, may give the impression that you are a 'cocky, patronising intellectual'!

Yet, there is often no reason to make the distinction between the project that you formulate in your department necessarily being the *real* one and that which you describe to outside gatekeepers and community members as being a more *tactical* version. The point here is that each account can be a simplification of a larger project, and/or can be the latest version of a project which is changing as you go along, and/or can be a version of a project which you are determined will gain its shape from what the people under study might want it to be about. Thus, the ways in which dissertation students can negotiate these kinds of ethical, political and practical problems can, again, be very different, depending on the circumstances.

In amongst all of these ethical, political and practical decisions, the two most important things which you will need to take into consideration are, first, whether your role can or should be 'overt, providing a full explanation of (your) role, or covert, concealing (your) purpose and identity' (Fyfe, 1992: 131); and, secondly, to what degree your research can or should be 'participatory' (being fully involved in the workings of a community and learning its culture through 'living it') and/or 'observational' (being less involved in a community and learning its culture through 'watching it'). Again, these will rarely be simple decisions as so much depends on the context. So, to give some impression of what is both possible and desirable under different circumstances, Box 10.4 illustrates how Bob, Zoe and Catherine settled on their roles in these respects.

In terms of deciding how to settle on particular presentations of yourself and your project which are ethically, politically and practically sound, there are no easy answers. Indeed, the deception which is so often practised through participant observation methods – even the most overt where you may get excited by

BOX 10.4

Three examples: adopting covert and/or overt roles in participatory and/or observational research

The extracts below show how Bob, Zoe and Catherine have described the roles that they took up in the communities where they did their research. The terms identified in the title of this box have the following meanings:

1. Overt observer: someone who *does* tell the members of the community being studied that they are *watching* them for their research.
2. Overt participant: someone who *does* tell the people that they are *living* and/or *working with* that this involvement is for research purposes.
3. Covert observer: someone who *does not* tell the members of the community being studied that they are *watching* what they do for their research.
4. Covert participant: someone who *does not* tell the people that they are *living* and/or *working with* that this involvement is for research purposes.

Sometimes, these terms in isolation are too simple to describe a researcher's role and, in practice, people using participant observation techniques move between these roles as their research progresses, as they talk to, work with and watch the activities of different people.

Box 10.4 continued

Bob's role was perhaps the most straightforward as he was very much a *covert participant* in the McDonald's crew among whom he did his research. However, at the same time, he also took on the role of a *covert observer* of other workers and customers in the store. Thus, as he wrote in his dissertation:

> '. . . the majority of my research was undertaken in the months of July and August when I became a full-time crew member at a McDonald's restaurant. I, however, had an ulterior motive, to covertly become an ethnographer constructing his field work data by taking on the method of participant observation. My work thus became my field work, so I was always researching and decoding what I did, what I saw and what I felt. This method, I felt, was particularly suitable because nothing was hidden or edited which probably would have happened if I had interviewed or questionnaired people. Very few people knew what I was doing, which was more practical because no one treated me or behaved differently. More importantly, I did not want managers knowing what I was doing because they seemed to portray a sense of loyalty towards McDonald's . . . As far as management were concerned, I was working for them and for the situation to remain this way, I was very discreet in what I said.'

Catherine, however, took on a more even-handed role *both* of a *covert participant* – because she had borrowed her friend's child and pushchair to take part in a 'research journey' which the people around her were unaware of – *and* of a *covert observer*, in that, in doing this, she also spent time watching how other women with children in pushchairs negotiated the city centre's paths, doors, phone boxes and so on. As she described this, then:

> 'Much of the study will centre around practical ethnography in the form of a participant observation research where I travelled around the town of Hull with a child in a buggy. By doing this, I aimed to understand this part of the world as I experienced it through the eyes of a carer with a child. Gaining access into town was extremely easy with it being a public environment and, because I was walking along with a child in a pushchair, I did not look particularly out of place . . . By doing this I got right into the situation in hand and observed the difficulties of [others] travelling with a pushchair, and I faced the problems with access and facilities directly.'

Finally, in Zoe's case, we have already seen how she adopted the role of the *overt observer* when talking to the security guards at the festival whom she told exactly what she was doing and, thereby, became identified by one as 'the girlie doing the project'. However, when she later entered the festival site on the days when the bands were playing, she was able to melt into the crowd as both a *covert observer and a covert participant* just like any other festival-goer. On top of this, however, she was both an *overt observer of and an overt participant with* a number of people whom she had told about her dissertation research, and whom she had arranged to meet at the festival through contacts which she built up through the fanzine which she had got involved with.

the 'juiciest' information which is often divulged in the most unguarded moments – is something which may cause you great anxiety. From the very beginning, it is necessary to consider how research is always bound up in networks of power and knowledge and is, therefore, inherently political. Many writers have argued that this is something that the researcher should tackle head-on, rather than simply deny through sheltering behind the traditional veil of 'objectivity'. If, in your research, you find yourself suspended between differently empowered groups, your roles and responsibilities may have to be compromised (Wade, 1984). Moreover, in situations where you might study powerful élites, on the one hand you may be seen as a threat through having the power to open out these people's lives for ridicule or ruination by others (Cook, 1995; Johnson, 1992) yet, on the other, these are also the people who usually have the power to bar your access in the first place (Cassell, 1988). So, in terms of gaining access to and establishing a role in your research community, not only must the significance of your position and apparent intentions be considered but so too must your responsibilities over how the people being researched will be represented in any account produced, how this will be circulated, and the impact that this might have on their lives in the future. Catherine was fairly comfortable with the compromises that she reached in this respect and did not consider them worthy of much comment. However, as illustrated in Box 10.5, Bob and Zoe felt that it was essential to discuss the ethical and political compromises that they had made in their research to justify what they had done.

In dealing with the ways in which you end up playing your roles as participant and/or observer in your research, many aspects of your identity will inevitably end up being played off against each other in various contexts as your appearance, body language, ideas, intentions, feelings, politics, ways of doing things, and so forth (have to) change through the experience of setting up and seeing through your project (see Parr, 1998). Through initial conversations and particularly through sustained periods of interaction, you can, first, learn which aspects of your identity allow you to be more or less acceptably placed in the world-views of both your key informants and the community under study and, secondly, thereby establish how any common ground might be found. And there are some key ethical questions which you will need to think about here: e.g. if you are expecting the people you live and/or work amongst to

BOX 10.5

Ethical conflicts in participant observation research

In his dissertation, Bob argued that his 'undercover' studying in a powerful corporation which appeared to be working in an 'unethical' manner did not oblige him to act 'ethically' as a researcher by telling his bosses what he was doing. In her completely different circumstances, however, Zoe felt that the way in which she made contacts in order to access and to take part in the activities of the community which she wanted to study was 'unethical' as she had befriended these people solely so that her dissertation could be made out of this friendship. As she hinted in the excerpt of her dissertation quoted in Box 10.2, going along with the

Box 10.5 continued

standard protocol of participant observation, she had gained access to a community through developing relationships with certain 'gatekeepers' in order to use them for her research. They both described the ways in which they had resolved such conflicts in their dissertations.

Bob:

'. . . before I set out exactly my intentions for this research, I would like the reader to read the two following quotes and ask what interpretations, and meanings you have encoded from them . . .

1) "There's nothing quite like a McDonald's" (McDonald's company slogan).
2) "I want every McDonald's worker to stand up for their rights, which is why I am backing this campaign 100%. In this way, [one worker's] death will not have been in vain" (McLibel leaflet).

I would now like to take this quote from the McLibel leaflet called *Do You Work For McDonald's?*: "35 workers will in October give evidence of what really goes on inside the stores. This will benefit all McDonald's workers by enforcing the company onto the defensive". What I propose to do in this dissertation is to take on the role of the 36th worker giving my evidence (field work research) of what went on inside the McDonald's store which I worked in.'

Zoe:

'Although I already vaguely knew [the two Kingmaker fans I had arranged to meet at the festival by letter], it was through my research that the friendship developed. Once friendships and contacts had been established, I was really doing no more than I would normally do in the course of my everyday life, except that I was writing about it and the people I encountered . . . One of the most common ways in which the participant observer approaches the problem is by identifying individuals and attaching themselves to them. It was during the early months of the research, when I was desperately trying to gain contact with people that I started to use [these two people] as key informants for my research. Looking back on the situation, I do now feel guilty that I was using them in such a way . . . For me, one of the most rewarding aspects of my research has been the development of relationships in the field. Since my area of study was so closely related to my own interests, I was lucky that I was coming into contact with people who shared similar interests. [One person], who was initially one of my key informants when studying the festival, has now become a close friend . . . A few weeks ago when I met up [with her], I was actually considering using the opportunity to interview her in depth about the festival – "I could [get] enough information from a single interview with her to write half the dissertation", thought I. I (luckily!) changed my mind about this, but instead I was able to just ask her questions about the festival at appropriate times during conversation. And I like to think that I was asking her these questions more because I was interested in her as a friend and thus genuinely wanted to know what she had done at the festival, rather than asking her questions specifically for the dissertation.'

be frank about their opinions and experiences, should you do likewise in order to foster the development of a genuine even-handed relationship? Or, should you step back, at least for a while, observe, ask innocent questions, and be careful what you reveal about yourself? How long should you spend skipping between different members of the community before relationships can develop in which all concerned can trust each other enough to 'open up' to share (often private) experiences and frankly argue out the issues which each thinks are important? If you come to form an opinion about the people you have been working with, should you present this to them to see whether this gels with their experience, or should you preserve the perhaps delicate nature of the relationship by keeping quiet until either the closing stages of your fieldwork or, indeed, the writing-up stage when it can, perhaps, be most carefully worked out and handed back for comment? Or, finally, if you have promised confidentiality to your informants, can you ask questions of members of one part of the community based on information gleaned from members of another? (Johnson, 1992).

Most researchers make uneasy and improvised compromises about such things as their work progresses. Some find themselves in situations where they are trusted with extremely private and/or damaging information, which they feel should not be written about, even in the most carefully anonymous account. They can also feel shocked, disgusted or threatened by some of the opinions that certain community members hold dear and/or act upon (and, no doubt, the opposite may also be true, e.g. Keith, 1992; Nast, 1994). At the same time, though, it is not uncommon for people under the researcher's gaze to feel self-conscious or threatened knowing that anything they say may be 'written down and used in evidence against them'. But it is a good idea to keep in mind the fact that few people, including yourself, are ever 100% (dis)honest, earnest, flippant, sure what they think, consistent in what they say across all contexts or anything else. And it can take quite some time before you come to understand these kinds of subtleties and to respond to them appropriately. First, second and third impressions can often be wrong because members of the research community may well be just playing on their expectations of your expectations to wind you up, to provoke a reaction and enjoy themselves at your expense (Taussig, 1987; Whitehead, 1986). So, you should always be suspicious, then, of why you understand what you understand within the contingent, intersubjective, time/spaces of your fieldwork (Crick, 1992).

Finally, in amongst all of these concerns, for many researchers the ideal role to adopt is that of 'an intelligent, sympathetic, and non-judgmental listener' to all of its members (Cassell, 1988: 95). Yet, there can be problems here because this approach can, in practice, make you stand out in that few, if any, members of a community take up such a role themselves. On this note, Jacqueline Wade (1984: 219) has argued that:

> *To present oneself as an unalterably 'neutral' character in the course of the subjects' life events courts an impression that the (researcher) is gullible, amateurish, inane, or uncommitted (or some combination of these) and, thus, unworthy of subjects' attention and time. Furthermore, such a stance could convey to subjects that the (researcher) has,*

> *in truth, a negative regard for their inner workings, thereby potentially causing inimical involvements in future areas of field relationships.*

At the same time, though, an entirely partisan, single-focus stance would preclude the possibility of critically understanding the meanings of particular situations or problems from the perspectives of differently positioned people who struggle over these with each other in the course of their everyday lives. What is important to understand from the outset of your research, then, is that there are no easy or final answers to such questions. All that I can do here is to raise these issues and suggest that they may only be precariously resolved at any given point in your research project. Once access has been gained to a community, you will hardly if ever simply blend into it via an uncontentious process of 'role-playing'. So, with so many factors being played off against each other in the field, any student's first stab at participant observation research is almost bound to take unpredictable twists and turns which are alternately fascinating, disturbing and challenging (Crick, 1992). All that you can do is, again, to be prepared and flexible enough to make the best of the situation and to think of yourself as the research tool. This means, as Zoe put it in her dissertation, that 'participant observation relies heavily on the social skills of the researcher, it is a method that requires considerable amounts of introspection, questioning and self-doubt'. So, the question becomes, what kinds of information can be constructed out of such field interactions to base your dissertation upon?

c) Constructing information

In terms of thinking about what kinds of 'data' you can construct from your participant observation research, the most important place to do this is in your *field diary*. Kept, at worst, every few days during your 'fieldwork', its purpose is to keep some kind of record of how your research has progressed in terms of what you have participated in and what you have observed, day by day, and what you have come to (mis)understand as a result. In such a diary, it is advisable to be sensitive to a number of different factors: for example, how you were able to access the community which you ended up studying; how your understandings have been affected by your developing role in the community; what power relations can be discerned in this; how your expectations and motives are played out as the research progresses; what you divulge, why and to whom, and how they appear to react to this; how various aspects of the research encounter make you 'feel' and how this affects what you do; what you dream about; what rumours have come back to you about yourself and the reasons for your presence in the community; if you took pictures, sketched maps and/or recorded a conversation, the kinds of contexts in which this took place; the kinds of places in which certain interactions occurred; who introduced you to whom and how they described and/or reacted to you and your purpose; what your immediate impressions were and how they changed; and so on.

As mentioned, one of the most important things which can make you different from the people that you are working or studying with is that your role in that place must include a writing function. If this is known by community members, it is likely to become a distinctive part of your identity because you are observing, writing things down, and forming opinions about 'them' (Miles and Crush, 1993). As a result of this, for instance, Joan Cassell (1988) found that the surgeons who she studied were very perturbed by her taking out a notebook and jotting down what they were saying and doing. Therefore, she adjusted her writing strategy so that: 'Eventually, I put {it} away . . . and carried 3″ × 5″ cards in the pocket of my white coat or operating room scrub suit; I took as few notes as possible, scribbling a few words every once in a while on the white cards, using them as mnemonics for each night's session at the word processor' (Cassell, 1988: 96). Other researchers have made sudden and frequent trips to the loo to write things down – symptomatic of the so-called 'ethnographer's bladder'. Again, in the note-taking experiences of Bob, Zoe and Catherine, much depended on the context of their work, the roles which they took on and certain practicalities which went with this. In Bob's case, for instance, the fact that McDonald's uniforms have no pockets meant that he had nowhere to keep his notes, even if he had found a time and a place to write them!

Having perhaps noted down important things to remember on index cards or in your research diary as you go along or, perhaps, like Bob, having no time to do this, they then have to be written up as an account of the participant observation experience – making a story of what you learned out of the fragments you have at the end of the day. Some people find this kind of account is best written as a stream of consciousness which may better set out, and allow them to think about, the developing connections between different aspects of their work and relationships. Others, however, prefer tape-recording their thoughts to be sorted out later. Whichever way you choose to construct these stories, it is better to do this while things are fresh in your mind because even eight hours later your recollections may have become more blurred. Having said this, first-time ethnographers often report how surprised they are at how much detail they *can* remember at the end of the day. It is the flow of observation and participation which is important, though, and this means that going, one by one, through a checklist of questions which you must answer – very much like the one presented above – is highly inappropriate. What this constructed account should convey is the detailed and vivid impression of your 'being there' in the community under study so that people reading it can imagine themselves in your place. Take for instance the kinds of accounts in Box 10.6 which Catherine, Zoe and Bob put together in their dissertations.

Finally, it is important to consider how what you present about your research process and what you can claim to understand from it could, first, be represented more vividly and, second, be complemented by other research methods. First, to make the accounts which they had constructed more vivid for the reader, Box 10.7 describes how Catherine, Bob and Zoe included photographs and Catherine and Bob included maps to illustrate their arguments. Secondly, alongside your participant observation work, you might also consider using other complementary research methods to deepen and/or broaden your understanding of the community under study. So, for instance, conducting a brief survey of the community in order to glean some basic information concerning its composition by age, gender, occupation, education, income, life-course,

BOX 10.6

Painting a picture of the community: blending participation with observation

The following excerpts are taken from the accounts which Catherine, Bob and Zoe constructed from their notes and/or memories of their research experiences. Here, I have divided the first two into accounts of a more participatory nature followed by those of a more observational nature. The final account, however, illustrates how these two aspects of a researcher's role in the field area can, in practice, blur into each other and therefore can and should be represented as doing so to give the reader a vivid impression of not only what was learned but how this was learned through the research process.

Catherine:

(a) participation:

'We moved along quite easily, meanwhile Victoria was beginning to get restless in the buggy and I thought that maybe it was time to stop for lunch in a while. Soon though we passed the new "Mothercare" on our left which had only just opened a few weeks ago . . . We turned inside using the automatic doors. The clothes were all close together which was surprising, and more than once I knocked some clothes off a display stand. Also, the clothes were displayed very close to the automatic doors, so while I was standing looking at the clothes, the doors kept opening. This was, when I thought about it later, quite worrying as children could easily have run out of the shop without their parents noticing. The talking tree, however, kept Victoria amused for a while.'

(b) observation:

'On our travels up this part of the road, I spotted a lady attempting to use the phone but, as she had a pushchair, she was finding it difficult as the pushchair would not fit into the telephone box so the buggy had to hold the door open for her.'

Bob:

(a) observation:

'The staff are paraded in their nice clean uniforms wearing the trendy modern cap, which straight away creates an image of them (customers) and us (staff) . . . the staff all work in a team which creates a professional ordered image in a friendly happy environment where there is no end to the amount of smiling people you can see, who all perform their jobs in an environment that becomes a joy to produce and to consume in. The individual talent or ability of crew members enables them to earn stars which are displayed to customers on our badges so they can see if we are good or not . . .'

(b) participation:

'The back room . . . is where all the food is kept. The main tasks to perform in the back room were to rack fries, take anything into the kitchen they (shouted) asked

Box 10.6 continued

for such as meat or cheese, so basically I became a "Gopher" – go for this, go for that. I only worked in the back room twice when I carried out my field work and one of my diary extracts reads: "There should have been two of us working back room today. However, Tracy did a no-show and when I approached the managers and asked if they were going to get anyone else in, they said no. I therefore ended up doing the job of two today and, seeing as I was constantly busy, it did not help when people kept on poking their heads around the corner asking for food such as a box of regular meat when, if they walked another five metres, they could have got it themselves. Stress would be an understatement to describe how I felt. I was snapping people's heads off, all of which culminated when I punched a hole in the side of a fry box in the freezer. After that, I felt a lot better".'

Zoe:

(a) moving from observation to participation:

'Before Radiohead were on, we were lucky to find a space to sit down in, fairly near the front, at the edge of the main crowd. Most people around us also sat down, so it seemed much less intimidating and claustrophobic than it had done last night. This side of the crowd was definitely a better choice as I could actually see, and hear, the music at the same time!! As the singer started a solo bit and attempted to show off his vocal skills, one of the lads in the group stood in front of us shouted "stop your fucking whining". He then plonked himself down mumbling "why doesn't he give up and go home?" On seeing that we had found his comments quite amusing, he turned himself around and asked me the time. After a few minutes, the group of friends that he was with also sat down and turned round to talk to us. He was from Plymouth, and at the festival with a group of about 15 others from home, but had split up from them, and tagged himself onto those others who were from Oxford. They were miffed that we had backstage wristbands, and wanted to know what it was like backstage. Being totally honest, I admitted that it was painfully dull, and not as full of stars as one would expect.'

social networks and any number of other factors appropriate to its specific membership can be useful. Alternatively, or in addition to this, you might consider conducting interviews with key members of the community before, during or after your participant observation work. Adopting these and other methods in combination can serve a number of purposes: first, if they are undertaken early in the work and are limited to fairly innocuous questions, they can serve as a means of introducing yourself and your project to community members in a relatively unthreatening way; secondly, they can generate descriptive statistics which can be used in the write-up to outline the community; thirdly, they can be used to identify the community's key groups and networks within which you might wish to develop contacts; fourthly, if you are keen to develop a research project which will address concerns within the community, they may also serve as a means to gauge what, and how widespread, these might be; finally, if you develop close relationships within the community and then

LIVERPOOL JOHN MOORES UNIVERSITY
LEARNING SERVICES

BOX 10.7

Making your account more vivid through illustration

Catherine's participant observation account included, for instance, a photograph of the pushchair wedged in the door of the phone box mentioned in Box 10.6 in order to bring this scene to life for the reader. Similarly, Bob took photographs of the restaurant in which he worked on one of his days off (again doing this covertly) and he also included colour photocopies of McDonald's promotional materials, both of which made his discussion of these things more vivid. Zoe also took photos which illustrated, for instance, the creation of the festival 'scene' in a field just outside Reading through taking two of them from the same place before and after the festival-goers had been let onto the site. Similarly, in explaining the contexts of their research, two of the three included maps. Catherine did so to show, in the early stages of her dissertation, the route which she took through the city centre with Victoria and her pushchair and, towards the end, to represent the relative ease of travel through this space for people on wheels based on her own experience. Bob included his map to illustrate the various public and private zones of the restaurant in which he worked and to show what kinds of behaviours could be expressed where – in the back room, in the kitchen, at the counter and in the public restaurant space itself.

undertake a survey or some interviews, you may be able to make your enquiries more appropriate to its members' lives, and the information thereby generated can be used to position their (and, indeed, your own) understandings of the community as being from one or more perspectives. To illustrate the advantages of such multi-method approaches, we can turn to Box 10.8 where the ways in which Zoe, Catherine and Bob drew in other research methods to complement their participant observation work are outlined.

In a similar vein, it is likely that within and/or outside the community, someone, somewhere has kept relevant records. People take photographs, receive telephone bills, keep school reports, compile tax returns, keep time sheets, compile reports and inventories, file letters, memos and faxes, trace their family trees, listen to radio shows, watch the TV, the list can go on and on. You should perhaps have this in mind at all stages of the research and, when relationships are developed sufficiently with people who might have access to such things, you can ask to see them and copy down or otherwise record them. Such records may also be in the public domain, so a questionnaire administered by a researcher may not be necessary if the community under study falls within a census tract. Finally, you can keep your own, and/or ask community members to keep, tallies of various objects and movements in the form of tables and/or maps (Gregory and Altman, 1989). Whatever form these records take and however they are gleaned, it must always be realised that, just because they can produce numbers, networks, and stories, these are no more nor less 'objective' or 'subjective' than any other form of information brought back from the field. Such apparently 'cold', 'scientific', or 'unbiased' 'data' is just as much socially and culturally constructed as any other form of information, and has to be sensitively asked for and interpreted with this in mind.

BOX 10.8

Mixing participant observation with other methods

Part of Zoe's preliminary research involved her interviewing the lead singer of the band Kingmaker for a fanzine. This, however, did not eventually have a significant enough bearing for her to include this as a central part of her research. Instead, after putting together her own impressions of the Reading Festival in her participant observation account, she sent out a series of loosely worded questionnaires to others who had attended, asking them what their experiences had been. She then reported these responses person by person and word for word in her dissertation in a section called 'Different views about the festival'.

Catherine had a number of options which would have helped to make a broader sense of what she had studied. For instance, interviews with planning officers were undertaken to help her to get to grips with why urban planning appeared to be so insensitive to the group whose travels she had been studying. However, having plenty to write about based on her participant observation fieldwork and other literary sources (see later), she chose not to use these. As in Zoe's case, Catherine's main supplementary method was her use of short questionnaires 'sent to a local play group in the Hull area in which parents were asked general questions on how they found access in the urban area of the city to give a wider perspective of what was actually happening in the town with other mothers and carers'.

The covert nature of Bob's research, unlike Zoe's and Catherine's, meant that such overt supplementary research was not possible. However, given that it was his intention from the start to compare the public image of the McDonald's corporation with the private, behind-the-scenes world that he was used to as a crew member, his supplementary methods were much more in the form of visual and textual analyses of advertising materials, restaurant and uniform design, newspaper reports from his local area concerning the restaurant he worked in, and the comparisons that he could make between these and the materials generated by the McLibel trial.

Concluding comments

In this chapter, I have tried to outline the main justifications for, and means to complete, dissertation research based on the method of participant observation. I hope that the illustrations taken from the work of students who have successfully used this method in their dissertations provides not only some idea of the diverse ways in which this can work but also some inspiration that you could perhaps use these methods to do some research on something which is both interesting and important to you. In so many ways, though, it is difficult to be prescriptive about what this should and should not involve and what can and cannot be understood through doing this. But I want to leave you with a couple of key questions which you will continually have to ask yourself along the way if you choose to take on this method:

1. How can I vividly convey what I have learned during my fieldwork to those who will read the final dissertation? and
2. How can I integrate this kind of account with the kinds of theoretical, methodological and substantive discussions which are usually required in a dissertation?

As far as the first question is concerned, what your dissertation markers will be looking for is a kind of creative writing which will allow them to feel that they have 'stepped into your shoes' and 'been there' too. When reading the boxed extracts from Zoe's, Catherine's and Bob's participant observation accounts, you may have found that you could imagine yourself trying to get into the Reading Festival site, pushing a small child around a Mothercare shop, or working in the back room of a McDonald's restaurant. If so, this is the kind of reading experience that you should be trying to conjure up for your dissertation marker (the last thing that they will want in your participant observation account is a dry and dusty list in which 'this happened', then two minutes later, 'this happened', and so on). The best advice that I can give here is to read through some of the further readings suggested below and to follow up references to ethnographies in which participant observation has been used (however irrelevant these might seem to your own research topic) to get a flavour of how experiences 'in the field' can be effectively represented.

As far as the second question is concerned, the integration of your participant observation account(s) into the rest of your dissertation should, as with any other method, be preceded by a discussion of this as a 'scientific' method which can be done badly or well: like other methods, participant observation research does have general principles and problems which can be discussed; it is suitable for studying certain kind of processes and not others; and its main research 'tool' – i.e. you! – is 'calibrated' in certain ways which have to be noted. On top of this, you will have to write about your theoretical and substantial concerns and how they feed into and out of the fieldwork which you did. However, given that such 'fieldwork' can lead to shifting of the horizons or focus of the project which you initially wanted to do, there will inevitably have to be some flexibility in how you represent these shifting grounds in the final dissertation. Thus, depending on what happens, you may have to be quite inventive in how you set out the relationships between 'theory' and 'practice', chapter by chapter, in your final dissertation. By way of conclusion, then, what I would like to finish with is a final box (10.9) outlining how Catherine, Zoe and Bob constructed their dissertations to take this into account in a way that worked for them.

BOX 10.9

The final versions: outlines of the three dissertations

In the following, I have tried to summarise how participant observation accounts were fitted into what each of the three dissertations were about. Each adopted a chapter structure in which theoretical, methodological, political, ethical, and substantive concerns were combined in ways which were suitable to represent what they had done.

Catherine's dissertation was structured as follows:

1. She introduced her dissertation with the whys and wherefores of her research.
2. She moved on to provide her participant observation account of the journey taken through Hull's city centre.

Box 10.9 continued

3. She attempted to make a wider and deeper sense out of this with reference to the feminist and disability studies literatures, which explain why urban landscapes are not built for all people to use equally.
4. She looked at ways in which alternative planning literatures have been produced in order to make such environments more accessible to more people.
5. She concluded with her thoughts on what had been, and could be, learned from such research in terms of the social geographies of access to urban environments.

Zoe's dissertation was structured as follows:

1. She set out the whys and wherefores of her research and how this was so personal to her as to justify writing it as an autobiographical account of six months of her life.
2. She provided her participant observation account of the days leading up to and through the festival.
3. She talked about what she had done afterwards both to find out what had happened from other festival-goers' perspectives and to find other festivals in which the connection between music and resistant youth cultures worked more effectively.
4. She discussed how the subject of the research had changed from the relationship between music and resistant-youth cultures to the purpose of research in general.
5. She concluded by attempting to reconcile the unexpected, difficult and important lessons learned through the research process with an anarchist geography more suited to what she was trying to find out.

Finally, Bob's dissertation was set out as follows:

1. He set out the whys and wherefores of his research and described the various 'layers' of the story to be peeled back, chapter by chapter, in the rest of his dissertation.
2. First layer: an analysis of the public face of the McDonald's corporation through advertising and other promotional literature.
3. Second layer: a more observational account of how the restaurant in which he did his research worked on a day-to-day basis.
4. Third layer: a more participatory account of serving at the till and comparing this to how serving is represented in the restaurant's marketing literature.
5. Fourth layer: another more participatory account of working behind the scenes in food preparation and comparing this to the claims which the corporation makes about its food, and what has been 'thrown up' about this by the McLibel trial.
6. Fifth layer: placing the working conditions in this one branch in the context of those in the thousands of other McDonald's restaurants around the world, again as highlighted by the McLibel trial.
7. Final layer: an experimental collage of images which represent what McDonald's means after taking all of the above into consideration and intended to contrast with the squeaky clean and happy public image described at the start of the dissertation.

Acknowledgements

This chapter has been adapted from the section on 'Participant Observation' which I wrote for the 1995 CATMOG booklet *Doing ethnographies*. This booklet was co-authored with Mike Crang and I would like to thank him for allowing me to 'own' this chunk for the second time. For the transformation of that chunk into this one, I would like to thank Catherine Hawkins, Zoe Howse and the anonymous 'Bob' for kindly allowing me to sing their praises, and Robin Flowerdew for his helpful comments.

BOX 10.10

Further reading

For discussions of the politics, ethics and practicalities of doing/writing ethnographic research outside of geography (including but not limited to participant observation), the following edited collections are a good place to start.

Amit, V (ed.) 2000 *Constructing the field: ethnographic fieldwork in the contemporary world*, London: Routledge

Bell, D, P Caplan and W J Karim 1993 *Gendered fields: women, men and ethnography*, London: Routledge

Clifford, J and G Marcus (eds) 1986 *Writing culture: the poetics and politics of ethnography*, Los Angeles and Berkeley: University of California Press

Gordon, D and R Behar (eds) 1995 *Women writing culture*, Los Angeles and Berkeley: University of California Press

Nencel, L and P Pels (eds) 1991 *Constructing knowledge: authority and critique in social science*, London: Sage

For a starting point to explore similar work in geography, the following will also be worth a read:

Cook, I and M Crang 1995 *Doing ethnographies*, Concepts and Techniques in Modern Geography 58, Norwich: Environmental Publications

Cloke, P, I Cook, P Crang, M Goodwin, J Painter and C Philo 2004 *Practising human geography*, London: Sage

Eyles, J and D Smith (eds) 1988 *Qualitative methods in human geography*, Cambridge: Polity

Hay, I (ed.) 2000 *Qualitative research methods in human geography*, Oxford: Oxford University Press

Limb, M and C Dwyer (eds) 2001 *Qualitative methodologies for geographers*, London: Arnold

Professional Geographer 1994, 46(1), special issue on 'Women in the field'

Section C Analysis of data

This section deals with some very different forms of 'analysis', ranging from the quantitative approaches that are usually conjured up by such a title, including the use of geographical information systems, to the analysis of qualitative materials. The choice of appropriate methods here will be largely determined by the type of data that is available to the project, whether from primary or secondary sources. Data analysis methods require careful thought, as they can play a very significant role in moving a research project from a simply descriptive exercise to one that produces new insight and information. An extensive data collection exercise can be let down by poorly chosen or inadequate analysis, making it difficult to draw any clear conclusions from the project as a whole.

Chapters 11 and 12 provide a basic framework for the analysis of quantitative research data. They are not intended to be used as a 'cookbook' for direct application to particular data sets (there are references to the many statistical texts available), but rather to explain the underlying issues which should be considered when dealing with quantitative analysis, including how to decide which methods should be used. In Chapter 11 Stewart Fotheringham introduces the analysis of numerical data, dealing primarily with ratio or continuous levels of measurement. In Chapter 12 Andrew Lovett deals with analysis of data at the ordinal and categorical levels of measurement, although much of the material in Chapter 11 is of more general applicability. Fotheringham stresses the importance of thinking about analysis before actually collecting any data, as this can be enormously helpful in identifying what needs to be collected. The chapter deals with the use of simple descriptive statistics, the value of data visualisation techniques, and the move from exploratory data analysis to hypothesis testing and model formulation. There is a particular emphasis on the explicitly geographical aspects of numerical data analysis, which are frequently inadequately addressed, even by geographers! Ordinal and categorical data are frequently produced by questionnaire research, and the examples in Chapter 12 are drawn from the real questionnaire case studies used by Parfitt in Chapter 6.

Chapters 13, 14 and 15 deal with the analysis of qualitative materials. Mike Crang discusses the analysis of information that may be obtained from primary qualitative research such as in-depth interviews and observations and notes made by the researcher. Stuart Aitken focuses on the analysis of textual material – taking a broad definition of what may be treated as 'texts'. In Chapter 15, Stuart Aitken and James Craine develop this discussion by a concentration on visual media.

Mike Crang guides the reader through the process of analysing materials derived from qualitative research, again emphasising the importance of thinking about analysis early in your research plans. Materials derived from this type of research may be voluminous and the processes of transcription and coding are important in the refinement of the researcher's ideas about the subject. Analysis of qualitative materials can be aided by the use of appropriate computer software, but whether computer-aided or manual, the process will often be iterative, gradually moving from a focus on the detailed findings to concern with more abstract theoretical issues. An important distinction between this kind of work and quantitative analysis is that the codes assigned to research findings here are not intended to be subjected to statistical analysis, but used as aids to the development of theory. There are many possible approaches to this task, and the chapter illustrates three alternatives, each of which asks different questions of the material, and uses different forms of explanation.

Stuart Aitken in Chapter 14 considers geographical analysis of textual material, showing that geographical research can be done from the armchair and does not require wellies or a clipboard. In addition to literary study, he gives examples of how geographers have studied texts in other media, including travel writing, visual texts (examined in more depth in Chapter 15), music, maps and landscapes. He discusses theoretical approaches which can be used in geographical analysis of texts, based on the use of semiotics (the theory of signs), including discussions of hermeneutics, Marxist, feminist, psychoanalytic and post-structuralist approaches to geographical text analysis. Examples and suggestions are given about how students might deconstruct a text and develop interpretations of new and exciting geographies.

Aitken and Craine in Chapter 15 concentrate on visual media, and how geographical knowledge is visually conveyed. They relate visual images to economic and political power structures, emphasising through a discussion of Vermeer's painting *The Geographer* how messages can be conveyed powerfully through visual media without being explicitly stated. They go on to discuss geographical perspectives on visual images, concentrating particularly on art, photography, film and advertising. They relate their analysis to semiology and again to various approaches within social theory, including discourse analysis. They conclude with a detailed example of how films can be interpreted, using the American 1950s film, *Kiss Me Deadly*.

As geographical information systems have become a familiar feature of geography programmes, the numbers of students who want to use GIS as part of their research projects continues to grow. Inevitably, technical aspects of using the software will form part of the work undertaken in such projects, but if GIS are being used in the analysis of a particular substantive research problem, then they are best seen in the context of this book as an analysis tool. In Chapter 16, David Martin reviews the types of analytical processes that GIS can offer, and provides some examples of the types of problem to which they are most usefully applied, drawing on specific student projects. The relationship between GIS and the types of quantitative data analysis introduced in Chapter 11 are explored, and the special requirements of data supply and data quality are stressed.

11 Analysing numerical spatial data

A Stewart Fotheringham

Synopsis

For many students the analysis of numerical data will form an integral part of a research project. Regardless of the source of data, whether it be a population census, a remotely sensed image, fieldwork measurements, a questionnaire, or a medical registry, in order to extract useful information some analysis of the data will be necessary. This chapter discusses several questions that will probably arise in the course of data analysis and it is hoped that if these questions are thought about *before* the analysis takes place, some of the more obvious pitfalls of data analysis might be avoided.

Rather than attempting to provide a 'shopping list' of techniques, the chapter discusses broader issues that are likely to arise in the analysis of numerical data. Five major topics are examined: the appropriate time to start thinking about data analysis (*before* the data are collected rather than *after*!); why data analysis is important; how to decide on what technique(s) to use; the optimal use of *spatial* data; and the possible problems one might encounter, particularly in the analysis of spatial data. The intention is that the student will be sufficiently motivated to explore particular techniques in more specialised texts.

Preamble

Many undergraduates will find that the analysis of numerical data will form an important, interesting and rewarding part of their research project. There can be few academic activities that produce a greater sense of achievement than being able to support initial ideas or being able to generate new ideas through the analysis of data. To many students, however, data analysis is the part of a research project approached with the greatest trepidation. It is often the part of a research project most closely scrutinised and most easily criticised. In order to increase the positive aspects of data analysis and reduce the negative ones, there are several relatively simple questions to ask oneself. These form the structure of the remainder of this chapter and are also summarised in Box 11.1.

BOX 11.1

Questions to ask

1. When should I start thinking about data analysis?
2. Why should I undertake data analysis?
3. How do I decide what technique to use?
4. How geographical is my analysis?
5. What problems might I encounter?

Rather than attempt to provide a 'shopping list' of procedures and detailed descriptions of particular techniques, which would be impossible in the space available and can be found elsewhere (*inter alia* Bailey and Gatrell, 1995, Earickson and Harlin, 1994; Robinson, 1998; Rogerson, 2001; and Fotheringham *et al.*, 2000), the chapter contains a discussion of broader issues to consider in data analysis. It is hoped from this that the reader might understand, and therefore avoid, some of the more obvious pitfalls that can arise in analysing spatial data (data having spatial coordinates), without having to understand the details of specific methods.

When should I start thinking about data analysis?

As will be discussed below, data can limit the type of analysis you might be able to undertake in order to pursue your research aims. It is therefore important to think about analysis *prior* to data collection and a chat with someone in your department knowledgeable about these issues would be useful before you embark on the research project. Failure to do so may result in the situation where weeks or even months of data collection are squandered: collecting data on a few other variables or gathering the data in a different way might considerably broaden the scope of the analysis that is possible. There is nothing more frustrating and guaranteed to set off panic attacks than having just spent your summer collecting data and then realising you cannot perform an appropriate type of analysis because of some limitation of the data.

Having said this, it is not essential to have every step of your analysis mapped out prior to data collection. Sometimes the data themselves present ideas that you might not otherwise have had and the analysis you conduct might go in directions that were not anticipated at the beginning of the project. Such serendipity is part of the excitement of undertaking research and is not to be discouraged. However, it is recommended that you at least make sure that a minimum set of analytical techniques that will provide information on your research topic is feasible before embarking on the data collection stage.

Why should I undertake data analysis?

Numerical data are commonplace in any modern society and web-based access to data is increasing rapidly. Spatial data, whether they are derived from a

BOX 11.2

Reasons to undertake data analysis

1. To summarise large amounts of data through descriptive statistics
2. To explore data through visualisation
3. To infer aspects of the population from samples or processes from observations
4. To develop models of the real world

census of population or a remotely sensed image or from one of a myriad of other sources, can be accessed relatively easily and in large quantities. Spatial analysis is needed in order to make sense of these data and to identify trends and relationships within the data. Spatial analysis can play several roles in a research project depending upon the particular type of analysis undertaken. These roles are now described, and the key reasons listed in Box 11.2.

Summarising large amounts of data through descriptive statistics

Examples of commonly used descriptive statistics include the *arithmetic mean* and the *median* as measures of central tendency, and the standard deviation and coefficient of variation as measures of dispersion. It would be much easier, for example, to appreciate that a North–South gap in house prices exists if one compares the mean or median house prices of a sample of houses in both regions rather than trying to make sense of the masses of data on individual house sales. Generally, for house prices the median is preferred over the mean because it is less sensitive to extreme values. There might be one or two house prices in one or both of the samples that are much higher than the others and such values would distort the mean but not the median. The median is therefore known as a *robust* indicator of central tendency. As well as having some indication of central tendency to summarise the two samples, one might also want to know something about the variability of the data within the two regions to assess (a) how representative the measures of central tendency are and (b) whether there is more variation in the data within one region than the other. The standard deviation could be used for this former purpose although the *interquartile range* is a more robust estimate of variability. Even better, however, might be the *coefficient of variation* because it is independent of the units in which the data are collected and is therefore useful to compare two samples of data with different means. The spatial equivalents of the arithmetic mean and standard deviation are the *centre of gravity* and the *standard distance*, respectively. The former is a location that represents a set of points in space; the latter is a radius of a circle depicting the 'spread' of points in space. Both these statistics utilise the spatial coordinates of the data in the derivation of the summary statistics and are therefore rightly termed spatial statistics.

Exploring data through visualisation

A second use of data analysis is in an inductive framework to *suggest* hypotheses about the data and the processes underlying them. This is called *exploratory analysis* as opposed to the more usual deductive-based *confirmatory analysis* that is discussed below. Exploratory analysis encompasses a wide and increasing variety of techniques, all of which are designed to suggest aspects of data or a modelling technique that might be worth investigating in more detail. At a very simple level, a *scatter plot* showing the relationship between two variables can shed light on the nature of the relationship (Does it appear to be very strong? Is it linear? Is there some obvious non-linear relationship?) and can also highlight individual data points that appear to be exceptions to the general trend. It should be remembered that although one purpose of analysis is to examine trends and regularities in data, another equally valid purpose is to identify deviations from these trends. Other popular and simple exploratory analytical techniques include *box plots, stem-and-leaf plots* and *spin plots,* all of which are available in most popular statistical packages. Bailey and Gatrell (1995), Fotheringham *et al.* (2000) and Sibley (1988) provide descriptions of various exploratory techniques for spatial analysis.

The key to exploratory analysis is that data are generally presented in some visual format such as a pie chart or a histogram or a box plot. More advanced forms of exploratory analysis, and ones that are particularly useful to geographers, allow windows in a display to be linked so that actions in one window are reflected by change in another window. For example, consider two linked windows, one of which contains a map and the other contains a frequency distribution. The frequency distribution can be *brushed,* meaning that values are highlighted moving say from low to high across the frequency distribution. At the same time that the individual data points are brushed in the frequency diagram, their locations are highlighted on the map so that one can immediately see where the low and high values tend to be located. The advantage over simply plotting high and low values on a map is that you have control over what is classified as 'high' and 'low' and can alter the map with a simple keystroke or movement of the mouse.

This type of tool can be very useful and can obviously be extended much beyond the simple example described above. For instance, an obvious application is to examine the spatial patterns of residuals from a model (this is termed post-modelling exploration) to identify whether the general trend indicated by the model is equally applicable to all parts of a region. An extension would be to allow elements of the modelling procedure to vary (such as adding variables to the model or removing data points) and then to investigate the sensitivity of the model outputs to these changes. In this way, a picture can be built up of (a) the reliability of the model, (b) the reliability of the data and (c) the nature and stability of the spatial processes producing the data. Statistical packages such as R and S-Plus provide the functionality of linking windows and an example is shown in Figure 11.1.

The left-hand window contains a simple scatter plot of two variables, percentage black population and percentage white population, for counties in part of the Midwest in the USA. The scatter plot shows that as the percentage white variable increases, the percentage black variable decreases: an inevitable relationship caused by the figures for other racial groups typically being very small and the percentages for all racial groups summing to 100. However, it might be

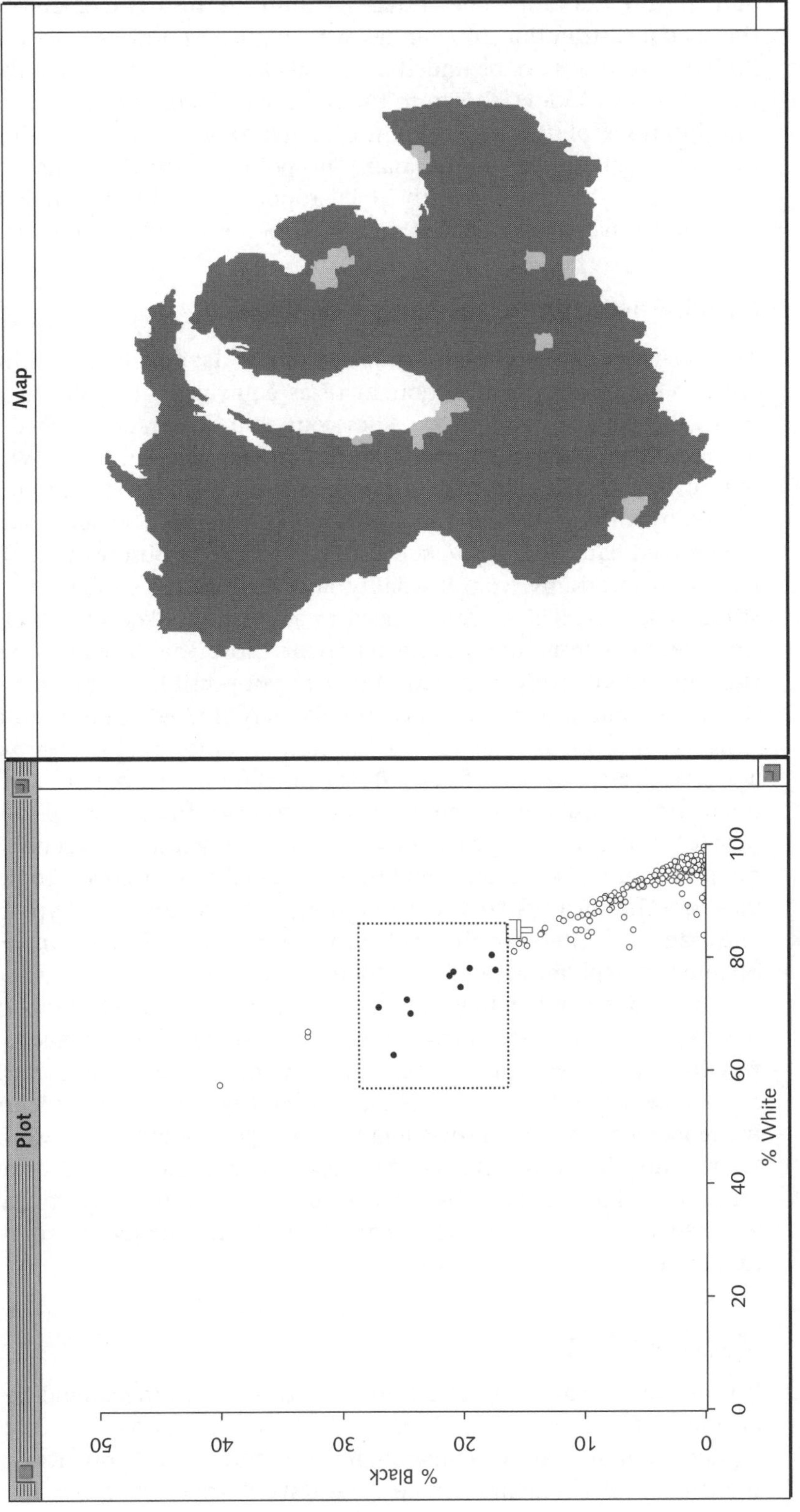

Figure 11.1 Example of linked windows in exploratory data analysis

of interest to examine some of the exceptions to this trend or simply to examine the spatial distribution of counties with high proportions of one racial group. Both of these tasks can be undertaken easily by linking the data in the scatter plot with another window, shown to the right, which contains a map of the region. The locations of the points in the scatter plot that are 'brushed' are automatically highlighted on the map. The points brushed in this example show that most of the counties with high proportions of black population coincide with large urban areas such as Chicago, Milwaukee, Indianapolis, and Cleveland.

Hypothesis testing

This is such a frequently encountered form of data analysis that the two terms are sometimes erroneously thought of as equivalent. Hypothesis testing often forms the core of first-year statistics courses in geography degree programmes and it still provides the basis for most geographical analysis. While there are several frameworks in which hypothesis testing can be useful, a fundamental use of hypothesis testing is to draw some inference about a larger body of unobserved data (the *'population'*) from a *sample* of observations. Typically, for reasons of economy, confidentiality and feasibility, our data usually do constitute a sample. Therefore the need to assess how likely it is that the conclusions reached from the sample represent those that would be reached from the population is very common. If it were not possible to draw inferences about the population, any analysis would have very limited application and use. For instance, suppose attitudes to the location of a proposed nuclear power station are investigated in an area of 10,000 inhabitants. How representative of the population in this area would be the responses from a sample of 50 or 100 people? What is the probability that the conclusions reached from these samples would also be the conclusions reached if all 10,000 people were interviewed? The answers to these questions are provided by hypothesis testing and statistical inference that fall into the category of *confirmatory analysis* (as opposed to exploratory analysis described above).

It should be noted that hypothesis testing has broader applications than making inferences about populations from samples. For instance, we might ask whether data have been generated by a random process. This forms the basis of a statistically testable hypothesis but does not rely on any sample-to-population inference: the data in question might form a population. What we would like to know is how likely it is that some observed distribution (say of disease incidence on a map) has occurred by chance. If it is very unlikely to have occurred by chance, then we could feel confident about looking for some alternative explanation.

Model building

It is perhaps debatable whether mathematical modelling should be classified as data analysis. However, modelling, which is often closely linked to the above types of analysis, can form an extremely important component of research and therefore merits a mention here. One test of whether we actually understand a spatial process, be it physical or human, is to develop a mathematical model that represents our understanding and to examine the extent to which this model can accurately represent that part of the real world in which we are interested. The calibration of models is a numerical exercise that can yield important information on the nature of relationships between variables. The estimation of parameters by regression, for example, is an example of model calibration frequently encountered in research projects. In such a model, a *dependent* variable is postulated to be a function of one or more *independent* variables and the objective of regression is to obtain estimates of parameters, each of which reflects the nature of the relationship between the dependent and a specific independent variable in the model. While regression models are typically written as additive combinations of independent variables,

$$y_i = \alpha_0 + \alpha_1 x_{i1} + \alpha_2 x_{i2} + \ldots \alpha_n x_{in} \qquad (11.1)$$

with y being the dependent variable measured at location i, x representing an independent variable and α being a parameter to be estimated, they can also represent relationships which are linear only after some transformation of the variables has taken place. A common example of such an *intrinsically linear* model is when a multiplicative model such as

$$y_i = \beta_0 \, . \, x_{i1}^{\beta 1} \, . \, x_{i2}^{\beta 2} \ldots x_{in}^{\beta n} \qquad (11.2)$$

is made linear in terms of its parameters by taking logarithms of both sides of the equation:

$$\ln y_i = \ln \beta_0 + \beta_1 \ln x_{i1} + \beta_2 \ln x_{i2} + \ldots \beta_n \ln x_{in} \qquad (11.3)$$

This type of transformation can be very useful in removing the problem of *heteroscedastic* (unequal variance) error terms (see Ferguson, 1977; Berry and Feldman, 1985; Schroeder *et al.*, 1986; and Berry, 1993 for fuller descriptions of this problem and regression in general) and also to produce parameter estimates that are scale-independent. That is, suppose the variable x_1 in equation (11.1) was distance and was multiplied by 1000 (for example if distances were measured in metres rather than kilometres), the estimate of the parameter α_1 would be reduced by a factor of 1000. However, the estimate of the parameter β_1 obtained from equation (11.3) would be unaffected by the change in the units of measurement.

It should be noted that the world is not always a linear place and that regression concepts can be extended to truly non-linear relationships with various methods of estimating such models being available in most standard statistical packages such as SPSS for Windows.

Models have another important use in predicting the outcomes of spatial processes. If a model replicates reality sufficiently accurately, it may be useful in predicting information for areas where data have not been collected or for future time periods. Models can also be used to ask 'What if . . . ?' questions where the impact of changes over time can be assessed in terms of the outcomes of a particular process. For example, a retail choice model could be used to predict what will happen to a supermarket's catchment area if one competitor opens a store of 3000 square metres two kilometres away and another competitor closes a store of 5000 square metres four kilometres away. Deviations from a model's prediction can also illuminate aspects of the underlying spatial processes that might otherwise not be noticed. Thomas and Huggett (1980) provide an overview of several different types of models relevant to geography and more specific examples are provided by Haynes and Fotheringham (1984) and Rogers (1985).

The analysis of numerical data can therefore play many important roles in research. It can be used to make sense of large amounts of data, to explore data as a means of generating hypotheses and understanding spatial processes, to infer processes leading to the observed data whether those data constitute a sample or a population, to assess the significance of findings, to help in model building, and in general to provide evidence for hypotheses about spatial processes. For many researchers, it is generally not sufficient to make statements about spatial processes without providing supporting evidence from the analysis of data. Such evidence is relatively value-neutral and its strength can be quantified. That is, through the analysis of numerical data, we not only have a mechanism for making a statement about the processes underlying observations but we can also measure how likely this statement is to be true. In some cases our evidence might be weaker than in others and we should be able to record such differences.

How do I decide what technique to use?

This is not an easy question to answer as there is a bewildering number of statistical techniques available, many of which can be applied to the same data. While there is no substitute for experience (again, try to discuss your research aims with somebody knowledgeable in this area in your department), there are two properties of data that limit the range of appropriate numerical techniques that can be applied to them. Again, this is why analysis should be considered *before* data collection. One of these properties is the number of samples; the other is the type or level of the data. An example of how the combination of these two attributes helps determine an appropriate statistical technique is shown in Table 11.1. This table includes several of the techniques with which an

Table 11.1 A typology of statistical tests

Type of data	*Task required*			
	Compare sample with fixed value	*Compare two samples*	*Compare more than two samples*	*Measure relationship between two variables*
Nominal	Chi-square one-sample test	Chi-square two-sample test	Chi-square *k* sample test	Coefficient of areal correspondence
Ordinal: weakly ordered	Kolmogorov–Smirnov one-sample test	Kolmogorov–Smirnov two-sample test	⇑	Kolmogorov–Smirnov two-sample test
Ordinal: strongly ordered	⇑	Mann–Whitney *U*-test	Kruskal–Wallis ANOVA	Spearman's rank correlation coefficient
Ratio	*t* or *Z* one-sample test	*t* or *Z* two-sample test	*F* test (ANOVA)	Pearson product moment correlation coefficient

BOX 11.3

Types of data

1. *Nominal*. Discrete (data can only take certain values); categories have no ranking (e.g. male / female)
2. *Ordinal:*
   (*weakly ordered*). Discrete; data are in categories which are ranked (e.g. strongly agree / agree / neutral / disagree / strongly disagree)
   (*strongly ordered*). Discrete; data are ranked (e.g. rankings of cities in terms of population)
3. *Ratio*. Continuous (e.g. rainfall figures)

undergraduate student in geography is likely to be familiar but it is not meant to be an exhaustive list of appropriate techniques. In order to utilise this table, it is necessary to identify both the number of samples you have and the level of data being analysed.

The number of samples refers to the number of 'frameworks' for which you have sampled data. For instance you might have data on a sample of house prices in one part of the country and want to examine whether this value is significantly different from the UK average. You have sampled data from only one part of the country (the UK average is considered a 'fixed' value in this context). On the other hand, if you have house price data from two regions of the country and want to determine if there is a significant difference between them, then you have two samples. Alternatively, you might want to compare farming practices in four different topographic regions: in which case you have four samples.

There are three levels of data that are commonly encountered: *nominal, ordinal* and *ratio* with ordinal data being further classified into two categories, *weakly ordered* and *strongly ordered* (see Box 11.3). Nominal and ordinal data are *discrete* (the data can take only certain values), whereas ratio data are *continuous*. Nominal data contain the least amount of information in that they are simply categories without any ordering. Examples are the division of respondents into male/female or the division of land use into arable, pasture, forest and moorland. The data consist of counts of observations in each of these categories. Data that are ranked either individually or in groups are referred to as ordinal. Weakly ordered ordinal data consist of ranked *categories* such as the common response asked on a questionnaire: strongly agree; agree; neutral; disagree; strongly disagree. Strongly ordered ordinal data consist of a ranking of individual data points; an example being when cities are ranked according to some attribute such as mean temperature or crime rate. Nominal and ordinal data are categorical and do not have metric properties which clearly limits the types of analyses that can be performed on them. For instance, it does not make sense to calculate a mean land use that might be somewhere between pasture and forest! Nor does it make sense to calculate a mean response to a questionnaire statement, somewhere between disagree and strongly disagree, for example, although this might be very tempting. The problem is that in calculating a mean from such data it has to be assumed that the perceived divisions between all the categories are equal when there is no guarantee that this is so. The statistical techniques that can be applied to nominal and ordinal data are discussed in more detail in the next chapter and form part of what are known as *non-parametric statistics* (see Chapter 12 and Coshall, 1989). Ratio-level data provide the most information since they are continuous and have metric properties. Examples include heights of individuals, stream velocities, clay content of soils etc. and the statistical techniques that can be applied to these data are referred to as *parametric statistics*. Data can always be transformed from ratio to ordinal to nominal but never the other way.

Knowledge of both the number of samples and the type of data being analysed can thus be a useful guide to an appropriate statistical technique. If the aim of the analysis is, for example, to compare two samples of strongly ordered ordinal data, then a Mann–Whitney *U*-test would be appropriate. If the aim is to compare one sample of nominal data with a fixed value, then a one-sample chi-square test is appropriate. An arrow in Table 11.1 indicates that a recommended methodology is to transform the data to the next level and use the appropriate test at that level.

How geographical is my analysis?

In conducting data analysis within geography, it is sometimes easy to lose sight of the geography of the problem and to concentrate simply on the analysis. Whilst the analysis is clearly important in improving our understanding of spatial processes, it is just a means to an end and the analysis should be linked to geography at every opportunity. For this reason, statistical packages that provide linked windows where data can be analysed and brushed in one window and then automatically displayed on a map in another window are very attractive. Being able to map data and results in increasingly interesting ways and to undertake numerical analysis are also attractive features of geographical information systems (GIS) that are considered in Chapter 16. Currently the analytical functionality of most GIS is rather crude but this is changing, and fairly soon most types of spatial analysis within geography will be possible within a GIS. Not only does a GIS provide an excellent medium for the display of spatial data and the outcomes of spatial analysis, but it also facilitates access to large spatial databases such as boundary files, road and river networks, digital elevation models and remotely sensed images.

The display of data or results on a map utilises the characteristic of spatial data that makes them special: *the data have spatial locations*. It is possible to distinguish two types of analytical research using spatial data on the basis of this property. The first, which I will term *weakly geographical*, uses data for different locations but effectively throws the interesting spatial component away and utilises the data as if they are simply lists of numbers that could be reshuffled without affecting the outcome of the analysis. Performing an ordinary least-squares regression analysis on spatial data is an example of a weakly geographical analysis. Unless the residuals of the regression are mapped, the analysis would produce exactly the same results no matter what the spatial arrangement of the data. The data become merely numbers that have been stripped of all their spatial information. Spatial analysis that is *strongly geographical*, on the other hand, explicitly uses the information on the spatial location of the data points. Examples of strongly geographical techniques include the calculation of spatial means and standard distances as spatial equivalents to the more usual mean and standard deviation (Kellerman, 1981), and the calculation of directional means and directional variances where the data represent movement in various directions (Gaile and Burt, 1980; Schuenemeyer, 1984).

One strongly geographical analytical technique involves the measurement of spatial autocorrelation that refers to the spatial patterning of data. Data where the value for one spatial unit is similar to those in its neighbouring units are said to exhibit positive spatial autocorrelation (most data exhibit positive spatial autocorrelation); data where the value in one unit tends to be dissimilar to the values in neighbouring units are said to exhibit negative spatial autocorrelation (Goodchild, 1987; Odland, 1988). An interesting aspect of spatial autocorrelation, and a feature of several other types of strongly geographical analysis, is that its measurement is sensitive to the definition of 'neighbouring unit'. Clearly many definitions could be used, ranging from a rather restrictive nominal contiguity criterion to a continuous distance–decay-based definition. However, this should not necessarily be considered a negative feature of spatial analysis; rather, the challenge is to examine how sensitive the calculation of spatial autocorrelation is to the definition of neighbouring unit and perhaps report spatial autocorrelation values for a large range of definitions. In this way, more interesting and complex issues in the way data are related in space might be illuminated.

Other strongly geographical techniques include various types of point pattern analysis (Boots and Getis, 1988) where the interest is to examine whether or not points in space are clustered. Clustering is particularly important in medical data where diseases might have some point source that could be identified from an examination of the spatial pattern of the disease itself. There are many ways of undertaking point pattern analysis; a common one is with nearest neighbour analysis, where average distances between points are measured and compared with expected values under a random distribution. Others involve placing a circle randomly within a region and counting the number of points that fall within the circle. If the observed number of points is significantly larger than expected given the population-at-risk within the circle, then the circle is highlighted. The process can be repeated many times until it becomes obvious where interesting clusters appear on the map. Clearly, circle size will affect the results and so this aspect of the analysis can also be made a random variable.

A final example of strongly geographical analysis is the movement towards producing local or 'mappable' statistics. Typically, in analyses such as regression, for example, and even spatial autocorrelation, whole-map statistics are produced. In the case of regression the whole-map statistics include the parameter estimates and the goodness-of-fit statistics. In both cases, these are clearly 'average' values that are assumed to apply equally to all parts of a region. However, it could be that the estimated regression model replicates data in some parts of the region more accurately than in others. Similarly, the general relationship between data points depicted in a spatial autocorrelation statistic is an average over the region that could mask important geographic variations in the relationships between data over space. For this reason, the realisation is growing that whole-map statistics are limited and that wherever possible we should attempt to provide local or mappable statistics. Developments such as the expansion method (Jones and Casetti, 1992), spatial association statistics (Getis and Ord, 1991) and Geographically Weighted Regression (Fotheringham *et al.*, 2002) reflect this trend and also provide greater links to geography and mapping.

What problems might I encounter?

It is perhaps true that all analysis can be criticised in some way and certainly the analysis of spatial data is no exception to this statement. There are many options available at various stages of a research project and often more than one option may be appropriate in a given situation. It is therefore necessary to make decisions and this lays the researcher open to criticism such as 'Why didn't you do this . . .?' For instance, it is possible to debate the type of data collected, the way the data were collected, the number of samples taken, the analytical method used, the assumptions embedded in the use of the method, the interpretation of results, etc. But before being paralysed into inaction, the appropriate questions to ask yourself are: Is the analysis useful? Has it provided some insight or evidence that I would not otherwise have? Clearly you do not want to commit gross errors that invalidate your analysis but the goal of producing a piece of analysis that cannot be criticised in some way is probably unattainable. More realistically, aim to undertake an analysis that is useful and which avoids the major pitfalls associated with spatial data summarised in Box 11.4 and described below.

BOX 11.4 Problems to consider

1. *The modifiable areal unit problem.* In the analysis of aggregate spatial data, the conclusions reached might be dependent upon the definition of the spatial units for which data are reported.
2. *Spatial non-stationarity.* Relationships might vary over space making 'global' models less accurate. Global models will also hide interesting geographical variations.
3. *Spatial dependence.* To what extent is the value of a variable in one zone a function of the values of the variable in neighbouring zones? Strong spatial dependence can affect statistical inference.
4. *Non-standard distributions.* Examine the nature of the data being analysed. Not all data are normally distributed! Consider experimental methods of statistical inference if data have an unusual distribution.
5. *Spurious relationships.* Make sure that any relationships identified are meaningful and are not caused by a variable or variables omitted from the analysis.

The modifiable areal unit problem

This problem occurs with aggregate spatial data: data that are reported for zones rather than individuals. The problem is that the results of the analysis can be sensitive to the way in which the zones are defined. For example, an analysis of the relationship between crime rates and unemployment rates using data for census enumeration districts may well produce a different set of results from an analysis performed with exactly the same data aggregated to the level of wards. This is clearly worrying especially if the results are sufficiently different to alter the conclusions reached regarding the nature of the relationships being examined. Fotheringham and Wong (1991) provide examples of such a situation in the analysis of census data for the city of Buffalo.

The modifiable areal unit problem can be separated into two components. The first, described above, is the sensitivity of results to scale and the numbers of spatial units into which a region is partitioned. The other is the sensitivity of results to the way in which space is partitioned even when scale is held constant. Even keeping the number of spatial units constant, for example at the enumeration district level, there is a large number of ways in which space can be divided into this number of units; the particular configuration of enumeration districts is just one of these. Of course, it could be that a particular partitioning of space makes more sense than others but in many cases the partitioning of space in which data have been collected is for administrative purposes that have nothing to do with the spatial processes under investigation. Unfortunately, there is little one can do about this problem except to be aware of its potential existence and to examine the sensitivity of one's results to variations in scale, and possibly partitioning, if the data allow. The modifiable areal unit problem is part of *the ecological fallacy*, the inappropriate application to individuals of results obtained from aggregate data.

Spatial non-stationarity

Whilst, as geographers, we fully expect attributes to exhibit spatial variation and we are used to seeing such variation displayed on maps, we all too often ignore the possibility that *relationships* might vary over space. Such variation is referred to as spatial non-stationarity. It is very common, for example, to assume that the parameter estimates obtained from a regression apply equally to all parts of the region for which the data have been collected when in fact they might vary in ways that can yield interesting insights into the underlying relationships. Equivalently, it is not unreasonable to expect the same model to replicate data in different parts of a region with different degrees of accuracy. Uncovering spatial non-stationarity as a means of exploring data and relationships in more detail would thus seem a reasonable goal of research and it is related to the move from whole-map statistics to local or mappable statistics. The technique of Geographically Weighted Regression (GWR) is one method of producing local statistics as a solution to the problem of non-stationarity (Fotheringham *et al.*, 2002). User-friendly software, GWR 2.2, is available from the author so that students could undertake a GWR very easily on their own data. More information is available at: *http://www.ncl.ac.uk/geps/research/geography/gwr.*

Spatial dependence

It was noted above that spatial autocorrelation is a measure of pattern in spatial data. If high and low values are found in separate clusters, the data are said to exhibit positive spatial autocorrelation; if high and low values are interspersed, the data are said to exhibit negative spatial autocorrelation. The measurement of spatial autocorrelation is thus a useful descriptive spatial statistic for which formal significance testing procedures are available. However, the presence of significant autocorrelation causes a problem for techniques such as regression where an assumption necessary for tests of significance on the estimated parameters is that the observed data are *independent* of each other. Clearly, if the data exhibit strong autocorrelation this is not the case because the value of an observation in one zone is partially dependent on the values in neighbouring zones. A good example of such a process is the spatial distribution of house prices where the mean house price for one street is probably a reflection of the mean house prices in neighbouring streets (as well as being a function of attributes of houses in that street). Hence, techniques such as regression should be adjusted to account for the lack of independence in the data and regression models designed to handle the specific problems of spatial data have been developed (Anselin, 1988). Intriguingly, a simpler solution to the problem might be to use the GWR technique described above and as shown by Fotheringham *et al.* (2002).

Non-standard distributions

The usual procedure in confirmatory analysis is to examine the sustainability of a hypothesis by comparing a calculated statistic with a critical value obtained from a standard reference distribution such as the Normal, t-distribution, F-distribution, chi-square etc. The use of any reference distribution makes some assumptions about the underlying data being examined that may not be met, particularly in the case of spatial data. In such circumstances, *experimental distribution methods* may be used to confirm or reject hypotheses. Rather than refer to some standard distribution, an experimental distribution can be created using the observed data. Although there are many ways to do this, the following serves as an example of the general principles involved. Suppose the correlation coefficient between two ratio-level variables distributed across 25 spatial units is calculated and we would like to test the hypothesis that $r = 0$ in the population (that is, that there is no correlation between the two variables in the population and that the observed correlation can be attributed to sampling variation). Rearrange the two sets of data randomly and calculate a new correlation coefficient. Repeat this procedure many times (say 1000 times) so that we now have a distribution of r values that have been obtained from a random pairing of the observed data and which can be used to assess the significance of the observed value of r. If the observed value of r exceeds x% of the values, then the null hypothesis that $r = 0$ in the population can be rejected with level of confidence x. Such a procedure can be employed not only when the data suggest that a standard reference distribution may be inappropriate but also for those statistics for which a standard reference distribution do not exist.

The identification of spurious relationships

Care must be taken in data analysis to avoid mistakenly attributing any importance to a spurious relationship. Such a problem can arise when complex relationships are measured in simple terms so that the underlying causal relationship between two or more variables might be very different from that which is suggested by the statistical analysis. A common example where spurious relationships can occur, for example, is in the interpretation of a matrix of correlation coefficients between several interrelated variables. Each correlation coefficient measures the strength of a simple linear relationship between a pair of variables and can give a very misleading impression of the underlying causal relationship which may involve several other variables. For example, suppose data are collected on crop yield, altitude, and rainfall within a region and it is found that the correlation coefficient between crop yield and altitude is positive. It might be incorrect to infer that the causal mechanism producing this statistical result is one where the crop under investigation has an affinity for high-altitude locations. This might be so but it might also be the case that within this region rainfall tends to increase as altitude increases and that crop yield increases with rainfall. The calculation of a simple correlation coefficient can therefore suggest a relationship that is spurious. A more accurate picture of the determinants of crop yield would be obtained by a multiple regression where each parameter indicates the relationship between the dependent variable and one independent variable, *everything else being equal*. A regression analysis, for example, might indicate that the relationship between crop yield and altitude is negative even though the correlation coefficient suggests it is positive. In much the same way, a simple scatter plot can give a false impression of the nature of the relationship between two variables. Partial regression or leverage plots which depict the relationship between two variables, *everything else being equal*, are much more reliable. These can be obtained quite easily from popular statistical packages such as SPSS for Windows.

Summary

The analysis of numerical data can be an extremely valuable constituent of geographic research. It drives data collection and data presentation and it provides a powerful framework within which our understanding of spatial processes can be tested and developed. This chapter has described how the analysis of numerical data can be useful in summarising large amounts of data through descriptive statistics; how relatively new, and primarily visual, exploratory techniques can be useful in both formulating hypotheses and examining results; how we can infer aspects of a population from a sample or a process from a set of observations; and how model building and calibration can be used as a test of how well we understand the real world.

One of the more confusing aspects of data analysis is deciding on what technique to use in a given situation. Whilst experience is useful here, and discussing your problem with a user-friendly statistician is highly recommended, this chapter also discusses how to narrow down the choice of technique fairly easily. It is noted that spatial data have certain properties that lend themselves to different types of analysis and many of these are highly visual which often makes their presentation in research projects more appealing.

There are pitfalls, however, and care has to be taken to eliminate problems that could invalidate the conclusions reached from the analysis. A number of potential problems that might be encountered or might need some thought are outlined. These include the modifiable areal unit problem, spatial non-stationarity, spatial dependence, non-standard distributions and the identification of spurious relationships. However, these problems are also challenges and should not deter students from incorporating data analysis as a central theme of their projects. With due thought, the analysis of numerical data can provide a pivotal contribution to research.

Acknowledgement

The author is grateful to Chris Brunsdon for supplying Figure 11.1.

BOX 11.5

Further reading

Concepts and Techniques in Modern Geography This series provides introductory reviews of a wide range of geographical techniques, mostly quantitative. There are approximately 60 volumes at the time of going to press.

Fotheringham, A S, C Brunsdon and M Charlton 2000 *Quantitative geography: perspectives on spatial data analysis*, London: Sage (especially Chapter 1)

Fotheringham, A S, C Brunsdon and M Charlton 2002 *Geographically weighted regression: the analysis of spatially varying relationships*, Chichester: Wiley

Robinson, G M 1998 *Methods and techniques in human geography* Chichester: Wiley

Rogerson, P A 2001 *Statistical methods for geographers* London: Sage

Schroeder, L D, D L Sjoquist and P E Stephan 1986 *Understanding regression analysis: an introductory guide,* Quantitative Applications in the Social Sciences 57, London: Sage

12 Analysing categorical data

Andrew A Lovett

Synopsis

Categorical data are encountered in many areas of research in human geography, notably those where questionnaire surveys are an important means of obtaining information. This chapter focuses on how to undertake statistical analyses of such data most effectively. It describes how to explore trends within sets of categorical variables and the use of non-parametric techniques (e.g. the chi-square test) to evaluate hypotheses. The chapter then briefly explains how generalised linear models can be used to investigate more complex relationships between categorical variables and discusses the circumstances in which these methods are likely to be of particular benefit.

Introduction

During the previous chapter there was some discussion of the differences between nominal, ordinal and ratio types of data. To recap, nominal variables are the simplest kind and comprise distinct categories that cannot be ordered (e.g. a classification of questionnaire respondents according to the supermarkets where they buy most of their groceries). Ordinal data consist of ranks. This means that items can be placed in a relative order (e.g. preferences for shopping centres), but the extent to which one is greater than another is unknown. Both nominal and ordinal variables are discrete (i.e. can take only certain values) and are types of *categorical data*. By contrast, ratio data are continuous and have metric properties. It is always possible to convert data from a higher level of measurement to a lower one (i.e. ratio to ordinal to nominal), but not in the reverse direction.

If you are undertaking a questionnaire survey as part of your dissertation, you may well find that much of the data generated is categorical in form. Examples of questions producing categorical data include those where a yes/no response is requested or where individuals are asked their opinion on a statement (e.g. strongly agree; agree; neutral; disagree; strongly disagree). Categorical variables may also be encountered in a wide variety of other sources (e.g. health service records or historical documents). Statistical analysis of such data need not cause you problems, but does require a little care. This is partly because there are many possible techniques, and their appropriateness depends on the manner in which variables are measured and the question being investigated. Some examples of the statistical tests suitable in particular circumstances are given in Table 11.1 of Chapter 11. You should also appreciate that most statistical packages or spreadsheets cannot distinguish category codes or ranks from ratio scale data. This means that such software will readily permit completely inappropriate analyses (e.g. the calculation of Pearson product moment correlation coefficients from nominal data). It is consequently important that results are not blindly accepted as correct just because the computer software produced them. If you are not sure about appropriate techniques after consulting this book (or other statistics texts), then it is best to ask your supervisor before proceeding too far with your work.

Another of your objectives in an analysis of categorical data should be to ensure that relationships between variables are examined in as thorough a manner as possible. It is therefore a good idea to undertake *exploratory* analysis of the data before using *confirmatory* methods to test hypotheses. During the latter phase of a research project it is also prudent to recognise that many phenomena of interest to geographers have multiple (and interrelated) causes. Exclusive reliance on statistical tests that evaluate associations between pairs of variables (e.g. chi-square) can easily lead to a situation where an analysis does not achieve its full potential or, worse still, results in misleading conclusions. This is particularly likely when *interactions* are present (i.e. the association between two categorical variables changes according to the value of a third factor). One example of such a situation would be where the relationship between social class and attitude towards a development proposal varies according to the area in which a respondent lives. In circumstances of this sort, statistical investigations of interrelationships are usually best undertaken through the use of generalised linear models (e.g. log-linear or logit methods). These techniques allow formal testing of the significance of particular associations or interactions between variables, and can also be used to assess how well the values of one categorical variable are predicted by a combination of other factors. It may be that such methods of categorical data analysis were not covered in any statistics course you have taken, but this should not be regarded as an overwhelming obstacle to their use in your dissertation. Recent years have seen implementation of the techniques in several widely available software packages (e.g. SPSS) and there are a number of excellent step-by-step guides to undertaking such analyses. The mark you ultimately obtain for your dissertation can only benefit if you first explain why more advanced statistical methods are necessary and then demonstrate that you can use them competently.

In the light of these considerations, this chapter aims to provide guidance on the statistical analysis of categorical data. The intention is not to discuss individual techniques in great detail (though references to sources of further information will be given) and instead the focus will be on broader issues that arise in exploratory investigations, hypothesis testing, and the construction of statistical models.

Exploratory analysis

Methods of exploratory data analysis place particular emphasis on the descriptive summary and graphical display of variables (for further details see Tukey, 1977; Marsh, 1988; Erickson and Nosanchuk, 1992). They can be used to examine trends in data and are also effective at detecting outliers (i.e. unusual values such as a category of a nominal variable with only a single observation in it). You should always check for any outliers early in an analysis since they may simply reflect coding or typing mistakes. Prompt identification of such errors can prevent much frustration and the need to repeat analyses. Even if outliers do not appear to be mistakes it is worthwhile to know that they exist because some statistical techniques are very sensitive to such values.

A good starting point therefore is to examine the distribution of values for each variable in the data set. With nominal data you can do this by using spreadsheet or other statistical software to compute the number of observations in each category of a variable, and perhaps also express these counts as percentages. This type of descriptive information (e.g. the percentage of questionnaire respondents who had visited a particular recreational facility) is invariably useful when writing up results, though it is often better presented in graphical form (e.g. as bar charts) rather than as text. Good discussions of how to implement these types of summary and graphical techniques using SPSS are provided by Bryman and Cramer (2001) or Miller *et al.* (2002). The book by Townend (2002) covers similar material with instructions referring to the Excel and Minitab software packages.

Associations between nominal variables can be investigated by *cross-tabulating* the data to form *contingency tables*. An example is shown in Table 12.1 using data from the recycling survey discussed in Chapter 6. In this survey, respondents living in three areas of Norwich were asked if they took materials to their local recycling centres. Scrutiny of the table suggests that participation in recycling varied between the three areas, a point that can be clarified by transforming the counts to percentages of the row totals (see Table 12.2).

A rather different view of the data occurs if percentages of the column totals are calculated. Table 12.3 shows this result and, for instance, indicates the percentage of all recyclers who lived in each area. It is therefore important to think

Table 12.1 A contingency table for area of residence and recycling activity

	Respondents undertaking recycling		
Area	*No*	*Yes*	*Total*
Costessey	64	152	216
South Park	16	85	101
Woodcock	42	58	100
Total	122	295	417

Table 12.2 The percentage of respondents undertaking recycling in each area of residence

	Percentage of respondents undertaking recycling		
Area	*No*	*Yes*	*Total*
Costessey	29.6	70.4	100.0
South Park	15.8	84.2	100.0
Woodcock	42.0	58.0	100.0
Total	29.3	70.7	100.0

Table 12.3 The percentages of all non-recyclers and recyclers who live in each area of residence

	Percentage of respondents in each area		
Area	*Non-recyclers*	*Recyclers*	*Total*
Costessey	52.5	51.5	51.8
South Park	13.1	28.8	24.2
Woodcock	34.4	19.7	24.0
Total	100.0	100.0	100.0

Table 12.4 A contingency table for area of residence and housing type

	Housing type		
Area	*Flat/terraced house*	*Other*	*Total*
Costessey	28	188	216
South Park	21	80	101
Woodcock	64	36	100
Total	113	304	417

about the question you wish to answer before deciding the direction in which percentages should run. If one of the variables you are cross-tabulating is thought to influence the other, then Marsh (1988) suggests that it is better to construct the contingency table so that the percentages sum to 100 within the categories of the *explanatory* variable. Assuming that area of residence influences propensity to recycle, then Table 12.2 illustrates this principle.

Tabulations of the type discussed above can be easily obtained using software such as Minitab or SPSS (e.g. see Bryman and Cramer, 2001). It is also straightforward to extend the approach and examine relationships between three or more variables. This is especially valuable given the difficulties with multiple causation and confounding factors that are frequently encountered in human geography. For instance, Table 12.4 indicates that recycling survey respondents in the Woodcock area of Norwich were more likely to live in flats or terraced houses than those from elsewhere. A question that therefore requires further investigation is whether the apparent geographical differences in recycling activity were simply a reflection of contrasts in housing characteristics (which

Table 12.5 A contingency table for area of residence, housing type and recycling activity

	Flat/terraced housing: undertaking recycling		*Other housing: undertaking recycling*	
Area	*No*	*Yes*	*No*	*Yes*
Costessey	9	19	55	133
South Park	5	16	11	69
Woodcock	31	33	11	25
Total	45	68	77	227

Table 12.6 The percentage of respondents using recycling facilities in each survey area and housing type combination

	Housing type		
Area	*Flat/terraced house*	*Other*	*Total*
Costessey	67.9	70.7	70.4
South Park	76.2	86.3	84.2
Woodcock	51.6	69.4	58.0
Total	60.2	74.7	70.7

might be interpreted as a general indicator of social status). This issue can be examined through a three-way cross-tabulation and Table 12.5 suggests that respondents living in flats or terraced accommodation were indeed less likely to recycle, but that the difference between housing types varied across the three survey areas. Another perspective on the data is provided by Table 12.6 where residence area and housing type define the classificatory framework, and the cell values are a descriptive statistic for a third variable (recycling activity). In this instance, the values are the percentages of respondents in each area/housing type combination who used the recycling facilities, but if the third variable was a numerical one it would be quite possible to calculate means or medians. Such a procedure is particularly useful when investigating associations between a mixture of categorical and numerical variables.

Appropriate exploratory methods for ordinal variables differ according to whether the data are *weakly* or *strongly* ordered. Weakly ordered variables (i.e. those consisting of ranked categories) invariably contain high proportions of tied ranks. In such cases it is usually best to treat the data as if they were nominal and apply the techniques already discussed above.

If the data are strongly ordered (e.g. contain few ties) then scatter plots are a useful means of exploring associations between variables. Such graphical displays may also be the most effective way of detecting any outlier observations (e.g. typing mistakes which stand out as a marked departure from the general trend). An example is shown in Figure 12.1 where two sets of ranks show a strong positive correspondence, but one observation is a distinct departure from the general trend. The use of linked window displays (as discussed in Chapter 11) provides a means of extending scatter plot analysis to consider relationships between three or more strongly ordinal variables.

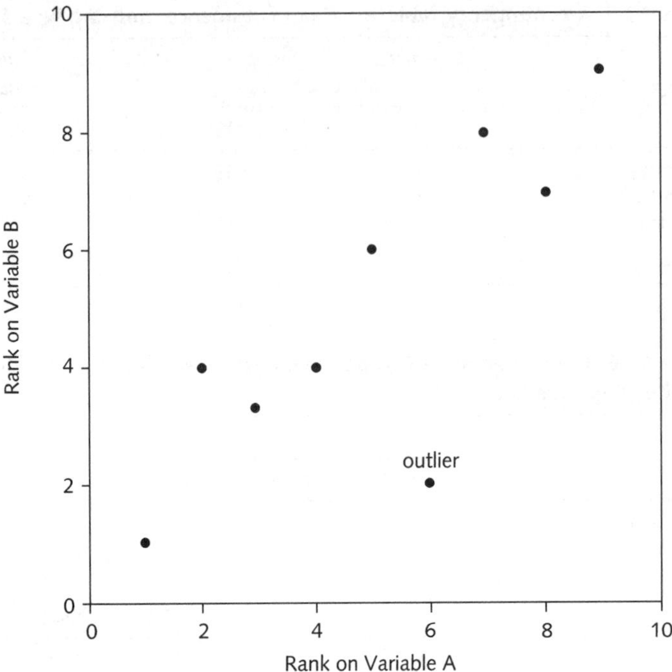

Figure 12.1 Example of a scatter plot showing an outlier

Basic confirmatory techniques

Once a set of data has been screened for outliers, and possible relationships between variables have been explored, you will often need to undertake some form of confirmatory analysis. Most commonly this involves the testing of hypotheses to determine whether trends apparent in a sample of data (e.g. a set of questionnaire responses) are also likely to exist in a wider population. A *statistically significant* result in this context is one where the trend in the sample data is sufficiently strong that there is only a small probability (conventionally under 5%) that it could have occurred by chance. When planning to undertake such tests, however, you should keep in mind that the size of the sample has a considerable influence on the outcome. In effect, the larger the size of sample, the weaker any trend need be to qualify as statistically significant. It is, therefore, vital to give some thought to the sample size required before embarking on data collection. If you do not do this, there is a real danger that your subsequent analysis will be handicapped by a limited ability to detect statistically significant trends or relationships.

It is impractical to provide universal guidance on the sample size needed for a credible confirmatory analysis as much depends on the actual statistical tests employed. In the case of nominal or ordinal variables the appropriate techniques are *non-parametric* methods and good reviews of these are provided by Siegel and Castellan (1988), Gibbons (1993) and Conover (1999). Non-parametric tests have lower statistical efficiency (i.e. the power to detect an

Table 12.7 An example of a chi-square test

Previous visit experience	*Willing to pay for flood defences*		
	No	*Yes*	*Total*
Day visit	36	46	82
	38.1	43.9	
Holiday	44	74	118
	54.8	63.2	
Never	64	46	110
	51.1	58.9	
Total	144	166	310

Observed cell frequencies are printed above expected values.
Calculated chi-square = 10.28.
Critical chi-square for 2 degrees of freedom at the 95% confidence level = 5.99.

effect) than their parametric equivalents, but they also make fewer assumptions about the parameters of populations from which samples are drawn. The latter feature makes non-parametric tests more robust and consequently there can be situations (e.g. where frequency distributions are heavily skewed) when they are the best option for analysing numerical data. There may also be instances where there are good theoretical reasons for treating apparently numerical variables as ordinal. Gilbert (1993) discusses the situation where one variable (years of schooling) is used as an *indicator* of a more fundamental but difficult to measure concept (education), and argues that as the relationship between these two cannot be precisely specified the former should be regarded as ordinal for the purposes of analysis.

Examples of the non-parametric methods that can be applied in different hypothesis testing contexts were given in Table 11.1. For nominal data the most commonly used technique is the chi-square test. Table 12.7 illustrates an application of this technique using data from a postal questionnaire survey conducted by Bateman *et al.* (1992). This research project sought to examine attitudes towards the protection of the Norfolk Broads and it was hypothesised that respondents would be more willing to contribute to the cost of flood defences if they had previously visited the area.

The chi-square test in Table 12.7 indicates a statistically significant result but, as is often the case with this technique, some care is needed in interpreting such an outcome. All it really confirms is that there is a difference in willingness to pay between the respondents classed according to their previous experience of the Broads. This is evidence of interdependence between the two nominal variables, but the chi-square statistic does not specify the nature of the association. To assess any trend it is necessary to examine the observed and expected cell frequencies in the contingency table. Scrutiny of Table 12.7 reveals that the observed counts for respondents willing to pay were greater than those expected if they had previously visited the Broads and less if they had not. This is in accord with the original hypothesis, but it is important to keep in mind that a similar chi-square statistic would have been generated if the relationship had been reversed.

There are several other cautionary points that should be considered when conducting a chi-square test. One is that the frequencies in the cross-tabulation must be counts, not percentages. A second issue is that the test will give less reliable results if several cells have low expected frequencies. It does not matter what the observed cell counts are but, as guidance, if there are expected frequencies of less than five it may be necessary to modify the contingency table (see Bryman and Cramer, 2001; Miller *et al.*, 2002; Townend, 2002). The most common solution in this situation is to combine categories, but sometimes amalgamation may be nonsensical or result in a considerable weakening of the hypothesis being investigated. You should also be aware that the outcome of a chi-square test might be different once the definition of categories has been changed (see Walford, 1995). The ideal, therefore, is to prevent such dilemmas arising by stratifying sample collection (see Chapter 6) to ensure sufficient numbers of respondents in any categories of particular importance.

When ordinal data are the focus of attention there are a wide variety of possible confirmatory tests. Selecting the best one requires consideration of whether the data are weakly or strongly ordinal (see Table 11.1), the nature of the samples involved (e.g. related or independent), and the relative power efficiency of tests. Good discussions of these issues, together with examples of how to undertake and interpret individual techniques, are provided in the references on non-parametric statistics given earlier, as well as books such as Bryman and Cramer (2001) or Townend (2002).

An illustration of a confirmatory test for a weakly ordinal variable is presented in Table 12.8. These data come from the survey of public knowledge and attitudes regarding HIV/AIDS discussed in Chapter 6. One of the questions in the survey was:

> *Some people have suggested that everybody in the country should be tested for HIV. Do you think this would be a good idea?*

and Table 12.8 shows some variation in the social class profiles of respondents who answered Yes or No. As social class is a weakly ordered variable, a Kolmogorov–Smirnov two-sample test is an appropriate means of assessing whether the difference between the two samples is sufficiently large to be statistically significant. The result, however, indicates that this is not the case and so the apparent contrast should be interpreted as occurring by chance.

Table 12.8 An example of a Kolmogorov–Smirnov two-sample test

	Approve of testing		*Disapprove of testing*	
Social class	*Number*	*Cumulative percentage*	*Number*	*Cumulative percentage*
I and II	20	11.9	33	17.9
III	53	43.5	65	53.3
IV and V	95	100.0	86	100.0
Total	168		184	

Maximum difference between cumulative distributions = 9.81%.
Critical value at 95% confidence level (two tailed) = 14.51%.
Cannot reject null hypothesis.

Table 12.9 Hypothetical summary of results from chi-square tests

Variables cross-tabulated	*Calculated chi-square value*	*Significant at 95% level*	*Hypothesised observed and expected differences*
Area and housing type	10.42	Yes	Yes
Area and car ownership	6.50	Yes	No
Gender and car ownership	3.20	No	Yes
etc. . . .	. . .	. . .	. . .
etc. . . .	. . .	. . .	. . .

Irrespective of the type of data and tests involved, you should pay some attention to the presentation of results from confirmatory analyses. Including the output from every test conducted (e.g. large numbers of contingency tables) is usually unnecessary and conveys the impression of a lack of thought. If, for instance, many chi-square tests have been conducted a suitable approach might be to discuss one or two in detail (to demonstrate appropriate competence) and then summarise the remainder in a list of the type shown in Table 12.9. You may also find scope for graphical displays – Pringle (1980) and Marsh (1988) provide examples using causal path diagrams. As a general rule, if you can represent information in a variety of formats it will help to make a dissertation more interesting to read.

Modelling more complex relationships

A limitation of the confirmatory techniques discussed above is that they are only suitable for investigating relationships between pairs of variables. This is a particular problem when the values of one categorical variable (e.g. participation in recycling or the extent of agreement with a statement) are thought to be dependent on those of several different possible explanatory variables. To analyse such situations effectively it is necessary to use types of *generalised linear models*. These have the advantage that they can be used to test explicitly for the presence of particular combinations of associations and interactions between variables, the ultimate objective being to identify the simplest model that fits the data. Gilbert (1993) describes this process in terms of comparing 'real' and 'imaginary' worlds, where the former is represented by the observed contingency table and the latter is derived by calculating the pattern of cell counts that would be expected if certain hypothesised relationships were present in the data. Clearly, many different 'imaginary worlds' might be generated (e.g. for a tabulation of three categorical variables there are nineteen possible combinations of associations and interactions), so *stepwise* procedures are commonly used to evaluate the goodness of fit for different models. The existence of any *design effects* must also be considered when analysing data that have been collected through schemes more complex than a simple random sample. For example, if respondents to a questionnaire survey were stratified by social class and area of residence, then the association between these variables would be fixed by design and allowance for this would need to be made in model construction and interpretation (see Wrigley, 1985; Skinner *et al.*, 1989; Bryman and Cramer, 2001).

Different forms of generalised linear model are suitable in particular circumstances. If the data are nominal or weakly ordinal, but cannot be divided into response and explanatory variables, then a log-linear model should be used. When a dependence relationship is hypothesised, a logit model is appropriate. This type of model would be suitable for undertaking further analysis of the data in Table 12.5 if the aim was to assess the combined influence of residence area and housing type on participation in recycling.

If the response variable is categorical, but a mixture of numerical and categorical data are being employed as predictors, a logistic regression model should be applied. This latter technique is most straightforward if the dependent variable is binary (e.g. Yes or No) and could, for instance, be used to extend the recycling participation analysis by including additional numeric explanatory variables such as household income or the distance to the nearest recycling facility. It is, however, also possible to undertake logistic regression analyses with dependent variables that contain more than two categories or are ordinal in form (i.e. consist of ranks). Examples of the former include studies investigating influences on modes of travel or shopping trip destinations (see Wrigley, 1985), while the latter often involve investigations into aspects of public preferences (e.g. Appleton and Lovett, 2003 used ordinal regression methods in research on factors affecting the rating of landscape views).

It is impractical to provide detailed discussions of generalised linear modelling techniques here and, indeed, to do so would simply duplicate information which is well presented elsewhere. Worked examples of analyses using software such as SPSS or GLIM are provided by Bowlby and Silk (1982), Wrigley (1985), Aitkin *et al.* (1989), Gilbert (1993), Miller *et al.* (2002) and Menard (2002). The books by Gilbert (1993) and Menard (2002) are particularly recommended for their clarity of exposition. Without doubt, it is now relatively straightforward to implement these methods and they certainly merit wider use in geography dissertations.

Summary

Categorical data are collected as part of many human geography research projects. Statistical analysis of such information can help you to identify trends and test hypotheses. You need to be careful, however, with the choice of appropriate techniques. Much can be achieved by analysing relationships between pairs of variables, but in some situations there are advantages in using more sophisticated generalised linear models. It is also important that results are presented in a succinct and interesting manner. Consideration of these issues will hopefully ensure that you gain maximum benefits from your analysis of categorical data.

Acknowledgements

I would like to thank Ian Bateman (UEA) and Julian Parfitt (Waste and Resources Action Programme) for supplying some of the data used as examples.

BOX 12.1

Further reading

A good introduction to exploratory techniques with particularly helpful discussion of contingency table analysis and presentation is:

Marsh, C 1988 *Exploring data*, Cambridge: Polity Press

Clear explanations of techniques for categorical data analysis (with many worked examples and instructions regarding the use of SPSS or Minitab) are provided by:

Bryman, A and D Cramer 2001 *Quantitative data analysis with SPSS Release 10 for Windows: a guide for social scientists*, Hove: Routledge

Miller, R L, C Acton, D A Fullerton and J Maltby 2002 *SPSS for social scientists*, Basingstoke: Palgrave Macmillian

Townend, J 2002 *Practical statistics for environmental and biological scientists*, Chichester: John Wiley

An extensive review of more advanced techniques is:

Wrigley, N 1985 *Categorical data analysis*, London: Longman

Good introduction to implementing generalised linear models with particular software packages can be found in:

Gilbert, N 1993 *Analyzing tabular data: loglinear and logistic models for social researchers*, London: UCL Press

Menard, S 2002 *Applied logistic regression analysis*, 2nd edn, Quantitative Applications in the Social Sciences 106, London: Sage

13 Analysing qualitative materials

Mike Crang

Synopsis

This chapter takes the student through the often initially alarming experience of being confronted with large quantities of material generated by qualitative research. It attempts to provide a range of ideas that could be used in making sense of such materials. Obviously, given the diversity of methods and objectives of research it should not be treated as a 'cookbook' of 'how to' interpret materials. Instead it suggests some preliminary steps that may help students find their feet in terms of managing the materials, and then goes on to illustrate a range of different approaches – broadly drawn from grounded theory, structuralist semiotics and narrative theory. Each of these is shown to allow a different handle on the material for different projects, but also to entail certain theoretical assumptions that need to be addressed. Some approaches emphasise the abstract, some stress the concrete, some the order and pattern of events, some their evolution and change. Hopefully, these approaches will act not as a shopping list of tools, but will encourage students to think more clearly about the materials they have collected.

Approaching materials

This chapter sets out to help you think through a problem that often strikes students rather late in their projects. If you are reading this before commencing your work, you may feel suitably self-righteous; if you are reading this after you have completed some fieldwork at least I can be sure of your avid attention. The problem is simply put – many books suggest the benefits and strengths of qualitative research and how to go about doing it, but rather fewer geography course books say what to do with the materials that are generated by it. If that sounds familiar to you, rest assured you are not alone. Discussions of analysis and interpretation have been rather like the dog's barking for Sherlock Holmes – remarkable by their absence. This chapter hopes to allay the anxiety this can cause, by providing a range of ideas and options as to how analysis can be conducted on a range of materials. Which is appropriate depends on what sort of topic and theoretical approach your project involves.

So let me begin with some caveats. Analysing qualitative material is not an ineffable and mysterious process but neither is it a case of painting by numbers. This chapter does not supply some formula that can be used to 'solve' projects, or produce answers. Furthermore this chapter deals mainly with verbal or textual material. Visual materials pose a different set of issues and often involve different approaches (outlined in Chapters 14 and 15). However, I do not want it to sound like these are mutually exclusive; projects can combine elements, say, from pictures and people's interpretations of them. Later sections will thus hopefully provide ways in which visual and verbal materials can be linked together and approaches to make their interpretation congruent to each other. Although focusing on the nitty-gritty of analysis, this chapter will draw in the theoretical implications of various strategies. This is important, since the separation of 'field' and 'theory' is more an artefact of books than a true portrait of research. No researcher refuses to think about the interpretation and significance of their research while they are doing it. Equally, detailed analysis can throw out new topics and questions that might need to be pursued with further 'fieldwork' if time and resources permit. Analysis should not be an afterthought, but needs to be included in early research plans, if for no other reason, because getting to grips with the materials properly can take as long as creating them in the first place.

Finally, then, come issues of what sort of materials the project has generated and what sort of approach the project takes. For this chapter, the term 'field materials' will refer to transcripts of 'interviews' or open-ended questionnaire responses, observations and notes made by the researcher, sketches, maps, itineraries, diaries, letters or memoirs forming a corpus of material generated by a project. Likewise, the project may be seeking very different answers from these materials – you might be researching the life-world of certain people, or changing patterns of employment, or perhaps the way work (paid or unpaid) is organised, or perhaps trying to sort out the meanings and significances attributed to particular places. The list could be endless. What this means is that different researchers will perhaps be asking very different questions of the same types of sources. This chapter is organised to start from processes that are common to most research projects – how to begin to manage all these materials. It then suggests some very common ways to develop preliminary ideas. From here I shall outline some more specific approaches, looking, first, to the most common 'grounded theory' approach, secondly to some alternative ideas of interpreting materials through the use of 'formal structures', and thirdly to the ways people make sense of things through narratives. These three are chosen partly to allow some congruence with approaches to visual media, partly for their popularity in geography and partly to illustrate different theoretical stances. Tesch (1990: 58) identified up to forty-three 'approaches' to qualitative analysis, so the three here cannot hope to be comprehensive. Instead the styles outlined are meant to show the differences between approaches and encourage you to develop a style that fits your own project. The chapter finishes with a section suggesting how interpretation cannot be divorced from the theoretical approaches adopted throughout a project – it is not simply a bolt-on addition.

Categories, contents and codes

The stage arrives in any qualitative project where the researcher has accumulated a voluminous collection of notes or tapes. To start making sense of diverse types of materials from diverse people on diverse occasions can seem a colossal task. The first step is to get them into a presentable, readable form. Thus tapes need to be transcribed and notes preferably typed up, so that they are at the least legible – an important point for those with handwriting like mine. Ideally this should be done as you go along, while your memory is fresh and you can add in or clarify any marginal notes needed – such as what acronyms might mean, places or people referred to, or so on. Typing up materials may be time consuming, but if done by the researcher, can allow a re-familiarisation which may pay off in the long run. The amount of detail in transcription really depends on what sort of study you are doing – if you are seeking information on who did what and when, for example, you can afford to miss out irrelevant portions, and confine yourself to typing up the substance of what is said. If you are looking at how and why people did things or made sense of them, you may need more detailed transcripts – including perhaps comments on tone of voice, hesitations, and the exact ways people put things (for useful ideas see Silverman, 1993: 118–120). In any case it is important to record your own actions at the time, to see how the research process influenced the reply. This is not to say you 'biased' the material but to become aware of how it was produced so you can use that in your interpretation (see Chapter 7).

Styles of transcript vary considerably between researchers and approaches and there are extremely detailed schemes that can be used – with notations for pitch, pause length and pronunciation (e.g. Fine, 1984). Similarly, if you are going to use computer software (see Box 13.1), you will find that each product has slightly different conventions for changes of speakers or topic, so it is worth looking these up in advance to avoid some tedious retyping. (It is possible to work from actual sound or video clips to record body language and intonation, in which case you need to think about the facilities available for recording and playing these back as digital files and then software for analysing these sorts of files – see Box 13.1 for some starting points.) As a rule of thumb, type the material up with plenty of room in the margins for you to make notes on. Most people type up dialogue so it appears rather in the form of a play, perhaps with the speakers' initials and then the rest of the material indented. Something which may save a lot of frantic work later on is clear identification of all the passages so you can find them again. Imagine having masses of papers swimming around your desk/room/kitchen – the potential for mixing things up can be huge. So it is a good idea to make sure that every page says where it came from (e.g., interview A, Joe's diary or whatever) in a running heading, so you can refer to your records and find out who 'Joe' was and put the material back in the proper place. Most times you will want to make sure the pages are numbered so

BOX 13.1

Software for qualitative analysis: a help and a hindrance

Before using software packages designed to aid qualitative data analysis there are a few preliminary considerations. Firstly, is it worth the time learning how to use one? If you have only a few materials and limited time, or plan only a broad-based use of the material then the answer is probably not. Secondly, they can cost up to £300 so they may be simply beyond your budget. However, your department may have such programs or one of them, which you may be able to use if you ask. Thirdly, individual programs have strengths and weaknesses (see Crang *et al.*, 1997; Fielding and Lee, 1991, 1998; Weitzman and Miles, 1995).

By and large what most of the programs do is cut/paste, allow you to code or annotate the text and retrieve information. That is, they help speed up sorting through your materials, so instead of piles of paper or coloured pens the computer keeps codes assigned to different sections of text. Having assigned these codes they then allow very quick retrieval of all sections labelled with a certain code. Moreover, they allow searches in a variety of permutations – for instance women talking about issue *x* or alternately occasions where both *x* and *y* were talked about. Different programs allow different types of searches to be made and present the results in different fashions. Some allow you to use prestructured questionnaires, or socio-demographic variables as well as less structured notes, and then these may be combined and exported in formats suitable for statistical analysis with SPSS (e.g. Textbase Alpha, Atlas-ti, HyperQual, Nud*ist, NVivo) or into spreadsheets (NVivo). Some allow you to create graphics or diagrams out of your codes (e.g. Atlas-ti, NVivo, Nud*ist, HyperSoft, HyperQual). Here are a few options and provisos if you are considering using these packages.

Firstly, any package takes time to learn and unless you have already done it as part of your course, this may be a barrier. It can be somewhat scary learning a new style of analysis and how to use a computer program at the same time. You might want to ask whether you need a specialist program. Some researchers have argued that using something as simple as Microsoft Word's table function, with text in one column and other columns for numbered codes can allow for chunks of text to be categorised pretty efficiently (La Pelle, 2004). Secondly, they all have different specific strengths – for instance some are very good for content analysis while at the other end of the spectrum, the program 'Ethnonotes' (Lieber *et al.*, 2003) specialises in cataloguing the multiple types of data and entries that comprise field notes and making them accessible among a team of researchers in multiple locations – and collates them into output that can be exported into some other types of analysis software. More subtle differences might be how the program relates to your material and method of analysis – for instance, Nud*ist uses a hierarchical coding system and offers a powerful range of search tools (see Peace, 2000), Atlas-ti allows you to annotate pictures, hypertextually link text segments into a sequence and visually manipulate code diagrams, as does NVivo through its modelling function that also has a merge facility for collaborative work and it easily assigns attributes to whole transcripts (such as the characteristics of the respondent). Most programs assume you are working with textual sources with perhaps some other media. However, if you have digital files (say taking interviews as .wav or mp3 or video as .mpeg) the imaginatively named Transcriber

Box 13.1 continued

software for instance allows the direct annotation of sound files as they play (useful if, say, intonation or pronunciation is crucial to you) or ANVIL allows the annotation of video footage (Loehr and Harper, 2003). You should really check that the program would help do what you want – so you control it and not vice versa (Hinchliffe *et al.*, 1997). Thirdly, they have differing hardware needs. You must consider what will be available to you – for instance some programs run on either MacOS or Microsoft, some have versions for either, and some allow you to shift between them. Fourthly, a common misapprehension is that in some way the programs do the analysis for you, or produce some sort of concrete results on their own. The programs do not produce magically correct results, or for that matter 'results' at all. The results are the connections, patterns and explanations *you* draw out of your materials – not something produced by the computer. However, appropriate software can speed up some of the routine work considerably, and you should speak to someone about your specific needs if you have the opportunity to use it. It may well be that using the appropriate software will help you manage the materials more effectively (I know that I for one flounder under mountains of annotated papers) and may give you some help in making sure your interpretations are thorough and clear.

you can find who said what in what order. If you propose to do a more fine-grained analysis looking at smaller sections it may also be a good idea to number the lines so they can be identified within a page quickly and easily. All these functions can be done on word processors with a minimum of effort and may save a lot of time later. If you are going to be using a software package for processing the material you should also check the format it uses – to check that it suits you and to save you having to reformat everything later. Finally, and especially if you will be physically 'cutting and pasting' as described later, it is a good idea to make a copy of the whole lot and put it in a safe place – disasters can happen with computer failures, or cars can be stolen, and in any case it will reduce the paranoia often felt when these materials are all you have to show for a lot of hard work.

Most researchers start by going through these materials a line or sentence at a time, trying to think what each one meant or what was being done and why, disciplining themselves to work slowly through the text. As ideas emerge about the topics in the material, they are noted down alongside the text. These do not have to be finalised or thought through, but are simply recorded as they strike you. This is known as 'open coding' and is based on getting as close to the material as possible, both to get a feel for it all and so nothing gets missed. So it is important to keep going slowly and thoroughly since there are often things that escaped your notice when you were actually doing the fieldwork. Gradually as these jottings accumulate, it is a good idea to write down what they mean or what ideas they sparked off in a separate set of memos – both so you can decipher them later and so good ideas can build up. These memos to yourself are often termed 'theoretical memos' and are worth keeping, and again need to be labelled so they can be linked back to the bits of materials they refer to. After finishing going through your materials once, reading these memos may help you clarify what were salient or recurring themes that might be worth pursuing.

Sifting and sorting: developing ideas

For now let us assume this is going to be done on pen and paper because some of the principles are most easily laid out that way. There are no hard and fast rules over the precise mechanics people adopt; it tends to be what they are most comfortable with or what fits their materials best. What tends to be done is that the researcher goes back over all those notes and jottings and starts to formalise them into categories or codes. So, say, each time people refer to a particular event, it is given the same code; each time they use a particular explanation of an event it might be given another. The codes are thus the abbreviations or acronyms put on similar segments. The segments coded can be of any length, there may be overlaps, or two or more codes may apply to the same bit of text. One way of dealing with this is to cut up the copy of the materials and put the coded sections into piles, so each code forms a pile. Michael Agar (1986) has called this the 'long couch or short hall' approach, where stacks and piles of material accumulate on significant themes, events or topics. What generally happens is that some codes 'break down', that is, although they apparently referred to the same thing, later on the researcher realises they actually have some differences or that they have been misunderstood. Then the researcher has to go back to that 'pile' to check and if necessary recode all those cases. Equally, there may be some very large piles that all generally refer to the same issue but contain shades of difference. For instance, in working on local senses of history in oral history groups, I developed a simple code referring to 'sense of loss' where people referred to places that no longer existed. This however, described a huge range of incidents, so going back through all these I divided them further by whether they were talking about total disappearance, or whether there were relics left, or whether they were even now being destroyed, and whether people were fatalistic or angry. In this way a large category is subdivided in what is sometimes called 'axial coding' where aspects and properties of each main code are teased out (Strauss, 1987: 32). Of course, these refined schemes can also break down and need modifying themselves. It is for this reason that Michael Agar (1986) has called the process 'maddeningly recursive', since it often develops through its 'breakdowns'. As a process, analysis tends to involve going through material, then going back to check or change in the light of later ideas. One of the important reasons for keeping theoretical memos is so you can trace back through these shifts, as well as recording what made you change your ideas. This in the end becomes one way of justifying your conclusions – by being able to show how they were arrived at (Baxter and Eyles, 1997).

You may be able to do this coding with coloured pens denoting the main axes or codes and subcodes in the respective colours, or you may wish to cut and pile sections of text. Other people work by setting up a card index system, with a card or set of cards per code, and different groups of cards for main codes, recording each occurrence on them. I would suggest using pens or cutting the material up, since this means you keep the text in front of you. In the next three sections I want to make some suggestions as to different ways of developing from this to a fuller interpretation, with varying theoretical presuppositions, about what theoretical 'status' codes might have, and how an analysis can be said to be reliable.

Building ideas, developing theory

It is important to note the different epistemology here from many quantitative projects. What is generally of interest is not so much the codes as the text they denote, not how often they occur but what is in them. The codes are not there to be rigidly reproduced, nor to be counted, but as an aid to the researcher in making sense of the material. They are not an end in themselves. Codes provide a means of conceptually organising your materials but are not an explanatory framework in themselves. Let me illustrate this by contrasting this with content analysis. In content analysis two or more workers adjudge each incident according to a set of predefined criteria, worked out so that any given incident will be assigned to the same code – an idea of 'inter-rater reliability'. Once this is achieved the incidents can be counted and used for statistical analysis – with all the rules of sampling and reliability implied – so that the analysis is done on the frequency and distribution of the codes, not on the material itself. This is definitely not what qualitative analysis is about. As you will have seen above, the codes here have emerged from the materials in an iterative process – going from the material to ideas, back to the material and so on, through what is called 'analytic induction' rather than 'enumerative' testing. These codes are not pre-given, and indeed they are probably not what another researcher would produce – they are in that sense creative (Bailey, White and Pain, 1999), since they rely on you making sense of material using the knowledge you have developed through research and reading literature. What the categorisation helps to do is to organise the materials so that interesting relationships can be seen. As a corollary the 'status' of codes needs a little clarifying. What I am suggesting is that there needs to be a little caution about what your codes do and do not mean. You might want to consider this in terms of the division of 'emic' (sometimes called *in vivo*) and 'etic' codes that I shall outline below.

The idea is to clarify the relationship between your codes and the materials with which you started. In this sense 'emic' codes are generally taken to be those used by informants themselves. Meanwhile 'etic' codes are those assigned by the analyst to describe events and attribute meanings and theories. A quick example may help illuminate this distinction. I did some participant observation with groups staging historical re-enactments. In the analysis, I was quite interested in ideas of realism in portrayal, so I picked up on how participants described some costume and props as 'authenty kit', meaning items of equipment that were replicas from actual historical sources. This I then used as an 'emic' category, compiling all the occasions when participants talked about this. Meanwhile I was also looking at issues of what effect was aimed for – which I was compiling under categories such as 'verisimilitude' (where it looked right), 'magic' (where people spoke about sudden insights), 'empathetic' (where people talked about identifying with the people they were portraying), and 'researched' (for finding historical sources). These latter we might say were then 'etic' codes, which could all overlap (or not) with incidents where the people themselves spoke of 'authenty kit'.

We can perhaps use this little sketch to exemplify a coding process. Obviously starting with hearing the particular jargon used by participants, I then set down to see what that term meant, and what it revealed about the views of the participants and the process of re-enactments. So we have a drift from a local term, an 'emic' category, towards more theoretically oriented, 'etic' ones responding to issues in the research. Moreover, the process involves the sort of axial coding mentioned earlier – where different sorts of realism were picked out as a key issue for the study. Then this one category is broken down into smaller parts so the 'etic' codes above provide a finer resolution in categorising events, and draw out all the 'dimensions' to a given statement, allowing the meaning of each to be compared against apparently similar ones to pick out the more subtle differences. However, we also need to question this division of code types. It is tempting to take 'emic' and 'etic' as a shorthand for 'insider's view' and 'outsider's view' but this is a little too glib since it is almost impossible to find terms which are not in some way a combination of the two (Agar, 1980). Thus for example, it was my theoretical interest in realism that led me to pick out 'authenty kit'. Likewise, I explained some of my ideas on 'verisimilitude' to some participants who broadly agreed – and added further details.

What the above sketch reveals is the need to think clearly about your epistemology (i.e. how you can claim to know something). Clearly I could not say that participants discussed issues in terms of the 'etic' categories. Instead what I could do is relate an incident – from the materials – and then unpack why I thought parts of it referred to 'research' based and 'empathy' based knowledge, or I could relate two incidents illustrating the differences of 'verisimilitude' and 'research' based realism. But the categories themselves worked as tools to help me do this, not as explanations. Merely coding is not the same as interpreting. Clearly these categories must have a certain level of robustness, that is they are not mere whims, but equally they may not be absolutes. One of the themes outlined so far is how to develop this robustness through the evolution of ideas and codes, so that in the iterative process of developing then redefining, categorising and re-categorising, the final categories are coherent and supportable. They should be developed along coherent 'axes' of interpretation, and develop 'dimensions' by constantly comparing the codes and their contents, new and old material, so all the themes and implications of the materials are drawn out. An excellent example of how codes are brought together and how they are interpreted is given by Jackson (2001: 206–209) on reader responses to 'lad mags'. He recounts the first coding of interviews then reading across transcripts to compile lists of themes, different types of readers associated with the magazines – and also 'silences' or topics that were not mentioned. From these he compiled 'discursive repertoires' or ways in which these magazines describe the world and 'discursive dispositions' of how the readers relate to them.

Above all, in thinking through the relationship of category and material, theory and evidence, researchers have to consider what sort of account they want to give. Thus, traditional qualitative research – ethnography in particular – has judged its accounts more reliable the closer they are to representing the world-view of the informants, 'to document the world from the point of view of the people studied' (Hammersley, 1992: 165; Katz, 2001, 2002). However, it can be critiqued as a tautological process – since it does not bring new understandings to bear on issues (Kirk and Miller, 1986). Likewise it also risks missing points where people disavow or conceal their interests. To take an example, Pierre Bourdieu (1984) studied how people described their taste in goods and art and found that, although everyone explained these tastes in terms of aesthetics and the attractiveness of the objects concerned, there were clear but unacknowledged links between people's class positions and what they found attractive. So it may perhaps be useful to turn to styles that pay more attention to the perhaps unsaid structures.

Formal relationships and structures

This style of analysis takes a rather different approach from that in the last section. It examines the categories for patterns and structures of meaning that may not be apparent to the informants themselves. As such it implies a search for meaning behind what is said. Before going further there are some implications of this that ought to be highlighted. First, there is a worrying tendency to position the researcher as knowing best and to play down the reflexive, thinking character of the people studied; raising not only epistemological but ethical questions that need to be taken seriously indeed – with arguments that structures do not operate 'behind people's backs'.

> *With a memorable phrase, Harold Garfinkel ridiculed social theories that presume that people, like cultural dopes, absorb perspectives through socialization and then apply them mechanically. The ethnomethodological call was to study not just how pre-existing values or interests are tailored to fit real life settings for action, but, more fundamentally, how the sense of situations is constructed in the first place. Ethnography has made some of its most professionally accomplished contributions to explanation by noting the situated character of social life. (Katz, 2002: 69)*

So if you have used, say, participant observation or interviewing (as outlined in Chapters 7 and 10) it is important to realise that it might be wrong to use ideas of 'formal structures' so as to downplay the way people make sense of their own lives. Secondly, there is a question about how such structures relate to real-life situations; they may account for regularities but they do not explain what actors think. Thirdly, concentrating on 'structures' may give the impression they were 'found' or that they owe nothing to the research process, thus losing sight of the positionality and politics of, say, interviews through which your data were created. Fourthly, this may lead to an implication that these structures act beneath the surface or are independent of interactions, and this may in turn lead to ideas of 'underlying' structures which downplay the fluidity and dynamism of situations. This is not the place to examine these issues but they do need to be thought through. There are various theoretical perspectives that might be used to develop ideas on how to handle this, for instance the work on structuration theory (Giddens, 1984; Thrift, 1983) or the work by Bourdieu (1990) which looks to structures embedded in actions and reproduced through them.

The type of analysis in this section will be developed from semiotics, which can be described as the study of signs and the construction of meaning. It is concerned with the way words, things, pictures and actions come to be 'signs', that is to convey meanings in particular times and at particular places. Although starting from this point I hope to suggest ideas that might help in other approaches and with media other than the solely textual. The emphasis in this section will be on ways of looking for patterns among the categories and codes you may have derived. I propose to lay out four possible devices or tactics, but these should not be taken as prescriptive – researchers continually develop combinations, refinements and different tactics for each project. I hope that these may serve to prompt your own thoughts rather than provide a model to be followed.

The first principle of semiotics is that meanings are relational rather than fixed. That is, signs derive their meaning from other signs and from the wider system of signs not from their actual form or content. An example might simply be a traffic light: there is no inherent meaning in red, amber or green; they come to signify 'stop, wait and go' through conventions in the society around them. For instance a celebrated case is a traffic light in Boston, Massachusetts, where the order of the lights had to be changed – because Irish residents refused to have orange (also a symbol of unionism) above green (symbolising republicanism). So it might be worth thinking through materials in terms of the signifier, in this case a coloured light, and the signified, the meaning it conveys. It is widely recognised that the latter can thus vary across cultures, groups and nations – nodding your head can be an affirmation in the UK while being a negation in India. It might be worth for instance exploring the gaps of denotive meanings and the connotations of particular signs – respectively, what they explicitly say and what implications others may read into them. More subtly, as a second tactic, the semiotician Charles Peirce divided the communicative ability of signs into three types – iconic, indexical and symbolic (cf. Parmentier, 1994). The iconic sign actually depicts what it means – thus a picture of a woman holding a child is an iconic sign conveying some meaning about that woman and that child. An indexical sign is part of a wider complex, as part of a whole, and standing for that whole – thus the same photo might also signify 'motherhood', with one particular instant standing for the whole. The symbolic level is where the relationship between the sign and meaning is purely conventional – thus the same photo could convey echoes of the Madonna and child, with all the values that places on certain models of femininity (cf. Hirsch, 1981). So an interpretation might look at how the same sign is used in different ways and at different occasions or places (Koskela, 1995) or, to put it another way, it

In Vivo terms	Connotation	Researcher Role	Gendering
The Unknown	Opposite to academia	Explorer	Feminised other
The Field			
Real world			
Substantive	Opposite to theory		'Masculine' role
Empirical		Bringing back data / Detached observer	
Observation	Activity of researcher		
Charting			
Original research			

Figure 13.1 Semiotic clustering

might look not for the 'right' meaning in a picture but for how it comes to mean specific things to specific people in specific places (Rose, 2001). The emphasis then is on how the context changes or stabilises the meaning of events, but it is possible to instead concentrate on the relationship between signs.

A third tactic might be to look for oppositions between categories, so that rather than looking for inherent meanings we see meaning created by the relationships between things. That is to say, things are often as much defined by what they are not as by what they are. Thus it might be possible to suggest some mutually exclusive categories instead of overlapping ones – cases where certain activities are grouped by opposed ideas, say, of masculine or feminine behaviour which might reveal the gender roles of the group. Equally there may be slippages between divisions in different times and places. In many cases groups of signs or events will emerge, whether or not they respond to some major binary division. One way of thinking through this has been termed 'semiotic clustering' (Feldman, 1995; Manning, 1987) which is a simple way of building towards theory. It is a variant on a tree diagram, where you might map all the meanings (or for that matter the uses/events or whatever you are studying) attributed to one thing. So in a column on the left side of a piece of paper you write all the ways a term or thing has been used. These can then perhaps be grouped in terms of whether they share characteristics or share oppositions. Thus, one might look at terms describing fieldwork in geography and chart their relationships rather like that illustrated in Figure 13.1.

I am of course being a little provocative in my choice of signs and meanings (but see Bondi and Domosh, 1992; Rose, 1993), but I hope what is clear is how organising the relationships between categories might help you develop ideas. It should be clear that these simple relationships by no means exhaust the possibilities. It would be possible to develop 'matrices' to help think through the possible permutations of relationships, for instance a range of sites on one axis and a range of activities on the other, to see if there is some consistent pattern of use (cf. Chalfen, 1987; Cook and Crang, 1995: 73–74; Jacobs, 1981). Such devices tend to be simple *aides-mémoire*, pointing out intriguing gaps or organising materials, but they can also be used to think through the relationships between categories – are they sub-categories, do they overlap, do they oppose each other? For instance, as a fourth and more formalised tactic, Andre Greimas (1985, 1987) looked at more complex cases than simple oppositions developing what he called a 'semiotic square' of rules. Thus the square organises incidents

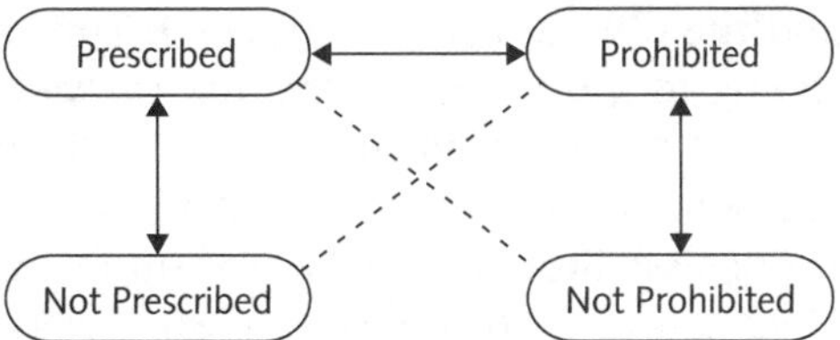

Figure 13.2 The semiotic square

into what is specifically included in a category versus what is specifically excluded, a simple opposition as above, but also incidents which are then not specifically prohibited and ones that are not specifically prescribed, as illustrated by Figure 13.2. So you might be able to look at the opposition of work and leisure and tease out which activities respondents feel are part of which (what is prescribed as work, what cannot be considered work), then which occupy the lower, more shifting positions in the semiotic square. What may then become of interest are the more permeable and shifting relations on the diagonals; what counts as one thing in one place but another in another place, say.

Now it is clearly possible to criticise this for making the interpretation over-rigid, by incorporating a formalist logic that makes (un)necessary distinctions out of grey-shaded continuums. That is not the point here. It is not necessary to use these particular devices; they are really examples to start you thinking about what might help your project and how it is possible to develop arguments from the relationships between your categories. For an alternative, you might wish to look at Jones's (1985: 60–62) reworking of clustering and oppositions in 'maps' of actors' beliefs about events and their descriptions of the relationships between them. Again these schemas serve to highlight the types of relationships shown in the materials. You should not then feel you have to cram any or all projects into these diagrams, nor feel worried if your material does not fit them. The diagrams should be interpreted as guides to thinking rather than something to be mechanically applied. The examples here show how conceptual diagrams may help clarify your thoughts and help you think through your project. However, all the examples discussed so far are 'static' so you might also use the common diagram I have not discussed – the flow chart, a diagram about change. It is the analysis of changing ideas and actions with which I want to deal next.

Narrative, plots and characters: stories we weave

Both of the last two sections actually share certain characteristics. Carr (1986) points out that the relationships discussed here are between terms at any given moment (paradigmatic), and not about how things change or are related over time and in sequence (syntagmatic) (cf. M Crang, 1994; Taylor and Cameron, 1987). The cutting and pasting often remove elements from original sequences, making it harder to chart the evolution of ideas or the reactions and changes of view that might take place. A simple example where this might be very important would be projects gathering material about how different groups of people saw an event unfold and the different stories they tell. Thus the third and final style of analysis I am going to introduce is narrative analysis.

Narratives are stories, and stories might be defined as having beginning and end 'states' and showing some change in the middle linking them. If you have the time, there is a considerable literature looking at different narrative theories (e.g. Budniakiewicz, 1992; Carr, 1986; Ricoeur, 1980, 1984). Common occurrences in qualitative methods are where people are reconstructing life histories or interpreting the competing versions different groups have of events (e.g. Lawson, 2000; Miles and Crush, 1993; Geiger, 1986). Narrative analysis suggests that 'belief systems' cannot be seen as static or external to events, but instead only exist in the way people 'emplot' events (Ricoeur, 1984), that is make them into stories. The key argument is that people tend to make sense of things as stories, comprising events, imputing motives, agency and roles, rather than in terms of static characteristics. This linking of people, places and events into stories presents a rather different slant than the previous two sections.

So how might this sort of analysis work? Perhaps by thinking of the materials you have as scripts or as dramas, and your work as retelling them (Revill and Seymour, 2000). You might begin to think about the story enacted on several levels. At the 'grand' level, you might try and think whether this is an ascensionist tale or a declensionist one, that is whether it talks of progress or decline. Think of your average movie, say a Western. We can quickly think of how the ascensionist story might involve the taming of the 'Wild West', the wagon trails and settlers bringing new land into cultivation and bringing order out of chaos. Equally, we might think of a declensionist story of indigenous peoples being intruded upon, having their lands seized, their sources of support removed and which might perhaps end with shots of European hunters slaughtering thousands of bison or beavers, bringing them to the edge of extinction, or of the dustbowls of the thirties due to inappropriate cultivation (cf. Cronon, 1992). In each case we see broadly the same history given totally different meanings by the direction of the stories.

On a more detailed level it might be possible to look at how events themselves are defined as significant or not, indeed are defined as 'events' by a wider story. Thus it might be possible to ask why certain events are picked upon, why they are defined as 'turning points', indeed how certain actions come to be seen as singular events. For instance, what in one story can be a range of events (say, thinking of a project title for your dissertation, getting it approved, planning the research, doing it, writing it up) could be interpreted as a single thing in another story (coming to university, doing a foundation course, choosing options, doing your dissertation, taking finals). This is not to say either is wrong, but the levels of detail and interpretations tell us something about the importance assigned to each element and thus perhaps the perspective of the teller of the story.

Moreover, the position of the teller, the actors and roles can be thought about. So it may be helpful to think how the story is narrated – in the impersonal third person or more personally? Is the narrator a character? If so what role do they play? What perspective have they got – all knowing? ironic? – in short, from what vantage point does the teller relate the events? Then think through people or places described. Are they being given characters or roles? Is someone being made into a hero or a villain? More subtly, who or what is seen as making the running, controlling the events, or being on the receiving end? Are places just scenery, or do they gain connotations from particular events, or indeed are they seen as actors themselves? Are landmarks or places used to help tell stories?

On a final level, though not strictly narrative, plenty of analysis has been done at the micro-level using the ideas of drama and scripts. The point is that most communication works through intentions. So when you say something to someone, you intend a certain message or maybe you adopt a certain role. Of course, the other person may misinterpret that message or want to play a different role or follow a different script. And these situational roles can vary over space and time – for instance a professed interest in geography may be a performance for a tutor, which would be treated very differently were it repeated in a student bar. So again the situation and what actors are trying to achieve need to be remembered. This is again not to say the material is flawed or biased, but rather to encourage you to look and see what 'scripts' are being used, to see how people expect the interaction to go, to look at what stories are being used to organise and make sense of the world.

From 'data' to theory and back again

In this chapter I have tried to outline some ways of developing ideas from qualitative materials. The range of styles shown here is not exhaustive but should have made it clear that there is no single or mandatory way that materials have to be understood or used. I hope they suggest ways of thinking about your materials that help you develop your own approach. By and large I have suggested that analysis tends to move from a detailed concern for the materials themselves through to interpretation of more abstract relationships. All the approaches maintain a commitment to the rigorous and thorough interpretation of materials and it is important to realise that they all take time to work with and develop through from the initial materials to a finalised argument.

I wish to end by reiterating some questions raised throughout the chapter. The approaches not only ask different questions but also use different forms for explaining your materials. Each one treats categories, codes and their relationships slightly differently. You have to decide what status is appropriate for the relationships or patterns you are going to talk about. In your final written account you may decide to organise sections according to these patterns, according to relationships and differences your analysis has highlighted. Proceeding as suggested here will give you the security of being able to justify these divisions and your interpretation but you should be clear whether you see the relationships of codes to materials as your own interpretations, as structures underlying the issues you studied, or as relationships your informants used more consciously. There are obviously grey areas between all these and your position will depend on a variety of theoretical and ethical considerations, which will vary for every project and researcher. If you keep this in mind and make clear what you are doing you should be able to justify your interpretation to the reader.

BOX 13.2

Further reading

Feldman M 1995 *Strategies for interpreting qualitative data*, Beverly Hills: Sage

Jones S 1985 The analysis of depth interviews. In R Walker (ed.) *Applied qualitative research*, Aldershot: Gower Press, pp. 80–87

Riessman C 1993 *Narrative analysis*, Beverly Hills: Sage

Strauss A 1987 *Qualitative analysis for social scientists*, Cambridge: Cambridge University Press

14 Textual analysis: reading culture and context

Stuart C Aitken

Synopsis

This chapter attempts to answer two questions: what are texts and how are they analysed? It is divided into two sections, each of which deals with one of these questions. In the first section, I introduce some different kinds of texts geographers study including art, music, film, maps, landscapes as well as novels. In the second section, I approach the world through textual metaphors, encountering a myriad of different theoretical and methodological perspectives including semiotics, hermeneutics, Marxism, feminism, psychoanalysis and post-structuralism. None of these, on their own or in combination, provides definitive answers. There are no cookbook recipes and menus in this chapter. Instead, students are offered a window to an exciting way of thinking geographically about the world.

Armchair geography, or how to curl up with a good book and call it research

Once, when presenting a paper on film texts at a British University, I was quite taken aback when a senior member of the geography department asked: 'How can you call this geography unless you go out into the field and validate the sense of place exhibited in these films?' The question led me to reflect on how much of students' experiences of geographic methods are focused upon being in the 'field' without any discussion of what constitutes the 'field'. Cultural and physical geography students alike are impelled to 'get your wellies out and dust off your clipboards'. With such pressure, it is not surprising that many students feel decidedly motion-sick and, perhaps, a little agoraphobic or provincial at the prospect of 'getting out there'. Given the need to 'do time in the field', it seems to me that students new to geographic research often miss wonderful opportunities in their own armchairs to read about culture and its geographic contexts. Students in my seminars on geographic methods chuckle and raise their eyebrows when I suggest that data they need for an original research project might well be found at their local video store, or between the pages of a novel. Only recently has the prospect of geographic study 'in armchair with popcorn' gained the respectability of geographic study 'in wellies with clipboard'. Today, armchairs need not be filled with senior faculty members dreaming up exotic theories of space. Rather, they can be filled with students like you intent on mapping new and exciting geographies through art, films, videos and books. This chapter is about some of the possible ways of constructing and interpreting these new geographies.

What I would like to do first is to outline some of the texts that geographers choose to study and why I feel insights from these texts constitute important geographies. In many ways, you can consider this section to be a concise literature review. Next, I trace the ways geographers analyse and interpret texts. In this second section, I describe some of the theoretical windows that aid interpretation of texts focusing particularly upon semiotic, Marxist, feminist, psychoanalytic, post-colonial and post-structural critiques. These are all heady names for different ways of approaching texts. The ability to interweave these theoretical perspectives around dissertation questions, and essentially use theory as method, is extremely important. These theoretical perspectives may, at first cut, appear complex and cumbersome but they are really about situating yourself in the field.

As a point of departure it is worth noting that methods of study often reflect changes in the way academics view the world. It may be argued that textual analysis gained importance with the recognition that representations and images dominate our culture. Many scholars argue that contemporary culture is an imaginative collage constructed through mixed-media video–audio texts. Methods of study in geography are influenced today by this flexibility and creativity. Textual analysis defines a field in geography, which finds form in large part through imagination. That said, textual analysis is not for the lazy, who view an evening on the sofa as an escape from thought and action. Nor is it for the faint-hearted who wish to avoid the realities of lived experiences. Don Mitchell (1995) cautions that focusing on cultural texts may lead towards long-winded idealism and away from concerns with the material world and how it can be bettered. Rather, he argues that through an analysis of texts we refocus our intent upon theorising the workings of power, and this is a goal with which I wholeheartedly agree. Texts are inescapably political, and an engagement with them is about effecting change, perhaps through elaborating new meanings or perhaps by representing resistance to dominant narratives.

Flights of the imagination

So, how are texts defined? In its most traditional form, a text is that written material that occupies anything from a newspaper article to a volume sitting on the shelf of a library. It is books, tomes, manuscripts, memos, archives, documents and treatises. These written materials collect dust on impressive floor-to-ceiling oak shelves or in metal filing cabinets that squeak when their drawers are pulled open. Over the last three decades, the text as a concept has broadened to include other forms of cultural production. The French philosopher Roland

BOX 14.1

Putting butter on your popcorn

There has been some serious questioning in geography about what constitutes the 'field' and the practice of 'fieldwork'. What kind of field does an armchair and a good book constitute? In 2001, the *Geographical Review* published two issues, comprising 56 short essays, on the subject of 'doing fieldwork' (DeLyser and Starrs, 2001). Titles about feeling and reading, visual literacy, digital worlds, place writing and so forth suggest that textual analysis is an important way of 'doing geography'. Heidi Nast (1994: 57) argues that the 'field' does not need to be thought of as 'a place' or even 'a people' but rather it can be read as a political artifact. Admonitions such as these parallel a growing respect for methods that are less constrained by 'field-oriented' empiricism and the 'rigour' of science, and are more about how we situate ourselves politically. The influence of perspectives from the arts and humanities gained credibility over the last two to three decades as the questions posed by geographers changed to incorporate knowledge about a world that could not be tied to the seemingly solid foundation of science and empirical research. The realisation that our gaze as researchers was from a vantage point which had no secure foundations led to methods that express, among other things, humility and self-reflection. Humility is not paralysing, nor is it self-deprecating. Self-reflection suggests that our voices, as students and researchers, should not be silenced and that it is important to appreciate our own fallibility. The relations between the student and the data created from field experiences – from the quadrat to the digitising table, or from the pages of a novel – are now equally suspect and subject to critical appraisal. Rather than magical methods that 'substantiate' the internal and external validity of a research project, many geographers are now concerned with their work's 'trustworthiness' and their own relations to progressive social change.

Barthes (1972) expanded the notion of the text to include many other communication systems such as fashion, etiquette and urban design. The Italian semiotician Umberto Eco (1976) further broadened textual analysis to include such diverse subjects as zoology, medicine, proxemics and musical codes. Today, geographers 'read' images such as paintings, maps and landscapes along with social, cultural, economic and political institutions as texts (Barnes and Duncan, 1992; Benko and Strohmayer, 1997). With this expansion of the idea of texts to include the reproduction of things other than written materials, the assumption is made that these productions exhibit text-like qualities and are interpretable using textual methods. These methods are gleaned in large part from literary studies. The task is to look for text-like structures and connections, to appraise grammars, syntaxes, inter-texts and subtexts, and to elaborate the relations between authors, texts and readers. In what follows, I provide a sample of the kinds of texts that geographers study and some brief notes on different methods of study. These differences in methods are taken up later in the chapter in the discussion on possible theoretical perspectives. Woven through what follows are cautionary notes on the misuse of textual metaphors.

Literary texts

Literary texts are compelling. I think of the hours I have spent reading literature and poring over maps, and I wonder if the former may even supplant the latter in engaging my geographical imagination. In 1947, the President of the Association of American Geographers, J K Wright, suggested that geographers should begin to explore the *terrae incognitae* of the mind. He noted that travelogues and diaries provided important data for geographers. Some early studies used novels as sources of geographical 'data' (Darby, 1948), but Wright's use of the term 'geosophy' to describe the study of common everyday texts never caught on, and the next two decades witnessed a certain disdain for textual methods as geographers embraced science and quantitative methods. But with the waning of science as a methodological panacea in the 1970s and 1980s, literature was rejuvenated as a respected focus of geographic inquiry and that tradition continues today. Initially inspired by the humanistic tradition in geography, researchers focused on poetry, novels and other fiction as representations of the essence of our lived world. Creative literature embodied the geographic imagination because of the various places and spaces occupied or envisaged by writers (Porteous, 1985; White, 1985). For many humanistic geographers, sense of place provided an organising concept for the life-world and concepts such as 'embeddedness', 'at-homeness', 'alienation' and 'belonging' were explored through literary texts. An early volume edited by Douglas Pocock (1981) emphasised this form of humanistic interpretation. A later collection of essays entitled *Geography and Literature*, edited by William Mallory and Paul Simpson-Housley (1987), focused upon 'the appraisal of literary places' although, for the most part, it sidestepped any detailed account of textual analysis or literary methods. In the preface to this book, the editors specify the geographer's method as 'factual description' and the writer's method as 'flights of imagination' (Mallory and Simpson-Housley, 1987: xii).

By the early 1990s, geographers were taking issue with the 'factual' prerequisite of their field. They began to note that texts and representations were extremely important to understanding lived experience and their neglect was due to geographers' misguided emphases on the physical conditions of social life wherein representations of the world are subsidiary to factual descriptions of a known reality. Attention was given to the sociological and geographical imaginations of writers (Schmid, 1995) and the place of literature in the production and consumption of power and cultural differences (Sharp, 1994; Phillips, 1997). Illustrative of this latter trend is a recent volume entitled *Lost in Space*, which includes a number of essays that probe the relations between forward-looking science fiction and contemporary power relations (e.g. Morehouse, 2002; Taylor, 2002). As with other writing, science fiction often reveals more about the culture of the author than that of the people and places represented, but it is also about the creation of new environments. With the genre, as science fiction writer Michael Marshall Smith (2002: xi) points out in the book's foreword, 'new places are wrought and telling futures conjured, and within both we hold up a mirror to ourselves'.

As issues of writing and representation came to articulate a theoretical core for the new cultural geography (Cosgrove and Domosh, 1993), the relationship between literary theory and geography became less distinct. Increasing attention was given to space and power by those working in cultural criticism, literary theory and elsewhere (Jameson, 1981; Davis, 1987). Concerns for the political inscription of cultural texts and how they reflect dominant ideologies turned geographers' attention to issues of resistance, contestation and subversion. For example, an essay by Stephen Daniels and Simon Rycroft (1993: 461) on the novels of Alan Sillitoe not only describes how Sillitoe imagines the region around Nottingham at a particular time in its history, but also takes care to consider geography and literature 'as a field of textual genres . . . with significant overlaps and interconnections'. This merging is significant because it suggests a new way of looking at the world, which incorporates more than just a description of the geography within a text but also the relations between authors, texts and landscapes. Daniels and Rycroft document how Sillitoe's vision of Nottingham competes with and contests official and academic geographies of Nottingham. In terms of social and spatial relations, they also paint an intricate picture of the relations between locality and citizenship. Similarly, Tim Cresswell (1993) explores the subtleties and ambiguities inherent in a geographical reading of *On the Road* by American counter-culture writer Jack Kerouac. Focusing on the geographic concept of mobility, Cresswell notes that this theme destabilises some traditional American values while contriving an ambiguous relation to gender and sexuality. It is important to note here the importance of taking critical and political standpoints. This contrasts with the earlier humanistic interpretations, which emphasised how writing mirrored the material landscape.

Timothy Oakes (1997) advises geographers to turn to literary texts so that the concept of place may be reinserted into discussions of societal transformations through modern times. He argues that in these discussions place often gets conflated with a nostalgic community of traditional values. One important source for a place-based understanding of societal transformations is from travel writings and diaries. John Pipkin (2001) takes Oakes's advice in a study of the writings of mid-nineteenth century transcendental theorist and nature writer, Henry David Thoreau. Pipkin (2001: 525) argues that a study of Thoreau's 'rich literary contrivances' provides a keen elucidation of 'written landscapes, inscribing and erasing places in varied ways, expressing the contradictions of early modernism'. Thoreau was wildly averse to travel, and viewed the transforming modern world from the local vantage of Concord. On the other hand, travel writing and diaries are hugely useful sources for textual analysis of societal transformations and the systems of 'othering' travellers bring to a place, and the geographical differences they discover upon arrival.

Travel writing, diaries and other anthropological texts

With increased geographic interest in colonialism and post-colonialism, research on travel writing attempts to elaborate imaginative geographies in the context of 'othering' and masculine heroic projects (Gregory, 1995; Phillips, 1997). From the early writing of explorers such as Richard Burton and Captain Cook (Phillips, 1999; Clayton, 1998), to the exotic travels of Gustave Flaubert (Gregory, 1995) and the contemporary travelogues of globe-trotting teenagers (Desforges, 1998), it is possible to glean how sexuality, race and power are geographically elaborated.

At its best, this work questions the politics of representation and elaborates the contested spaces of globalisation. In their introduction to an edited collection on travel writing, Duncan and Gregory (1999: 2–3) note that travel and its cultural practices are now critically located within larger formations of power and privilege.

The influence of anthropology is significant for geographic analyses of biographical texts and diaries. Much contemporary concern derives from the silenced voices of women and from post-colonial sensitivity to the ways Western anthropologists and geographers wrote about 'exotic people and places'. For example, Lydia Pulsipher (1993) uses diaries and oral histories to describe and analyse the matriarchal focus on gender relations in traditional West Indian houseyards. Similar ethnographic methods are used by Patricia Sachs (1993) to describe changes in household and community relations in a coal-mining community at a time of disinvestment in this activity in West Virginia. Both these essays are important because they describe shifts and changes in the global political economy that affect local areas and the everyday lives of women and men. In a collection of essays which study women's diaries, art and literature in the American Southwest, Vera Norwood and Jan Monk (1987) highlight forcefully that the geography and culture of this region are neither monolithic nor static in character. Many of the essays are an appropriate contestation of past writing, which celebrated masculine stoicism, dominance and control.

Rather than focusing on diaries and other cultural artifacts, anthropologist Clifford Geertz (1973) spearheaded a revolution in the social sciences by proposing that culture is a text, which may be read by ethnographers similarly to reading written materials. His basic notion was that culture could be read like a book: it has an author or multiple authors; it has a structure, a grammar, a syntax; it connects with other texts; it elaborates certain kinds of meanings if you understand the basic rules; there are subtexts and other things that are not written in the culture but nonetheless understood; there are multiple possible readings and so forth. By making this connection, Geertz opened up the possibility of using textual methods from theology, the study of comparative literature, and rhetoric to better understand culture. Although Geertz's pioneering use of textual metaphors spawned a plethora of studies in many disciplines, Mitchell (1995) cautions that a focus on culture as a text misses the workings of power in the way societies reproduce themselves. He argues that although there is a complex politics of reading and interpreting cultural texts, reducing lived experience to language and the politics of language sidesteps the issue of what culture actually is. Culture is an idea that orders the world, and by focusing on how culture functions as a constraint and context as well as a textual metaphor, we can better understand how powerful groups operationalise the concept to, for example, elaborate cultural differences to aid colonisation (Mitchell, 1995: 113).

Visual texts

Visual methodologies are addressed in Chapter 15, but it is worth noting that geographers use textual methods to help read paintings, photographs, advertisements, television shows and movies. I noted earlier that literary texts might well supplant maps in engaging the geographical imagination. Whether this is true or not, there certainly is some methodological tension between the use of textual and visual metaphors to understand the contexts of the geographic world. Cresswell and Dixon (2002: 5) argue that the textual metaphor predominates because of our limited ability to read vision. A sophisticated repertoire of methods from literary analysis propels textual understandings, but we do not have the same kinds of methods to help us understand what we see in its own terms. Here I focus on textual metaphors, but in Chapter 15 Jim Craine and I elaborate some of the other ways we come to know the visual world.

Several collections of essays by geographers are particularly noteworthy for their focus on the production of political culture through visual texts. In *The Iconography of Landscape*, a collection of essays edited by Denis Cosgrove and Stephen Daniels (1988), there are several papers that describe the confluence between art, geography and dominant ideologies. In the essays in *Place Images in Media*, edited by Leo Zonn (1990), the predisposition is towards the production of meaning in terms of the portrayers and creators of texts about places. The intent of this collection is to provide an understanding of the processes through which information is conveyed by various media texts, with a specific focus on representations of place. *Power, Place, Situation and Spectacle* (Aitken and Zonn, 1994) is a compilation of work by geographers focusing exclusively on film texts. In this volume, Jeff Hopkins (1994: 49) notes that, although other texts such as literature and painting may 'blur our sensibilities, . . . film is peculiar because of the semblance of actuality attributable to the film image and the obscurity of its very production'. Literary theorist and postmodern philosopher, Fredric Jameson, provided an impressive rendering of cinema and space in *The Geopolitical Aesthetic* (1992) which brings together Marxism (see discussion below), postmodernism and methods from cognitive geography. Jameson's project encompassed a cinematic vision of the world system, and helped reinvigorate the study of popular culture. It also moved the study of film beyond textual analysis, which, some argue, is not sufficiently flexible and mobile to accommodate such a dynamic visual medium (Cresswell and Dixon, 2002: 4).

Musical texts and performances

The meanings and values embodied in musical texts are understudied topics in geography. Influenced by humanistic concerns for sense of place, a collection of essays edited by George Carney (1994) focuses upon American folk culture and popular music. Elsewhere, Jim Curtis writes about Latin and folk music (Curtis, 1976), Larry Ford (1971) documents the geographic evolution of Rock and Roll, Warren Gill (1993) uses structuration theory to analyse music from the Pacific Northwest, and Lily Kong (1996) focuses on local Singaporean music. Although music is clearly an important cultural text (Eco, 1976), these geographers analyse music in terms of regional differentiation, cultural diffusion, and perception of places rather than anything specifically textual. Alternatively, Andrew Leyshon and his colleagues (1998) use post-structural methods that explore issues of economy, society, polity and culture in 'local' and 'global' music.

Of late, geographers have turned to the study of 'sound politics' or the emergence of an understanding of music as texts, performances and products of material struggles. At one level, music is popularised through political struggles over whose music shall prevail. The last few years witnessed an expanding literature by critical geographers exploring music as it intersects with cultural texts, performance, place, nation, gender, identity and so forth (Ingham *et al.*, 1999; Gibson, 1999; Revill, 2000; Aitken and Craine, 2002). Smith (2000: 617) argues that the most promising aspect of this work is not understanding music in relation to other things, but rather conceiving music as a text and a performance through which social relations exist and change. It is not enough to specify the place of music in society but we must further understand through music, the place of individuals and societies. This is not just about reading lyrics, but about the physicality of music and its call to action.

Reading maps

It is perhaps not surprising that textual analysis in geography found great potential in cartography. Maps may be described as visual texts and, accordingly, they are more or less amenable to the forms of analysis accorded photographs, art and cinema. Brian Harley's (1989) 'Deconstructing the Map' is one of the first attempts to bring cartography into a wider theory of representation. Drawing on the work of Foucault and Derrida, Harley deconstructs the alleged 'truth' and 'impartiality' of maps. He notes that maps are cultural texts embodying a collecting of codes, few of which are unique to cartography. Deconstruction is needed to unpack the many alternative meanings of maps. Harley also outlines the power of maps, a topic that is detailed more fully by John Pickles (1992) who uses hermeneutics to get at the intention behind the construction of maps, and how maps are used. Pickles's work is not only a useful deconstruction of propaganda maps, but also an excellent description of the hermeneutic method, which I shall draw upon a little later. Denis Wood's *The Power of Maps* (1993) is another important contribution to the deconstruction of maps. Drawing on the communicative power of maps and using many familiar examples, Wood highlights the importance of understanding the signs and myths which are embodied in cartographic codes. More recently, Matthew Sparke (1998) focuses on the embodiment of male colonial power and the context of cartographic recognition in nineteenth-century Canadian maps. Students interested in geographical information systems (GIS) should note that research on the textual power of maps and cartographic representations is extended to the visual authority of GIS in the work of Gregory (1994b: 60–69), Curry (1995), Aitken (2002), and a collection of essays edited by Pickles (1995).

Landscape as text

Contemporary textual work on landscapes is often traced from Carl Sauer's (1925) reading of the origins and development of past landscapes in order to describe the morphology of contemporary landscapes. Sauer inspired much of what constitutes cultural geography in North America, which led to a variety of work on how to read landscapes. Concern for the 'atheoretical' nature of the Sauerian school of landscape interpretation led to new forms of textual analysis in cultural geography (Price and Lewis, 1993; Duncan, 1993; Mitchell, 1995). Although it may be misguided to characterise a 'new school' of cultural geography, it is clear that cultural geographers today are very open to literary post-structuralism, sophisticated social–theoretical constructs and perspectives on textuality offered by hermeneutics, post-structuralism, feminism and post-colonialism (Cosgrove and Domosh, 1993). Three collections of essays represent well the evolution of how we use textual metaphors to understand landscapes. Cosgrove and Daniels's (1988) *Iconography of Landscape* has been discussed already for essays focusing upon landscape painting. This collection is also an attempt to apply fairly sophisticated literary methods in order to interpret symbolic imagery. Barnes and Duncan (1992) focus on discourse, text and metaphor, and how geographers self-reflexively insert themselves and their culture into what they write. Focusing more broadly on representation, Duncan and Ley's (1993) anthology builds on the previous two works with sections on conceiving residential landscapes and political institutions as texts.

Mitchell (2000: 270) is encouraged by these building blocks for landscape studies that may help provide 'a theory equal to the world we live in'. His optimism is couched specifically in understandings of the relations between the rendering of landscapes (as both texts and material contexts for our lives) and the emergence of capitalism. The point he makes is that we have come a long way from Sauer's (1925) atheoretical reading of landscapes as 'naively given'. Instead, argue Mitchell, Cosgrove, Domosh, Duncan and a host of other geographers interested in landscape, places are actively produced and struggled over, and it is the politics of this struggle that ought to engage those who adopt textual analysis.

BOX 14.2

Texts as metaphors for lived experience

Can textuality be applied as a metaphor for all aspects of our lives? Many scholars think so. In the early 1970s, Paul Ricoeur suggested that social and political institutions inscribe values, meaning and behaviours in much the same way as texts. From this, many geographers point out that the life-as-text metaphor is applicable to all social and cultural productions whether they be movies, maps, landscapes or institutions. If we accept this postulate then texts may be used as a metaphor for lived experience and all life is amenable to textual analysis. Certainly, textual analysis provides interesting and useful metaphors which make the unfamiliar familiar, which deconstruct and unpack the complex and help us to understand more fully social and cultural productions. Textual analysis probes the communication of meaning about places and the political manipulation of space, but can we take it too far? What if we abstract our analysis far beyond the material conditions of life? Of what concern are textual readings of social life to the lives of people at the everyday level? Neil Smith (1993: 97) notes that the metaphorical uses of space are so fashionable in literary and cultural discourse that there is a danger of losing sight of material conceptions of space. If we are concerned with everyday geographies and academic practice that is critical and political then we should take care not to be constrained by textual metaphors that discourage political insight.

Making sense of texts and textual metaphors

The previous section covered some of the texts and textual metaphors that geographers study. In what follows, I turn to how we come to understanding texts (and things we want to think of as texts) and how we situate ourselves as readers. Theories help reconceptualise specific texts and representations with reference to comparative concepts and a broader literature, and by translating specificities into more general (but not too general) frameworks, debates beyond the immediate text can be opened up (Duncan and Duncan, 2001: 402). Perhaps more than anything else, critical theoretical perspectives enable students to situate and position themselves in a way that brings the text alive and opens its political subterfuge.

Allow me to begin with what might seem like a step backwards, and let me assure you that it is not. The chapter synopsis admonishes you not to look for cookbook approaches to textual analysis, but I want to start this section on theoretical perspectives with some simple how-to steps, which may or may not be taken in order. My point here, which hopefully will become clearer as you read through to the end of this section, is that every decision you make about your research is steeped in your politics. So read the following list with that in mind:

1. Find something about which you are passionate.
2. Decide whether a textual analysis is appropriate. Box 14.2 suggests that just about everything may be likened to a text and analysed as such. The question you must ask is whether a textual analysis is appropriate for you and what you are studying.
3. Read the text, watch the movie, contemplate the painting, observe the landscape.
4. Ask yourself why the author(s) created this text. Does it support larger dominant groups and ideologies (white men, colonial powers, multinational corporations)? Does it contest and resist these dominant forces? If so, who is contesting and resisting?
5. Read everything you can find on the topic, and its historical and geographic context. Try to fathom the relationship between the author(s) and the text. Note where your sources are situated politically.
6. Do 3 again.
7. Look for structure in what you are studying: grammar, syntax, rules of engagement. What is outside the structure? Is there part of what you are studying that defies the relative fixity of textual analysis?
8. Determine how your text relates to other texts.
9. Look for important subtexts or missing/hidden texts. What is your text silent about?
10. Do 6 again.
11. Make connections to larger geographies and societal processes.
12. Create your dissertation text with awareness of how you situate yourself as a researcher and writer.

The adoption of textual analysis and the reading metaphor has given rise to a methodology that is often thought of as being largely implicit and derived from years of apprenticeship and practice. There is no magic recipe for success in the above list. I cannot tell you what you should be looking for or how you will find it. This makes things a little difficult for the beginning researcher who has little training and experience in interpreting texts. One aspect of contemporary cultural geography is an attempt by researchers to make explicit their philosophical and theoretical underpinnings. Just reading some of the essays and papers already discussed in this chapter reveals a myriad of appropriate and usable methods. The treatment of any kind of geography as a text that can be read implies that, at least at the moment of reading, the text has some meaning which may be grasped. This meaning derives from a body of knowledge. What I would like to do now is briefly outline some of the theories, perspectives and ways of knowing that make these bodies of knowledge more accessible. A number of useful theoretical perspectives help us focus on the at once structured and flexible context of meaning systems. Such theories are more for orientation than for generation of hypotheses in that they help the researchers situate themselves and their way of thinking in the research. As students, we must also ask ourselves how theory enables a greater sensitivity to the data at hand, and how the data in turn help us to rethink theoretical constructs.

My discussion neglects computer-based methods, but there are a number of textually based software packages that may be used to analyse texts. I particularly like QSR NVivo as a computer aid to textual analysis because it is derived from grounded theory. Interested readers are referred to Chapter 13 by Mike Crang for a discussion of grounded theory and computer software. Crang's chapter also contains a discussion of semiotics.

Semiology: a science of signs

Semiology is intent upon uncovering the meaning of texts by dealing with what signs are and how they function. Language is a system of signs and, like language, social and cultural products are not simply objects or events because they have meaning: they are also signs. Swiss linguist Ferdinand de Saussure (1966) indicated that any sign could be divided into two components: the signifier (a sound-image) and the signified (a concept). To take an example from film, the image of pages dropping off a calendar (signifier) is commonly thought to represent the passage of time (signified). Saussure noted that the relationship between the signifier and the signified is arbitrary, unmotivated and unnatural but it may be fixed for a certain time by the nuances of political culture. The point that makes the analysis of texts interesting and problematic is that there is no logical connection between the signifier and the signified. If the relationship between signifier and signified is arbitrary then the meaning of texts has to be learnt. Just as there is a learnt grammar for writing and speaking, there is also a learnt grammar (or a set of codes) for different kinds of texts. Generally, as in speaking, people are not consciously aware of the codes and cannot articulate them even although they respond to them. Codes are highly complex patterns of associations that are common for a particular society at a particular time. They affect the way we interpret signs and symbols found in all forms of texts. To be socialised and given a culture means, in effect, to be taught a number of codes. Many codes are specific to a person's social class, geographical location, race, ethnic group and so forth. Semiotics tries to decipher these codes so as to bring them to the forefront of consciousness.

Intertextuality, metaphors and metonyms

Important concepts of semiology include intertextuality, metaphor and metonym. Intertextuality refers to the conscious or unconscious use of previously created texts. Readers must be familiar with the original text to fully appreciate the new text. For example, a full appreciation of the work of Irish songwriter Mike Scott (founding member of *The Waterboys*) requires knowledge of the texts of other Irishmen such as Van Morrison and W B Yeats. He pulls from other texts to arrange his lyrics and melodies. Some lyrics focus on the landscapes of Ireland accompanied by traditional melodies and uilleann pipes while others pay tribute to the life of Jimi Hendrix accompanied by wailing rock guitar solos. Nonetheless, all Scott's songs are recognisable as work of *The Waterboys*.

Metaphor and metonymy are two important ways of communicating meaning. In metaphor, the relationship between two things is suggested through analogy. The relationship in a metonym is based upon association of names. Metaphors are used to give objects and events different identities. For example, an enduring metaphor of contemporary cultural geography is 'culture is text'. The twin towers of the World Trade Center in New York were a powerful metonym for globalisation, trade and the evil centre of capitalism for the terrorists who destroyed them on September 11, 2001. Another example of a metonym is the colour red at the end of David Lynch's acclaimed *Twin Peaks* TV series. The colour encompasses the protagonist as he finally encounters the focus of the series, Laura, in a dreamlike hell. The colour symbolises the hell of the place, but it also suggests the passion that has pervaded much of the TV series. That hell and passion are both represented by the colour red is understood only within a specific set of cultural codes.

The tension between different frames of reference that may be found with metaphors and metonyms enables a certain amount of clarity or obfuscation depending upon the intent of the author. This connection of two seemingly unrelated concepts may, as Anne Buttimer (1982: 90) points out, touch a deep level of learning and intelligence. At one level, it is a way that the unfamiliar may be made familiar. Whatever their function, the use of metaphors and metonyms implies the existence of codes in people's minds that enable them to make connections.

Hermeneutics: a method of interpretation

Hermeneutics is often thought of as a self-reflexive form of semiotics. It has much in common with semiotics but its base is in literary criticism rather than linguistics and its scope is broader, embracing methodological principles of

BOX 14.3

Firing the canons

A rigorous hermeneutic interpretation of texts revolves around five canons.

1. The integrity of the text must be preserved in such a way that meaning is derived from, not projected into, the text.
2. The interpreter has the responsibility to bring themselves into a harmonious relationship with the text. As such, any critique must be rooted in the claims, conventions and forms of the text.
3. The interpreter must give an optimal reading of the text and of the meaning the text had for those for whom it was written, and show what the text now means in the context of contemporary views, interests and prejudices.
4. The whole must be understood from its parts, and all parts must be understood from the whole (the hermeneutic circle).
5. The interpreter of ambiguous texts must make explicit what the author (or subsequent readings) left implicit (Pickles, 1992: 225).

interpretation which incorporate the context of the reader as well as the context of what is written. John Pickles (1992) is of the opinion that hermeneutics is one of the most appropriate methods for reading texts and his application of the method to propaganda maps is an excellent example of how it works. Pickles points out that hermeneutics usually begins with establishing the authenticity of a text, noting if it is a coherent whole and ascribing an author or authors to the text. It is also concerned with understanding the meaning of the text, how it is related to its own contemporaneous world and, in turn, how it is related to our present world. Accordingly, hermeneutics establishes the several layers of meaning which accrue to a particular text.

Both semiotics and hermeneutics are criticised for having no theoretical base that, some argue, leads practitioners to invent underlying structures rather than discover them. Many geographers argue that theories that derive from a particular 'standpoint' such as Marxism and feminism offer a more pointed interpretation of texts.

Marxist analysis: the production and consumption of texts and 'the political unconscious'

Elspeth Graham provides an excellent discussion of the general method of Marxism in Chapter 2, but it is also a powerfully suggestive way of interpreting texts. In that semiotics and hermeneutics may be applied only to spaces and texts already produced, Marxist analysis is concerned with the production as well as the consumption of texts. Several important questions concerning the production of texts are opened to the student through a Marxist perspective. For example, what social, political, and economic arrangements characterise the society whose texts are being analysed? Who owns, controls and operates the media? How are writers, actors, artists, architects, and other creators of texts affected by the patterns of ownership and control of media? For Gerry Macdonald (1990, 1994), for example, a geography of or from film goes beyond the study of artistic or technical production to the social and political ramifications of its consumption. He is intent upon examining the tension created between the cinema of radical politics (Third Cinema) and the Third World condition.

If a classical Marxist analysis is called for, then we assume that beneath the superficial randomness of things there is a kind of inner logic at work. Everything is shaped by the economic system of society, which affects ideas and consciousness. In short, the 'base' of any given society is the economic system, or mode of production, which influences, in profound and complicated ways, the 'superstructure', or the institutions (texts) and values (what is encoded in those texts) of a given society. Of particular concern for Marxists is delineation of the precise mechanisms that lead from base to superstructure. How exactly does economic organisation affect superstructural levels, which cannot be directly related to it? A secondary, but no less problematic concern, is how the economic base can be defined without having recourse to the categories which are themselves superstructural.

The most traditional form of Marxist analysis presupposes that an analysis of the economic base will enable us to read off the elements of the cultural superstructure. But if this position is adopted, then all cultural texts end up with the same content, and analysis continually reveals the same messages, which are based upon economic rather than cultural terms. That is, there is simply an endless recoding of property relations that reveal class conflict. Your results are a foregone conclusion, your theory a self-fulfilling prophesy. In the early 1980s, Fredric Jameson (1981, 1984) offered a way out. Trained as a linguist and literary analyst, Jameson was particularly concerned with specific and small variations in meaning. Jameson's form of Marxist analysis respected minute differences rather than collapsing them into undifferentiated reflection upon the mode of production. To do this he assumed that the relation to the economic is a fundamental element within the cultural text to be analysed, not in terms of the economic processes within which the cultural object takes form, but in the psychic processes which engage in its production and consumption. Jameson (1981) developed his theory with a reading of nineteenth century and early twentieth century fiction writers such as Balzac and Conrad, suggesting that these writers help students understand more fully the joint effects of colonialism and imperialism from an aesthetic and psychic standpoint. Primacy is still given to economic forms of organisations and institutions, but a nuanced understanding of these forms needs the light of analyses produced within cultural texts. Jameson's *The Geopolitical Aesthetic* (1992) takes these methods further by developing Kevin Lynch's (1961) notion of cognitive mapping as an important psychic and aesthetic tool. Although consciousness is socially produced, it always filters through the minds of men and women and, as such, the dominant ideology can always be contested through appropriate cognitive mapping. This reformulation of theory and method enables students to respect levels of textual and cultural differentiation within a Marxist tradition. These differences become a primary element in the analysis of new social and economic relations of late capitalism.

Jameson's Marxist-inspired approach to the study of everyday texts and media reinvigorated cultural studies through the 1980s and 1990s, but more recently an extension of these ideas from feminist, psychoanalytic and post-structural frameworks has had a great impact on geographers' engagements with texts.

Feminist and psychoanalytic criticism

An important part of feminism addresses aspects of women's geographies that have heretofore remained silent (or been silenced). The work of Vera Norwood and Jan Monk (1987), described earlier, is an attempt to look at the texts produced by women in the southwestern desert environments of the United States. The authors in this edited collection weave stories about women/environment reciprocity and renewal rather than male stoicism; they talk about personal vulnerability rather than dominance and power. Other early examples of this kind of analysis include Domosh's (1991) and Katz and Kirby's (1991) reconstruction of the texts of explorers in order to rewrite a feminist historiography of geography and person/nature relations respectively.

Another important aspect of feminist criticism is focus upon the gendered power relations inherent within texts. Gillian Rose (1993: 99) argues that masculinist perspectives – particularly the new cultural geography's preoccupation with landscape as text – are enabled by the textual metaphor. She asserts that if we accept that landscapes are objectively fixed like written texts (cf. Barnes and Duncan, 1992: 6), then we must also accept the possibility of a certain (male) knowledge about the landscape. Rose suggests that if we move to such an impersonal objective then the geographer as an author himself is also removed from the narrative: 'This removes the geographer from the interpretive rules that he applies to the texts of others, and renders him invincible as an author – all-seeing and all-knowing' (Rose, 1993: 100). Alternatively, we should always try to see how our gendered values and mores are inserted and incorporated within readings. If we do not do this, then textual analysis becomes yet another way to deny the phallocentrism of the geographic gaze. Spurred by feminist concern, work by geographers has focused upon the ways masculinity and femininity are inscribed upon texts (Lukinbeal and Aitken, 1998; Sparke, 1998).

Often as an extension of feminist theory, geographers use psychoanalytic theory and psychotherapeutic practice to help analyse and understand texts, images and representations (Pile, 1996; Bondi, 1999; Nast, 2000). Lacan's (1978) linguistic reinterpretation of the Freudian theory of 'othering' has proved a particularly important perspective for textual analysis. Lacan suggests that sexual oppositions develop at the 'mirror stage' through a process of subtraction similar to that found in Freud's Oedipal stage. Identity formation in a subject is a mirage, which finds form when the subject fashions an image of itself by identifying with others' perceptions of it (Lacan, 1983: 5). The important point of Lacan's work for textual analysis is that this identity crisis comes with the acquisition of language and the appropriation of a socially and culturally delimited system of signs. With language, we form an image or text of ourselves that takes the place of the *real* us, but we are always trying to rediscover the *real* (Aitken and Zonn, 1993: 198). Our sexuality, then, is social and significatory rather than instinctual and hereditary.

There is some criticism that psychology's focus on interiority detracts from enduring (larger, exterior) political and economic problems (Hamnett, 1999; Martin, 2001), but Pile's (1996) defence of psychoanalytic theory highlights its focus on the 'politics of the subject'. The importance of bodies, emotions and sexualities as ascribed in texts and engaged to larger social struggles also suggests an important politics of scale that relates to individual psyches as a partial product of globalised forces (Smith, 1993; Lukinbeal and Aitken, 1998).

Is a post-structural reading of texts possible?

The previous discussions of feminism and psychoanalysis resonate with postmodern critiques of the 'factual basis' of everyday lived experience and post-structural critiques of the structures that are assumed to hold the world together. Post-structural textual analysis takes a lead from hermeneutics and problematises the storyteller as well as the receiver of the text, but it also jettisons the rigorous structure of hermeneutics. By treating texts as social products that are dynamic, and culturally mediated by discursive practices, post-structuralists reject the quest for meta-narratives found in early semiological and Marxist writing, as well as empiricist efforts to make 'facts speak for themselves' which presupposes some form of meta-narrative. Accordingly, everyday life is not immediately 'present', but is *re*-presented simultaneously through the contradictory texts that constitute our world and through the everyday filters and pretensions of our gender, class and racial/ethnic identities. Working within the anti-essentialist rubric of post-structuralism, students consider new ways of thinking about reading, writing and, importantly, authority and structure.

One aspect of problematising the storyteller that may seem paradoxical is postmodernists' preoccupation with the 'death of the author'. This perspective is sceptical about originality in that every text refers in some way to other texts. Meaning, then, is not an individual invention but a social product. The social production of a text does not mirror reality, nor is it about something more real than itself although it may have an intrinsic worth all of its own. A reading of James Joyce's *The Dubliners* does more than just provide a description of Dublin. The novel creates new images of Dublin. The student also brings something of themselves to this text, and so Dublin is once again transformed through a combination of Joyce's text and the reader's knowledge. It is fairly easy to argue, then, that the reality of Dublin, the reader's knowledge and the image Joyce created are interconnected in profound and sophisticated ways to the extent that it may be impossible to pull them apart. Students grappling with post-structural ways of knowing draw explicit attention to the context within which they are working, noting its contradictions, complications and complexities as insightful rather than obfuscating aspects of their positioning as readers, researchers and writers. At the level of textual representation, a post-structural perspective questions not only what is known, but also how it comes to be known. It is based upon the simple perspective that nothing in the world is fixed or immutable, that things are grounded on moving foundations. And perhaps most importantly, post-structuralism questions the basis of any method that assumes a structure of signification and understanding that is not politically based.

A concluding caution

The meaning of a map, a painting, a novel or a historical narrative is produced, at least in part, within a frame of reference that assigns patterned regularity to the flow of everyday existence. All forms of text are produced in large part through discursive practices, which structure flows of understanding. Discursive practices imply intertextuality rather than rambling chaos, but these practices are also always embedded in social relations of power and ideology that give authority to some texts while subverting others. Be wary of authoritarian texts, including chapters that attempt to teach textual methods in human geography. That I bring to this chapter twenty years of experience with textual methods means nothing if it does not fire your passion and speak to your interests. There is nothing here that is anything but a partial rendering of approaches to texts that have worked for me to some degree. Some I like better than others. None of the methods or theories described in this chapter, on their own or in combination, is infallible. Textual methods, like texts themselves, are social constructions and, as such, they take on the characteristics of their users. It may seem that there are almost as many ways to analyse texts as there are texts and students analysing them. There is some truth to this, but there is always a need to develop sensitive generalisations (but not overgeneralisations or essentialisms), either from a specific text and/or from critical reflection about the worlds we experience through that text. By so doing, we may communicate more effectively with colleagues and supervisors who do not share our interests in the specifics of a text or place. Today I accept my foibles and the fallibility of my texts, and I try to produce work that is honest and trustworthy. That is all that I ask of you.

BOX 14.4

Further reading

The following anthologies are useful collections of research by geographers who practise textual analysis:

Aitken, S C and L E Zonn (eds) 1994 *Place, power, situation and spectacle: a geography of film*. Lanham, Maryland: Rowman and Littlefield

Barnes, T and J Duncan (eds) 1992 *Writing worlds: discourse, text and metaphor*, London: Routledge

Cosgrove, D and S Daniels 1988 *The iconography of landscape: essays on the symbolic representation, design and use of past environments*, Cambridge: Cambridge University Press

Duncan, James and David Ley (eds) 1993 *Place/culture/representation*, London: Routledge

Duncan, J and D Gregory (eds) 1999 *Writes of passage: reading travel writing*, New York and London: Routledge

15 Visual methodologies: what you see is not always what you get

Stuart C Aitken and James Craine

Synopsis

The world is a miasma of images, spectacles and visual representations. In this chapter, the student is introduced to a range of methodologies used in understanding how geographic knowledge is visually conveyed. The first two sections of the chapter cover geographic images and visual methodologies respectively. The third section provides an extended analytic example from film, and boxes are used throughout the chapter to provide illustrative examples. The first section discusses geographic visualisation from maps to GIS, and then elaborates the importance of understanding the spaces of representation inherent in painting, photography, film and advertising. In the second section we focus on how to critically examine the visuality inherent in images. Lastly, these critical methods are illustrated using a segment from the apocalyptic *film noir* movie *Kiss Me Deadly*, which is unpacked for its geographic and cultural significance.

Seeing is believing

We are told that of all our senses, the visual is our most useful and, by extension, our most acute. Sight deprivation is one of our more debilitating sensory losses. What we see is important, perhaps more important than what we hear, smell or read. Television producers and advertisers understand that carefully constructed images sell products. Many filmmakers with box-office successes know that their craft turns on how they transform narrative into visual spectacle. Curiously though, until quite recently academics have been silent about how to study images, and about how to apply and develop visual methodologies to the sights that bombard our senses and sensibilities. This is hardly surprising because, despite its all-encompassing immediacy, there is a lot of confusion about what actually constitutes visual imagery.

We know quite a lot about the structure of vision and how it works: from understanding the structure of lenses and corneas to the firing of synapses. Scientific studies abound on what people attend to when they look at things.

Questions on what different people look at first and for how long have been studied for decades. Cartographers, geographic information scientists and cognitive researchers focus specifically on how to make better images, ones that are more responsive to the needs of map-readers, data-miners and globally positioned subjects. But there is still huge debate on how, precisely, images work! Do they work on our senses as narrative-like texts? Can we make distinctions for visual images, as in linguistics, between *langue* (the formal structure of a language) and *parole* (the freer, more informal and creative use in speaking)? In education, there is a hue and cry against visual illiteracy as teachers ask academics to devise methods for teaching visual grammar in a similar way that children are taught to understand and be critical of texts. As students of visual imagery we ask colleagues and supervisors for methods to help us decipher our world of representations, but there seems little in the way of cookbook methods for this kind of analysis.

Rather, in what follows, we offer a brief introduction to visualisation methods by highlighting past geographic work and offering our own examples of visual interpretation. We then discuss some of the theoretical positions that help geographers to situate their work, and again we offer examples of visual interpretation to illustrate the ways theories help ground interpretations. In everything that follows we weave our enduring commitment that no visual method can escape consideration of the politics through which the images are created and wielded.

Visualisation: image, power and representational landscapes

It may be argued that visualisation is at the heart of geographic practice. The study of landscapes and their portrayal in maps, paintings and photographs require visual knowledge. But landscapes are more than depictions of scenes. Recent understandings of geography through film and art suggest the creation of new cultural landscapes. In addition, the penchant for more sophisticated technological representations of places, patterns and processes through geographical information science clearly focuses on visual acuity. More often than not, these are not simple representations of a lived reality. Since the 1990s, the traditional visuality of cultural geography has been questioned by a forceful critique that highlights the neglect of *power relations* that are imbedded within maps, landscapes, paintings and movies (Aitken and Zonn, 1994; Robins, 1996; Fyfe, 1998; Crang *et al.*, 1999; Cresswell and Dixon, 2002). There are many ways to study power relations in images, some of which we detail later in the chapter. To whet your appetite and as a basis for some of the discussion to follow, Box 15.1 offers a way of looking at Vermeer's famous depiction of *The Geographer* that draws on issues of discourse and power.

The use of visual representations as a pedagogic tool in geography can take many different forms and can offer great insight into the power relations between individuals and social institutions. Raymond Williams (1993: 6) notes that '[e]very human society has its own shape, its own purpose, its own meanings. Every

BOX 15.1 Image power

Johannes Vermeer's *The Geographer*, an oil-on-canvas painted in 1669 provides a poignant and interesting example of power imagery. It presents a male figure, seemingly endowed with intense energy, looking out of a window at some unseen exterior landscape. This may well be the very stuff of geographical practice, but let us look a little further. First, consider what we do not see. That the male figure is intent and purposeful stands in stark contrast to Vermeer's portraits of female figures, which in his other paintings are represented as enervative or, at best, introspective. A feminist critique of *The Geographer* would note this distinction, and note further that the energy of the male's gaze outside suggests contemplation of an exterior landscape, perhaps directly outside or, more likely, further afield. Detailed study of the canvas reveals that the geographer originally looked down at the table, with his dividers also pointing down. Adjusting the composition to align the man's face and the dividers with the flow of light gives further energy to the movement of light across the canvas. The flow of light from left to right activates the canvas. It is accentuated compositionally by the massing of objects on the left, that represent maps, sea charts, books and a hastily thrown cover. The light spills on to an open area on the right, casting a powerful series of diagonal shadows that highlight a globe on the cupboard. Vermeer adjusted his initial depiction of the figure to provide a more active stance, but it also activates the purpose of the geographer, his craft and his instruments. This may well be the epitome of the patriarchal geographical gaze so ably critiqued by Gillian Rose (1993). A dissertation on this topic might focus on Vermeer's art and his times, as well as his known connection with the scientific community. The painting opens up the world of mid-seventeenth century Dutch science and commerce, and the place of geography in these early global discourses.

human society expresses these, in institutions, and in arts and learning.' Geographical shapes, meanings and arts do many things. They are used as mimetic devices to represent real world places and people; they provide sites that permit the exploration of social issues ranging from gender and sexuality to spaces of resistance and contestation; and, perhaps foremost, geographical images create *landscapes*, the traditional domain of geographers, that allow the investigation of dominant ideologies and alternate forms of social contestation. You, as a geography student, can employ images to uncover patterns and relationships constructed and located within what we are calling *representational landscapes*.

Carl Sauer (1956: 400) once said that geography is 'knowledge generated by observation'. Since the publication of his *The Morphology of Landscape* in 1925, cultural geographers recognise landscape as a central concept and subject, and focus particularly on how landscapes reflect and symbolise the activities and cultural ideas of a place. Representational landscapes are viewed as metaphors that are interpreted and analysed in order to show the role of landscape in social and cultural creation and to help geographers understand space within a social and cultural context. In addition, you can also explore visual representations to determine the role landscapes play in these ongoing and always changing processes (Schein, 1997).

The human eye is bombarded almost to sensory overload with a continuous flow of representations and images directed through a wide variety of media. We 'connect' with these *representations* of social life via our 'gaze', thereby perceiving that social life as *experience*, and so we are able to become a part of spatial relationships contained within the spaces of the images/representations. Expanding on Shiel (2001: 5), we define these spaces of representation, noting how they function as a spatial form of culture that affects the organisation of space as follows:[1]

> ***Space in images:*** *the space of the image; the space of the narrative setting; the geographical relationship of various settings in an image or a sequence of images; the mapping of a lived environment onto an image.*
>
> ***Images in space:*** *the shaping of lived spaces by images as a cultural practice; the spatial organization of the various media industries at the levels of production, distribution, exhibition, and consumption.*

The first important point to underscore is that representational spaces – the social and mechanical processes whereby visual variables (cf. Bertin, 1983) such as colour, light and shadow, and dynamic variables (cf. DiBiase *et al.*, 1992) such as duration, rate of change, and order are used to represent another time and space (cf. Hopkins, 1994) – are shaped and created by life experiences. The second important point to note is that the opposite is also true: conditions within material society are most certainly shaped through their visual representations. Thirdly, it is important to understand, as we study the examples below, that geographical images are not objective representations of the material world

[1] Shiel primarily talks of 'filmspace' but his explanation of that space is equally applicable across many representational mediums.

but are socially created under the influence of particular ideologies; thus, meanings are social constructions.

Gillian Rose (2001: 14–15) clearly articulates the nature of the power relationships within social projects that account for how places, people and events are made and constructed:

1. Visual representations have their own effects and can be used by many people for many different reasons.
2. Ways of *seeing* are geographically, historically, culturally and socially specific; how we look at an image is not natural or innocent and is always constructed through various practices, technologies, and knowledges.
3. Visual representations both depend on and produce social inclusions and exclusions (what is seen and what is hidden).

The interpretation of geographical representations provides insight into how society and space are ordered and how the construction and the representation of that space is manipulated by powerful groups through cultural codes that promote dominant ideologies. Uncovering the shared systems of meaning provides clues to the social, economic and political circumstances of the societies that produce the visual image being analysed. In what follows, we look at some of the ways geographers have come to understand images with a particular focus on art, photography, film and advertising. This brief exposition is elaborated through boxed examples of images from the American West that help us to illustrate social, political and economic contrivances.

Space in images; images in space

The visual methodologies we introduce in this section explore four types of representational images that have been studied by geographers: art (specifically in the form of painting), photography, film, and advertising (a unique yet extremely important genre of visual and cultural representation). We do not explicitly address television because it has received scant attention from geographers despite Jacquelin Burgess's (1990) advocacy of research on television as one of the most important underpinnings of contemporary culture.[1] Nor do we discuss cartographic images or landscape interpretation, although both embrace a rich tradition in geographic visualisation.

Art and photography

Much of the early research by geographers on art and photography derives from the humanistic tradition, and involves highlighting the ways that landscapes are depicted. One of the best examples of a humanistic interpretation of visual texts is David Seamon's (1990) phenomenology of the New York photographs

[1] An interesting article by Paul Adams (1992) attempts to reconcile the place-based nature of television by focusing on its ability to constitute a 'gathering place' for social interaction. See also Robins (1996).

Figure 15.1 W H Simmons *Royal Sports on Hill and Loch* 1974. Engraving after Sir Edwin Landseer, 1850

of André Kertész. He uses Heidegger's (1962) concepts of *immersion-in-the-world*, and the distinctions between *readiness-to-hand* and *presence-to-hand*, to interpret the stories behind Kertész's photographs. Peter Jackson's (1992) work on Edward Curtis's pictorial portrayal of North American Indians is also noteworthy. Jackson's perspective differs from Seamon's phenomenology because he feels that the interpretation of pictorial texts involves political as well as aesthetic judgments. Jackson's method focuses upon Curtis's contrived images of cultural 'otherness', which he shows are replete with demonstrable falsehoods, misrepresentations and deceptions.

In *The Iconography of Landscape* (1988), Dennis Cosgrove and Stephen Daniels introduced several papers that not only addressed art and geography, but also their relations to dominant ideologies. For example, Trevor Pringle (1988) describes the appropriation of Scottish history and geography through the commissioned landscape paintings of Sir Edwin Landseer (Figure 15.1). By positioning Queen Victoria and members of her court in deliberate ways and painting them with particular demeanours, Landseer represents the loyal Scots nobility and a royal appropriation of the Highlands. Landseer's visual texts tell a story of loyalty and subjugation at a time when the English crown needed to consolidate its territorial possessions for imperialist expansion. We return to Pringle's (1988: 146) discussion of Landseer's paintings of Queen Victoria in the Highlands of Scotland as an example of semiotics later in the chapter, but it is worth noting here that the concept being signified is the 'royal imposition and appropriation of the Scottish landscape and Scottish history'. Similarly, Box 15.2 elaborates the creation of a mythic landscape designed to perpetuate imperialist spaces with representations of the American West.

In Boxes 15.2 to 15.5 we use examples of images from the American West to elaborate ways to get through and below visual images. The important point we want to make with each example is that they provide a site that can embody

BOX 15.2 Nineteenth-century representations of the American West

It may be argued that the nineteenth-century American West was a landscape waiting to be occupied by the imperialist forces of manifest destiny. Albert Bierstadt's representations of the American West (*Mt. Whitney* [1875, above], *The Yosemite Valley* [1868], *The Great Trees, Mariposa Grove* [1876], and *Giant Redwood Trees of California* [1874]) are problematical for their overt representation of the pristine (and unpopulated) nature of the West. As part of an imperialist dialectic, the paintings invite settlement and conquest, calling to would-be pioneers with their idealised wide-open spaces and unoccupied lands. Herman Melville (1857) of *Moby Dick* fame expressed the propagandist nature of this type of work: 'They look not only for more entertainment, but, at bottom, even for more reality than real life itself can show.' Lee Clark Mitchell (2000), discussing Bierstadt and his influence on the Western film genre, states more succinctly that '[p]art of what critics have persistently thought of as the "problem" of Bierstadt derives from his socially transgressive, metaphysically transcendental yearning – the unappeased craving.'

many different scales of spatial relations and meanings (cf. Elwood and Martin, 2000). All are capable of showing and revealing the complex nature of landscape. These geographical representations are viewed as cultural images that represent, structure, and symbolise our surroundings – the images become *places* that can be analysed to better understand lived experience. The images are important cultural signifiers that, when interpreted, reveal social attitudes and material processes where ideologies are transformed into concrete material forms (Duncan and Duncan, 1988).

Mitchell (2000: 47), noting the integration of the geographical image into the promotion of capitalist culture, argues that 'the obvious commodification of these cultural forms, their clear complicity in the maintenance of capitalism (after all, mass cultural forms circulated precisely by making money, first and foremost), and their singular ability to push out more local, less "mass" cultural forms, made artifacts of mass culture objects of ready suspicion'.

BOX 15.3

Twentieth-century representations of the American West

In the 1940s, Standard Oil of California commissioned some of the most prominent photographers in America to provide spectacular photographs of the American West. These photographs, including a number of works by Ansel Adams, were then incorporated into a series of promotional give-aways at Standard Oil service stations located throughout the Western states. An accompanying text (some penned by *Perry Mason* author Erle Stanley Gardner) described the wonders awaiting those willing to venture into the still relatively wide-open spaces of the West. Thus, Bierstadt's mythical landscapes of the nineteenth century in Box 15.2 are updated through the medium of photography – however, the same imperialist dialectic can nonetheless be located within the represented landscape.

Film and advertising

It was not until the early 1990s that geographers began to look at power and representations in film (Aitken, 1991; Natter and Jones, 1993; Rose, 1994; Benton, 1995). *Power, place, situation and spectacle* (Aitken and Zonn, 1994) is a compilation of work by geographers focusing exclusively on film landscapes, with the intent of looking specifically at the power of film and its geographies in contemporary culture. Interest in geography often revolves around the power of film to shape contemporary culture. Drawing on the work of Guilliana Bruno, David Harvey (1989, Chapter 9) suggests that Ridley Scott's *Blade Runner* and Wim Wenders's *Wings of Desire* are parables in which postmodern conflicts are set in a context of late capitalism with its flexible accumulation and time–space compression. To take another example, the appropriation of Scottish history and geography by English and American filmmakers provides a basis for showing how the films of Scottish director Bill Forsyth (*That Sinking Feeling, Gregory's Girl,*

BOX 15.4

The American West in twentieth-century film

Elaborating the importance of the medium of film, we can once again return to Monument Valley as a representational site created to enforce spaces of dominant ideologies. The western films of director John Ford, most notably the 'Cavalry Trilogy', consisting of *Fort Apache* (1948), *She Wore A Yellow Ribbon* (1949), and *Rio Grande* (1950) and starring John Wayne, form sequential narrative representations that invoke the righteous conquest of the West by the white patriarchal social order (cf. Coyne, 1997). The destiny of the true America is fulfilled (although not without noble sacrifice and loss of life) and the colonisation of the West becomes memorialised upon landscapes of new capitalist mythologies. Wayne (although he chose to sit on the sidelines during World War II) assumes the status of archetypal American icon and appropriates both meaning and geography through Ford's cinematic representations.

Local Hero, Comfort and Joy) subtly subverts and transcends dominant discourses of Scottish culture and politics (Aitken, 1991). How close to the real William Wallace did Mel Gibson get in *Braveheart*, and how does his representation of a brave but failed hero reflect on contemporary Scottish politics (see also Gold, 2002)?

The tradition of highlighting power, ideology and discourse continues with Cresswell and Dixon's (2002) edited collection of essays by geographers, which focuses on film as a mobile medium that elaborates identity in ways different from other media because it is not fixed (like photography and other art pieces). The focus of this collection is to construct an understanding of the relations between film and identity that is not tied to essential notions of either; 'films are no longer considered mere images or unmediated expressions of the mind, but rather are the temporary embodiment of social processes that continually construct and deconstruct the world as we know it' (Cresswell and Dixon, 2002: 3–4).

Perhaps one of the most influential forms of imaging to emerge in modern times is advertising. Henri Lefebvre (1984) notes that everyday life in late capitalist society is in a state of psychological terror because it is constantly under attack from advertisements and television commercials. These terrors include pressure to look a particular way or consume specific products. Under this barrage of signs, Lefebvre argues, we are often unable to articulate our feelings, and this hesitance leads to alienation and anxiety. In an apparent contradiction, advertising provides momentary gratification for the alienated spirit because it stimulates desire, but this then leads to us working harder and getting further alienated. Those who control advertising have learnt to fuse sexuality onto commodities and thus gain greater control of the conscious and subconscious aspects of our lives. Important questions revolve around the spatial ramification of this commodity fetishism. For example, Caroline Mills (1992) focuses on how postmodern fetishism sells houses and lifestyle landscapes while Peter Jackson (1991, 2001) elaborates the world endorsed by news-stand magazines that target male audiences.

LIVERPOOL
JOHN MOORES UNIVERSITY
AVRIL ROBARTS LRC
TITHEBARN STREET
LIVERPOOL L2 2ER

BOX 15.5

The American West in post-WWII advertising

In the post-WWII American West, thousands of servicemen, having passed through California on the way to the Pacific war, returned, families in tow, to provide the labour force for the new booming defence industries occupying the Fordist landscapes of Southern California. New enemies, the heathen (and thoroughly unChristian not to mention anti-capitalist) Communists from the Soviet Union, having the temerity to steal our atomic secrets, now threaten the very fabric of the American Way. The Southern California landscape is filled with newly constructed

Now you can protect precious lives with

An all-concrete blast-resistant house

Here's a house with all the advantages of any concrete house—PLUS protection from atomic blasts at minimum cost.

A firesafe, attractive, ***low-annual-cost*** house, it provides comfortable living—PLUS a refuge for your family in this atomic age.

The blast-resistant house design is based on principles learned at Hiroshima and Nagasaki and at Eniwetok and Yucca Flats. It has a reinforced concrete first floor and roof and reinforced concrete masonry walls. The walls, the floor and the roof are tied together securely with reinforcement to form a rigidly integrated house that the engineers calculate will resist blast pressures 40% closer to bursts than conventionally-built houses.

Anywhere in the concrete basement of the house would be much safer than above ground but a special shelter area has been provided in this basement to protect occupants from blast pressures expected at distances as close as 3,600 feet from ground zero of a bomb with an explosive force equivalent to 20,000 tons of TNT. This shelter area affords protection from radiation, fire and flying debris as well. And the same shelter area also can serve as a refuge from the lesser violence of tornadoes, hurricanes and earthquakes.

The safety features built into this blast-resistant house are estimated by the architect and engineer to raise the cost less than 10%.

Concrete always has been known for its remarkable strength and durability. That's why it can be used economically to build houses with a high degree of safety from atomic blasts.

Like all concrete structures, blast-resistant concrete houses are moderate in first cost, require little maintenance and give long years of service. The result is ***low-annual-cost*** shelter. Write for folder.

PORTLAND CEMENT ASSOCIATION
Dept. A6-9, 33 West Grand Avenue, Chicago 10, Illinois
A national organization to improve and extend the uses of portland cement and concrete through scientific research and engineering field work.

Interiors of a blast-resistant house have all the charm and livability of conventional houses.

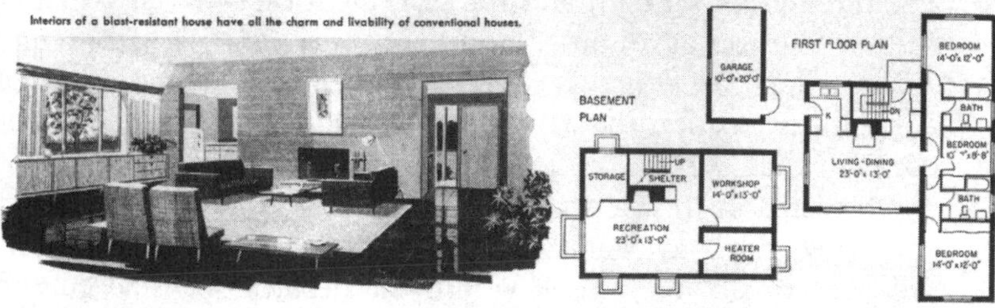

Portland Cement Association, 1955

LIVERPOOL
JOHN MOORES UNIVERSITY
AVRIL ROBARTS LRC
TITHEBARN STREET
LIVERPOOL L2 2ER

Box 15.5 continued

suburban housing developments containing houses that have been stylistically and symbolically linked to mythology of the American West (Ovnick, 1994). These *Ranch* homes embody the transitional process from frontier settings to urban settings and from the open spaces of the American West to the more confined settings of the American city. Containing a rustic 'den' area for our new Cold War foot soldiers and an 'all mod cons' kitchen designed to provide women (now returned to their previous place of employment after being unceremoniously 'let go' from their wartime jobs) with everything needed to maintain a happy (not to mention productive) homestead. The imperialist patriarchal mandate is restated in the 1950s and the conquest of the West is complete.

Visual representations of the landscape, according to Rosalyn Deutsche (1991: 18), are only rescued from idealist doctrines and seen as social if they are recognised, as other cultural objects have been, as representations. Neither autonomous nor social because they are produced by external society, representations are not objects at all but social relations, themselves productive of meaning and subjectivity. It is important for you, as a student, to recognise that the four types of images illustrated here – painting, photography, film and advertising – are more than just passive visual images, they are symbolic representations of a dominant order and organisation. When subjected to a visual analysis, it becomes clear that their production and consumption are socially mediated. But what kinds of visual analysis should be applied and under what circumstances? What we offer now is clarity on some visual methodologies that help make sense of images in our quest to reveal larger social and spatial concerns.

Making sense of images

Before choosing your method and beginning your analysis, you need to first read through and around the context of your image. It is important to know something about all aspects of the image that pique your interest. Knowledge of the painting in Box 15.1, for example, enabled us to contextualise *The Geographer* with other works by Vermeer that depicted women rather than men. Writing by art critics gave us a sense of how Vermeer portrayed his subject through use of light and the placement of artifacts. We also gained knowledge on Vermeer as an artist, and his connections with contemporaneous scientific practices. Other knowledge about his times and place – exploration, the enlightenment era and the 'golden age' of Dutch capitalism – enabled us to comment on what we see as the quiet, introspective arrogance of 'the geographer' depicted in the painting.

Secondly, to avoid what Rose (2001: 29) calls 'analytical incoherence', you need to situate yourself within the debates of visual culture and embrace the 'modalities' that you think are most important for interpreting your chosen image. In our boxed examples of the represented landscape of the American

LIVERPOOL
JOHN MOORES UNIVERSITY
AVRIL ROBARTS LRC
TITHEBARN STREET
LIVERPOOL L2 2ER

West, we choose to focus on a dialectic that elaborated imperial – white, male – domination. It is important to understand the theoretical position that you bring to your analysis and that the methods you choose carry with them political baggage that is also part of your analysis. There are many different ways of understanding visual imagery and different theoretical positions have different methodological implications, and oftentimes you need to be clear on your standpoint before embarking on a piece of research. That said, a theoretical standpoint often brings clarity to your methods and, ultimately, your image. What follows is a discussion of the use of semiotic, feminist, psychoanalytic and discourse analyses as visual methods. Again we use illustrations to highlight how these methods may be applied to your work.

Semiology: a method and theory of signs

Chapter 14 elaborates the textual basis of semiotic analysis and discusses its origins, but the technique is also widely used as a visual method. Part of the popularity of semiotics is attributable to its ability to take images apart and then focus on how they relate to larger systems of meaning. Semiotics can be used to interpret images by identifying signs and then analysing what meanings those signs construct, especially in terms of the construction of social difference. As the basis of semiological analysis, a 'sign' is 'everything that, on the grounds of previously established convention, can be taken as something standing for something else' (Eco, 1976: 16). 'Signification' is the social process whereby 'something' (the signifier) comes to stand for 'something else' (the signified), and by questioning this process, semiotics provides a popular way of getting behind and through images (Hopkins, 1994: 50). Roland Barthes (1972) argued that when things gain power as myths, the sequence of signifier, signified and sign is reified in a second-order semiological chain (see Figure 15.2).

Returning to our earlier example of Pringle's (1988) work, Figure 15.3 articulates the connections between Landseer's paintings of Queen Victoria in the Scottish Highlands and the construction of myth using a semiological chain. As Pringle (1988: 146) points out, the process is one of depoliticisation, the privation of Scottish history and geography, and establishing the apparent naturalness

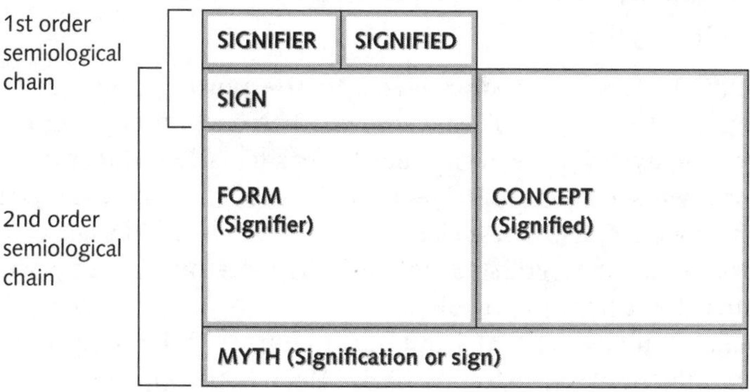

Figure 15.2 The nature of myth
Source: Barthes, 1972: 115

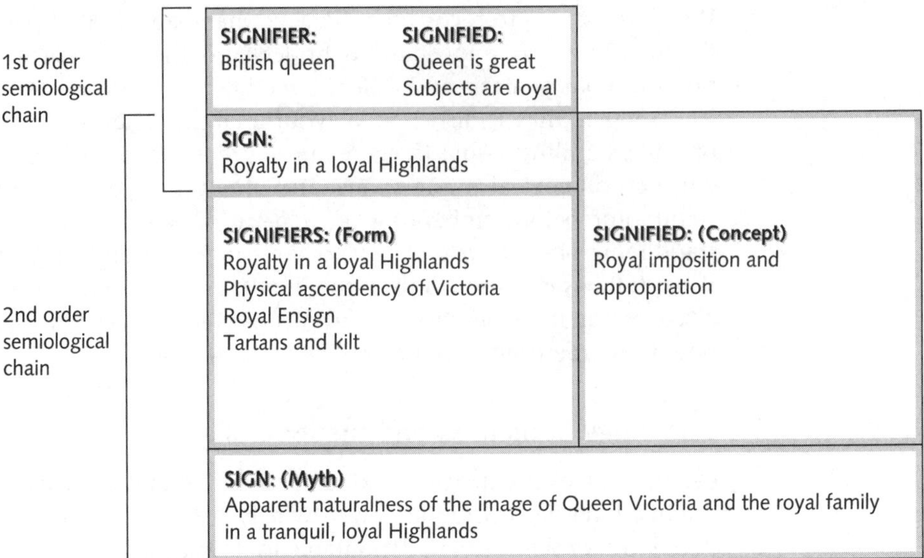

Figure 15.3 The Victorian Highland myth
Source: Pringle, 1988: 147

of royalty in the Scottish Highlands. The signification is the myth itself: the naturalness of the image of Queen Victoria and the royal family in a tranquil loyal Highlands.

Be wary of semiology's sophisticated theoretical terminology. The elaboration of second-order semiological chains, for example, often gets wrapped up in its own logic and neologisms when there may be simpler ways of stating your findings. Also, you should take care not to fall into the trap of making truth claims about what you find. Obviously there are other, perhaps more aesthetic, ways to evaluate Landseer's Scottish work that draw on his penchant for painting nature, particularly animals. Semiology does not articulate well diverse ways of seeing and knowing.

Psychoanalysis and feminist analysis: visual ideology and the male gaze

In Box 15.1, we introduced you to the power of feminist critique with our example of Vermeer's *The Geographer*. The encounter of feminist geography and psychoanalysis has been productive, particularly engagement with, and critique of, the work of Sigmund Freud and Jacques Lacan (Grosz, 1990; Bondi, 1999; Nast, 2000). Chapter 14 elaborates the ways that Lacan reread the linguistic/textual basis of Freud, so here we focus more on visual pleasure and patriarchal claims about image control.

Laura Mulvey's (1975) seminal work on visual ideology and the male gaze has been influential in geography (Rose, 1993, 2001; Aitken and Zonn, 1993). Mulvey problematises the viewer/reader as voyeur and the image as spectacle. She argues that voyeurism and the male gaze are part of the classic Hollywood script whereby 'the spectator looks, the camera looks, and the female character *is looked at*' (Saco, 1992: 28). In Freudian terms, women are never represented as *self* in these images, but rather as *other*, the dark continent, the love inspired in the hero, or the values from which the protagonist is trying to escape. Voyeurism involves a process whereby the male gaze seeks to exercise power over its subject by marking 'her' as the 'bearer of guilt' (Mulvey 1975: 11). Mulvey's militant stance against the male gaze offered one of the first feminist psychoanalytic considerations of the relations between readers/viewers, the producers of texts, and the narrative conventions of the text as set within the confines of a patriarchal discourse. Mulvey (1992) also notes the significance of space as a site of gender differences: a home or homestead as a signifier of stable space, femininity, and family, the scene of domestic space that is the location of narrative events. Opposed to this is the outside, the masculine space of adventure, action, and movement. It is through these codes that meanings are constructed not only through visual images but also through film's ability to control the dimensions of time and space through dynamic modes of cinematic production such as choice of shots, framing and editing.

More recent feminist geography focuses on how bodies are represented by suggesting a more complex framing of the gaze. In other words, mapping politics onto body space is affected by how the body is seen and recognised (Pile, 1996). Box 15.6, for example, takes one of Freud's early conjectures over men's concerns about emasculation and turns it, literally, on its head with a feminist reinterpretation. Bodies have become part of the social agenda of Western societies to the extent that Kristen Simonsen (2000: 7) calls them a 'cultural battlefield'. Their imaging in advertising and film is clearly at the forefront of this battle.

Discourse analysis

Recently, some geographers have distanced themselves from what may be called penetrative methods such as feminism and Marxism that try to look beneath

BOX 15.6

The Medusa's head

Freud (1955), in his famous essay entitled 'Medusa's Head', suggested that the terror of castration is linked to the sight of female genitalia and, similarly, 'the sight of the Medusa's head makes the spectator stiff with terror, turns him to stone'. At the same time, he argues, consolation is offered because 'the stiffening reassures him'. Teresa de Lauretis (1984), in her equally famous essay 'Oedipus Interruptus', asks the question 'how did Medusa feel?' She points out that when Medusa is looked at straight on she may not be deadly, but beautiful and laughing. Portrayed otherwise, de Lauretis suggests, the Medusa is 'death at work' not only for the male but also for the female because it contrives the conditions of vision and meaning production. The Medusa's Head, then, is how ways of knowing – in this case Freudian theory – would have us relate images to bodies. See Aitken (2001) for an elaboration of this example.

things to try, rather, to look beyond things. They explicitly reject classical Marxist claims that meaning may be reduced to our understanding of supra-organic and economic accounts of the system of production (see Chapter 14). Similarly, they are suspicious of classical feminist perspectives that reduce explanation to gender inequities. For some, discourse analysis – based on the writings of Michel Foucault – addresses these problems in other analytic frameworks because it looks specifically at the social consequences of difference through power in tandem with the construction of identity (the primary focus of feminism and psychoanalysis). Discourse refers to groups of statements and practices that structure the way we think about things. Rose (2001: 140) identifies two forms of discourse analysis as applied to visual images:

- Discourse analysis I pays attention to discourse as articulated through various kinds of visual images more than it does to the practices entailed in specific discursive formations and their productivity.
- Discourse analysis II attends to the practices of institutions rather than their portrayal through images. It focuses on issues of power, institutions and technologies.

Discursive practices – the ways meanings are connected through representations, texts and behaviours – become forms of disciplining and so discourses are also about power and knowledge production.

In this way it is possible to speak of a geographical discourse that refers to the special language of the discipline and how it is practised, with its focus on spatial relations, uneven development and the power of places. The discourse produces subjects (geographers, planners, environmentalists) and landscapes, but it also produces technologies like maps and GIS that enable particular ways to visualise the world (see Box 15.7).

BOX 15.7 Geovisualisation

Although geovisualisation is most often associated with GIS and cartographic applications, the basic principles of geovisualisation as a research methodology are equally applicable to uncovering and revealing geographic unknowns within the context of social theory. Geovisualisation, as defined by Buckley *et al.* (2000), focuses on visualisation as it relates to spatial data and its application to all stages of problem-solving in geographical analysis. DiBiase *et al.* (1992: 202) emphasise the importance of graphic representation in the realm of visual thinking by stating: 'Photographs and imagery, whose spatial dimensions correspond with those of the physical object being depicted, are more realistic than graphs, whose spatial dimensions represent nonspatial quantitative data or diagrams in which spatial relations are topological.' Geovisualisation can be used to determine and understand how social order is constructed. The combination of geovisualisation theory and conceptual thought with contemporary visualisation analysis creates a new and valid research process. The methodology may then be used to further explore spatial data and generate new geographic knowledge by expanding the domain of qualitative methodology (cf. Fairbairn *et al.*, 2001; MacEachren *et al.*, 1992).

Images can be deadly: a very brief case study combining methods

An important element of the visualisation of geographic space is the idea that the creator constructs a representation that is then seen and experienced by the viewer. Both participants add their social subjectivity to the interpretative act. Images of the landscape, for example, are communicated through visual signs and symbols thus making the representation a collaborative effort involving creator and viewer (Muzzio, 1996). Images are thought of as a spatial data set, organised into sequences that lead geographical exploration of the landscapes contained therein. It is a fictional landscape that is important in the production of geographical knowledge, particularly how an image presents a space disputed by opposing ideologies and identities (Said, 1993). As such, and to reiterate and underscore much of what we have been saying, qualitative visual analysis must be concerned with systems of meaning: especially how the landscape is viewed, experienced and created by those who populate it.

We close this chapter with a short analysis of a scene from the 1955 film *Kiss Me Deadly* to illustrate how visual methodologies are employed to interpret a landscape representation. The DVD version of the movie provides excellent contextual analysis of the movie.

Our semiotic analysis of *Kiss Me Deadly* focuses on the images contained within the film, using the sign to explore social conditions and social effects coded into the movie. Signs work together in relation to other signs and the signs contained in a film gain extra meaning because the film is an animated

BOX 15.8

DVD: an ideal spatial data set for analysing movies

The DVD format provides an additional feature that offers researchers new insights into the textual metaphors and symbols found within the narrative structures of a particular film. Many DVDs and nearly all of those packaged in what are termed 'Special edition' releases include a separate audio track containing a 'Director's commentary', which allows the film's director to present his own comments and reflections on all aspects of the film. Although these commentaries are hardly objective (however, many do find the director being critical of his own work – or, as is more often the case, bemoaning the heavy hand of studio interference), the added information can qualify or enhance interpretations of particular facets of a given film. There is an inevitable degree of repetition and the tone can change from serious to joking depending on the number of artists present but, generally, people involved in the production and creative process have consistently interesting things to say. In addition, the researcher can find the unique situation of having a novelist and the screen adapter of his work hold an extended conversation on the record about the results of their work. As in any analysis of qualitative data, the researcher must consider the supplemental information in its proper context by taking into account that commentaries are provided months, and in some cases years, after the initial creation of the cinematic narrative.

and dynamic sequence of moving images, creating repetitions of constructed meanings that enforce the film's ideology. Our feminist psychoanalytical critique of *Kiss Me Deadly* centres on how the unconscious of patriarchal society is structured in film form. Our discourse analysis focuses on larger societal institutions and values that frame the movie.

Kiss Me Deadly is noteworthy for its striking use of Los Angeles location photography to provide a visual design that employs great depth of field, a naturalistic lighting scheme and disturbingly realistic depictions of violence. In many ways, the film offers what Telotte (1992: 7) terms an 'excess of distortion' that creates a sense of disorientation in both the film's protagonist and the viewer. The cinematic production process is thus made far more visible giving the film its highly constructed, almost fantastic nature in its depiction of modern life. More than just a highly stylised *film noir* detective film in the tradition of works like *The Maltese Falcon* or *The Big Heat, Kiss Me Deadly* is an examination of the strange sort of reality that characterises this examination of Cold War American life.

In this example (at 25:20 to 27:57 in the film), the main character, Mike Hammer, walks down a dark city street and realises he is being followed. To see his stalker, Hammer pauses to buy popcorn from a street vendor and, while doing so, casually surveys his surroundings. Hammer is poised in centre frame with a gleaming neon wall clock behind his head in the centre background. The initial rhythmic sound of his own and his pursuer's footsteps, his dialogue with the vendor, and a classical pattern of matching shots suggest that we have seen a brief continuous action lasting about 30 seconds of screen time. However, when Hammer first stops, the clock behind his head reads 2:10; when the vendor hands him his popcorn, it says 2:17; and when he turns away and continues walking down the street, it is 2:21. The very prominence of the bright neon clock in this dimly lit scene and its central placement in the frame ensure that we note the inexplicable temporal gaps, suggesting both the very *unreality* of this scene and its purposely constructed nature. Because movement is part of the everyday experience of life, Hammer's movement through space and time increases our sense of objective reality (Metz, 1974; Cresswell and Dixon, 2002). The unreality of the narrative and the carefully constructed space contained in the scene enables suspension of the laws of nature: space is distorted, our cognitive mapping process breaks down and our attempt to reconstruct space creates a discomfiting reality.

Many scenes in the film take place at night and director Robert Aldrich, using deep-focus camera lenses, transforms the dark into a shadowy, disconcerting landscape, populated by killers intent on hunting each other down in a futile attempt to reassert some kind of masculine superiority. The darkness engulfs the world, obscuring the landscape in an array of shadows cast by solitary street lamps (or neon clocks) that disrupt the clarity of the screen and swallow up the figures walking through these dangerous spaces. This is a world of paranoia; a coherent space now rendered confused and chaotic. Hammer walks through the night, the seeming confidence of his forward motion contrasted with the undoing of this confident progress by a paranoid looking back. The temporal displacement produced by the clock and the almost cubist fragmenting of the frame into several planes create a space that is artificially and explicitly constructed to the extent that the scene is blatantly obvious in its attempt to intrude upon our standards of reality.

Visual methodologies provide the tools to interpret this peculiar reality. The viewer is confronted with a less conventional real world, a fantastic, imaginary and nightmarish vision of the American landscape. *Kiss Me Deadly* is one of many post-World War II films that use urban space as a metaphor for the angst and paranoia of the Cold War period, and resistance to a larger hegemonic discourse of the McCarthy era (Wilson, 1991). Not a Marxist, but sympathetic to the communist cause (a 'fellow traveller' in the words of the McCarthyites), Aldrich creates a Cold War allegory that takes aim at the politics of greed disguised as patriotic fervour. As in most *films noirs*, women play central roles in the film, creating the tension that propels the narrative to its apocalyptic end. Mike Hammer, unable to restore patriarchal order by punishing the women who strive to take away his power, becomes 'petulant, temperamental and uncertain' (Wilson, 1991: 138). Davis (2001: 40, 43) labels Hammer 'a serial killer with a detective's license' and a 'primordial material boy, a sleazy hustler interested only in Jaguars and expensive gimmicks'. The end of the film reduces him to cringing in panicked confusion, clutching at his loyal secretary's blouse (a woman he nonchalantly pimps out in an earlier scene) as his world explodes around him.

According to DiBiase *et al.* (1992), time is a vector that can be divided into orderly intervals, intuitively meeting the expectations of the viewer. However, by reordering the attribute of time, the viewer is presented with meaningful alternatives – new spatio-temporal data that emphasises the relationships among the attributes of symbolised cartographic features. *Kiss Me Deadly* disturbs and confounds our normal film experience by distorting and violating realism, calling attention to what passes for reality. The viewer must reinterpret this new reality because, according to Fairbairn *et al.* (2001: 9), the 'experience with reality is indirect' and relies 'upon the cartographic transformations implicit in the representation'. Thus, visualisation becomes a constructive process, an active and creative act that emphasises the viewer's awareness of a greater need to understand their own cultural reality. As this short example illustrates, by using visual methodologies – in this case psychoanalytic, feminist, semiotic and discourse methodologies – film can become a valuable medium for the exploration of landscapes and other forms of visual data. The interplay of social constructions upon the landscape through space and time is a common thematic element of film and these cinematic representations can provide a site for geographers to uncover and define meanings and identities. Importantly, geographic critiques can be directly applied to the landscapes, spaces, and places of cinema without having to resort to incorporating a film criticism that relegates geography to secondary status.

Some concluding cautions

Images are an important part of geographic study, and they have been studied for years. In this chapter, we pulled knowledge garnered from some of these studies to highlight visual methodologies as they apply to peculiarly geographic endeavours. By utilising visual methodology, geographers can form uniquely valuable understandings of the geospatial environment. We highlight methods that show how to unravel the conventions that structure what is visible and what is invisible, with the caveat that there are many ways to unravel and much will always remain obscure. We explore how the content of an image is elaborated by narrative conventions, and how those conventions draw on past representations and cultural norms.

The appeal of visual imagery to geography students resides with the ability of all types of visual images to use the physical world as a metaphorical space for the representation of spatial and social relationships. We have tried to illustrate in this chapter how images produce socially mediated metaphors and systems of metaphors about landscapes of power, and how these landscapes can be interrogated with visual methodologies. Various visual methodologies help us to understand how particular spaces are produced by manipulating images of the landscape and the built environment, but there is no precise relationship between landscapes and images. In addition, the relationship of images to narratives and texts (such as the one you are reading now) may be metaphorical at best: we do not really know. Nor do we know with any great certainty what images are seen, how they are seen and by whom. There is no precise relationship between images and language structures, so perhaps there is some arrogance in our suggesting that we know how their geographies are put together. That said, it is for us one of the most endearing wonders of visual imagery that it resists any kind of fixity and structured knowing.

BOX 15.9

Further reading

With the increasing interest in visual methodologies, a number of new and extremely useful books have appeared. Some are general in nature, others specific to a particular visual medium; some are highly theoretical and conceptual, others more empirical – however, these are the ones we feel should be at the top of everyone's reading list:

Tim Cresswell and Deborah Dixon's *Engaging film: geographies of mobility and identity* (2002) is, as of this writing, the best introduction to the use of film in geography – in particular, Scott Kirsch's chapter on spectacular violence in *Pulp Fiction* and the excellent introduction to the volume.

Picturing place: photography and the geographical imagination by Joan Schwartz and James Ryan (2003) is an extremely valuable introduction to the relationships between image and place. Derek Gregory contributes a worthwhile chapter on the production of space in Egypt while Deborah Chambers' essay on family as place is a fine example of qualitative methodology using visual material.

Box 15.9 continued

Kevin Robins' *Into the image: culture and politics in the field of vision* (1996) covers the complete visual range from photography to cinema to television to video to virtual realities with a special emphasis on the relation of image to the way we interact with the world. It has the advantage of being post-Gulf War I but, unfortunately, has not been revised to include the events of September 11, 2001. However, this is still a very important look into the *sociological* aspects of visual culture.

Gillian Rose's *Visual methodologies* (2001) provides perhaps the best introduction to the *interpretation* of visual materials. The reader is exposed to semiology, psychoanalysis and discourse analysis and their use in the exploration of visual representations. For any researcher making their first foray into visual methodologies, the Rose book should be a constant companion.

Having to do more with visual *culture* and its place in geography, Irit Rogoff's *Terra Infirma: geography's visual culture* (2000) explores issues of signification in geography, especially 'the complexity of contemporary art's engagement with problematics of place and identity'. Very useful for its post-colonial stance, the Rogoff book makes the link between the image and a sense of belonging and will satisfy those with a more postmodern bent.

And, as a way to begin to understand the connection between visual materials and *affection* (a term becoming more and more a part of visual methodologies), William E. Connolly's *Neuropolitics: thinking, culture, speed* (2002) is a most interesting starting point and a most valuable source for future research. While not particularly geographical, Connolly nonetheless succinctly places complex theories of culture directly into the realm of visual methodology.

Finally, the *Journal of Visual Culture* (a Sage publication) is a remarkable resource for all things visual. Geographers may be underrepresented in the contents but the journal is extremely interdisciplinary and geographers looking for insight into any type of visual material, theory, or concept would do well to consult *JOVC* on a regular basis.

16 Geographical information systems and spatial analysis

David Martin

Synopsis

Geographical information systems can be used to support many different research projects in which there is a requirement to process geographically referenced data, particularly where a wide range of data sources are involved. GIS are characterised here as a 'toolbox' of spatial processing functions. In order to determine whether GIS might be appropriate for your proposed project, it is necessary to understand the basic principles of GIS use, and to weigh up the major strengths and weaknesses of GIS-based research as applied to your specific problem. This chapter explains what kinds of tasks might be reasonably undertaken using GIS, and the range of software that is available, with examples of different types of project. There is a discussion of potential sources and problems with digital geographical data, and an overview of the types of manipulation and analysis that can be performed.

Introduction

This chapter considers ways in which geographical information systems (GIS) can be used to help carry out a research project. GIS are widely used computer systems for handling geographical data (simply, information which refers to specific places), which have seen enormous expansion since the mid-1980s, and are now a familiar part of geography degree programmes. There are many research situations in which GIS software provides most or all of the tools required, in a single computing environment. For this reason, it is probably more accurate to think of GIS as a 'toolbox' rather than a single tool. If your research project is likely to involve the handling and analysis of moderate to large quantities of geographical information, especially if it is already in digital (computer-readable) form, then you should consider using GIS. In general, GIS are better at data manipulation than analysis and it will often be necessary to use other software as well. GIS are particularly widely used in research projects which require the integration of geographical information from a number of different sources, for example in the evaluation of site suitability or the overlay of primary research data on existing information. The advantages are usually in terms of speed of processing diverse, large, or complex geographical data. The aim of the chapter is to help assess whether GIS is likely to be the correct tool for your proposed research, and to give guidance concerning the best places to find further information and the kind of questions to ask.

A very real challenge when conducting a research project that uses GIS is to maintain a correct sense of proportion with regard to the technology itself. Using GIS will not in itself turn a poor project into a good one, nor will it add any real authority to the results. There is a large and important area of activity, both commercial and academic, whose objective is to conduct research *into* GIS – the development of new software and applications, the evaluation of its organisational implications and the policy environment surrounding it. Many student research projects have examined aspects of GIS implementation and development, for example within local government or business. These can make good projects, but they will involve the use of research methods dealt with elsewhere in this book. In the present chapter we are concerned only with research that *uses* GIS in order to investigate something else, such as the changing pattern of retail provision in a neighbourhood, or some aspect of its demography. A frequently encountered application is one in which a number of different data sets are to be brought together to find the most suitable site for a new facility, such as a hospital, incinerator or superstore. A GIS will allow many different combinations of information in these cases, for example, the overlay of different land use or planning zones, the computation of distances from the proposed sites to areas of population, and the creation of corridors of specified widths around major routes. In this way alternative real world 'scenarios' can be explored. The golden rule is to perform each task because it is a logical requirement of your research, and not just because it happens to be available in the GIS software! Students using GIS for their research project must face up to the need for an understanding of both the substantive research problem and the GIS procedures involved.

The rest of this chapter is divided into five sections. The first of these aims to place GIS in context, both generally and in relation to student research projects. This is not an attempt to reproduce in one brief section all the topics found in a GIS textbook, but rather to help you make sensible decisions about the use of GIS for your research project. The following section gives some help with regard to working out exactly what functions are available to you and how you might take advantage of them. We then move on to a consideration of spatial data, which is a particularly important issue in GIS-based research. The last major section takes a look at some actual student projects and encourages you to begin thinking geographically so as to make the best use of GIS in answering your research questions. We end with a brief conclusion, which raises some wider issues about the appropriate use of GIS technology.

GIS in context

In this section, we consider what makes a system a GIS, and take a look at the kinds of operations that it might be expected to perform. Be assured that in order to use GIS in an applied research context, it is not necessary to understand the details of how the entire system works. It is, however, important to know if the chosen system will perform all the tasks required in addressing your research question. Much of the research literature assumes a level of GIS familiarity which is unlikely to have been achieved in undergraduate GIS courses and will not be of much help to you at this stage. There are now many excellent general GIS textbooks that you may already have encountered as course texts, and these can be a useful source of ideas. See, for example, Burrough and McDonnell (1998); Heywood *et al.* (1998); DeMers (2000) and Longley *et al.* (2001). A massive collection of papers on more specialised topics will be found in Longley *et al.* (1999), while basic concepts particularly aimed at human geography are covered in Martin (1996). If it is really necessary to find out more about the algorithms and data structures inside the software, Laurini and Thompson (1992) is a good starting point, but contains detail which should not be necessary for the average geography student. As discussed below, there is a very wide range of software marketed as 'GIS' and the final choice will depend on what is available in your own institution, but some general principles can be given. It is worth finding out exactly which systems are available to you, and then casting a critical eye over the software and accompanying documentation before committing yourself to any particular course of action. Many of the GIS teaching materials available are designed to give you an introduction to the entire range of functions available within a particular software system. The more common systems are now well represented in the computer book market, and volumes such as Booth and Mitchell (2001) on ArcGIS or Daniel *et al.* (2001) on MapInfo may be particularly useful to you if you know that all your work will be conducted within those specific systems. Note that there are frequent revisions to these software-specific texts to accompany new versions of such software. Your project is likely to require detailed use of a few functions, while much of the system is never used, so you must be prepared to use the documentation selectively. As with all projects, it helps enormously if you set out with a clear research plan. Staff involved in teaching GIS will be much more receptive to an intelligent question such as 'do we have software which could be applied to this problem?' rather than a desperate appeal for help when it is already too late!

As already suggested, a GIS may be thought of as a toolbox of procedures that operate on geographical data. The toolbox nature of many systems means that it may be possible for you to create new functions by building sets of primitive instructions in some kind of macro or programming language. A wide range of operations, including essential data input, manipulation and display functions will be readily accessible, even to the inexperienced user, but construction of your own macros is not to be undertaken lightly without some prior programming experience. As discussed below, it may be necessary to use other software if more complex statistical analyses are required. In this case, it is advisable to find out how easy it is to transfer data between the two systems at an early stage.

Many authors characterise GIS as systems for the input, storage, manipulation and output of geographical information (Longley and Clarke, 1995a and Figure 16.1). There is considerable variety between systems in terms of exactly what

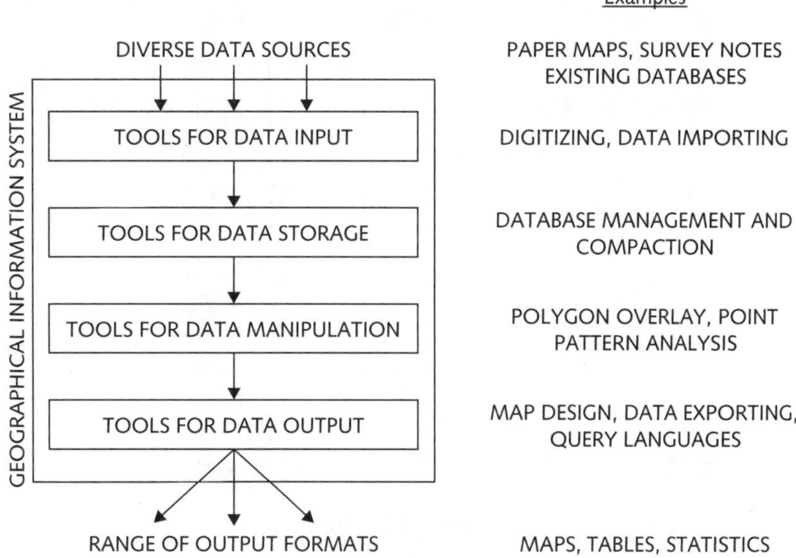

Figure 16.1 Typical components in the GIS 'toolkit'

tools are provided for each of these tasks. This means that you can use a GIS to handle pretty much all kinds of information for which you might think of producing a map. The geographical referencing of information is a fundamental feature of GIS, and is one of the things that distinguishes them from other kinds of computer-based information systems. Data issues are considered more fully below, but you will need fairly precise geographical locations in order to realise the benefits of GIS use.

Data input covers a variety of methods for getting geographical information into the computer. The most common data source was once paper mapping, although modern surveying instruments are capable of saving measurements in digital form, and global positioning system (GPS) receivers now offer a cheap and accessible method of recording your study locations for some types of application. There are sophisticated scanning and line-following devices available which permit the semi-automated capture of information from paper maps, but by far the most widely used input method is manual digitising, which is discussed below in the context of data sources. Many student projects will be able to take advantage of pre-existing digital data and will require very little direct data input, although access arrangements will vary and data sources are discussed in more detail below.

In commercial use, one of the most important roles of GIS is as an organising structure for extremely large quantities of operational information. These applications tend to require compact data storage and rapid retrieval, but do not place particularly heavy analytical or statistical demands on the system. For this reason, much of the development effort in GIS (and the corresponding literature) has been directed towards ways of holding and querying geographical data more efficiently. These discussions are rarely of much relevance in the

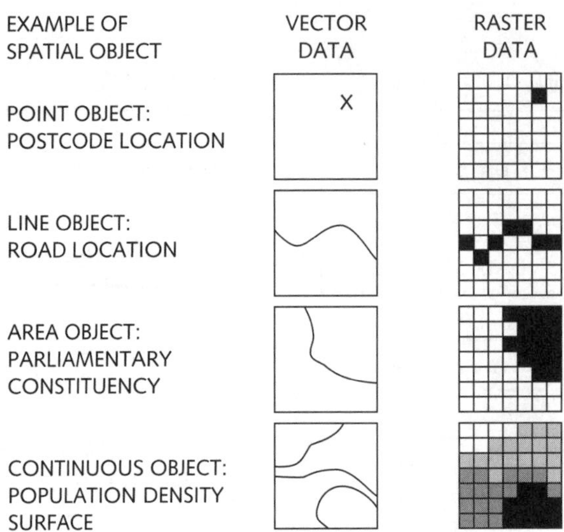

Figure 16.2 Vector and raster data structures

context of an individual research project. A commonly encountered division among strategies for data organisation in GIS systems is between vector and raster models (Figure 16.2), although these are not the only possibilities. In vector systems, location is recorded by using map coordinates, while in raster systems it is by row and column position in a grid. The most important differences are that vector structures usually include some kind of topological information (what is connected to what), and store geographical features (such as streets) individually. The attributes of geographical features (e.g. populations of zones, names of streets, sizes of retail outlets) are often stored separately from their locations, and held in tabular form. Raster systems are based on entire gridded layers (e.g. land use, population density); they do not include topology and there is no separate attribute information. Individual software systems may use vector, raster or both.

Probably the most important aspect of GIS use for your project will be the third component, which concerns the manipulation and analysis of geographical data. It is the presence of these functions that, for many people, define whether a piece of software is a GIS at all. It is worth noting that you may find your GIS system surprisingly lacking in basic statistical functionality, and you should check carefully at the outset that it will actually perform the tasks that you think you will need. Finally, you will probably want to output information from the system for presentation in a final report, either as maps or as non-graphical products such as tables of results, or the answers to specific queries. All the major GIS software includes tools for creating cartographic output and in some cases other statistical graphics of a standard appropriate for research projects. It should now be clear that GIS is not just computer mapping, nor is it necessarily the best way to produce maps with a computer. If your objective is primarily the display of geographical information, be sure to also consult Chapter 19.

What systems are available?

As with all the advice offered here, you will need to investigate the exact situation in your own institution, but some general indication can be given of what to expect from the available GIS software. Although there are many disparate software products on the market that claim to be GIS, most universities and colleges will provide you with access to at least one of the major systems. It is quite possible to achieve good results with the level of knowledge imparted in most introductory GIS courses, although those with some programming experience will be able to tackle more complex problems. Each of the proprietary systems use their own internal data storage formats, but usually provide tools for the import and export of data, either in commonly-used formats such as those used by ArcView or MapInfo, in generic data exchange formats such as GML or as plain text files: these abilities become important if you need to use existing digital data from a range of sources.

Many of the early distinctions between GIS systems have now disappeared but it is still important to recognise that not all GIS software will be able to do the same things! For example ArcInfo, which is widespread in UK higher education institutions, offers an enormous variety of functions and even a choice of interfaces but can be challenging for the novice user, despite its standard Windows interface. ArcView (also from the ArcGIS product range) is widely used in GIS teaching, but does not offer the same breadth of spatial manipulation functions. Other systems frequently encountered include the primarily vector-based MapInfo and the primarily raster IDRISI. Whatever software is available at your institution, check very carefully that the functions you need are actually available. If there are introductory GIS courses being offered as part of your degree programme it is strongly advisable to use the software on which those courses are based, as these will be the systems for which your local instructors and IT staff will have the greatest expertise and readily available training materials. Finally, it is worth noting that some statistical and spreadsheet packages are now being marketed with built-in mapping modules. If the emphasis of the work is in the statistical manipulation of the data rather than explicitly geographical analysis, these are worth considering.

The importance of data

Data availability will be a very major consideration in any research project using GIS. Walford (2002) addresses geographical data characteristics and sources in an international context, and in more depth than is possible here. Data comprise the largest single component of the financial costs associated with commercial GIS, and there is a whole industry concerned solely with digital data creation and supply. This commercial environment is important even to your project, because it determines the setting in which data are created and exchanged. It will most often be necessary to use existing digital data, and as no funding is usually available to you for data purchases, it is particularly important

Table 16.1 Potential sources of geographical data

Geographical data search resources	
http://www.gigateway.org.uk	A gateway site for searching for UK geographical information
http://www.geographynetwork.com	Resource for finding geographical data, including links to free data – particularly North American and world coverages
UK data available 'free' to academic users through national purchasing agreements	
http://www.census.ac.uk	ESRC/JISC 2001 Census Programme, providing links to registration system, data, boundaries, directories
http://convert.mimas.ac.uk	UK Postcode look-up tables
http://www.mimas.ac.uk/maps/barts	Bartholomew topographic map data
http://edina.ac.uk/digimap	Ordnance Survey Digimap products (currently Land-Line Plus; Meridian; 1:50000 Colour Raster; Land-Form PANORAMA; Strategi; Placename Gazetteer)
http://edina.ac.uk/ukborders	UKBORDERS range of census and administrative boundaries, including historical data
Census and neighbourhood data available free online to all users	
http://www.national-statistics.gov.uk	UK Office for National Statistics site, includes Neighbourhood Statistics and census key statistics
http://www.census.gov	US Census data and TIGER boundary files
Online mapping systems with international coverage	
http://www.multimap.com	
http://www.mapblast.com	
http://www.mapquest.com	
Visualisation tools	
http://www.mimas.ac.uk/descartes/	Descartes mapping software enabled with UK Census data

to know what data are available to you. Table 16.1 contains some suggestions regarding general purpose GIS data sources, more details of which are given below, and you should also refer to Chapter 5 for further discussion of secondary data sources.

Data input by digitising is suitable only where high-quality paper mapping is available as a source. This requires the use of a digitising table – rather like a drafting table, on which the map is fixed and its position registered – connected to a computer running GIS software. The required spatial information is traced off the map by the operator who moves a cursor over the map and presses a button to record each significant point. Most GIS products include modules for the control of a digitising table, and store the captured coordinates directly in the GIS database. Manual digitising can be time-consuming and error-prone, but is a necessary stage if you have paper maps containing information for your study (for example: school catchments, land-use zones, the route of a proposed road, locations of facilities). For most student projects, digitising is practical only for relatively small quantities of data. If it seems likely that you will need to input your own data in this way, remember to check at an early stage on the local arrangements for access to a digitising table. While GIS software may be universally available on your campus computing network, there may only be one or two digitising tables available, and they may be hidden away in departmental research laboratories rather than forming part of central computing provision. You should also note that much mapped material is copyright and digitising without prior permission is considered akin to illegal photocopying of any other publication. These issues are discussed more fully in Chapter 18.

An option worth considering for primary data collection is to use a GPS receiver, especially if working in an area without good map coverage. The network of navigational satellites now offer continuous grid coordinate readout typically of 5–15 metre accuracy to users of cheap hand-held receivers, and use is free apart from the batteries! Precision surveying really requires the use of a differential GPS system, in which both mobile and base stations are used to bring the locational information down to sub-metre accuracy, but this is unlikely to be a requirement of human geography research projects. Many geography departments will have hand-held receivers in their field equipment store, and these may be available for student project work. Given their low cost, you may also find that friends who are keen on hillwalking may have their own and be prepared to let you borrow them. Although designed primarily as a navigational tool, such devices provide an entirely adequate way of recording spatial locations as part of a field survey, and this may be of particular utility to projects concerned with environmental management or development issues, where there are no fixed features such as buildings at your study locations. Even the cheapest receivers offer the capacity to store each location with simple attributes, and many have associated PC kits that allow you to upload your information directly to a PC for input to GIS and mapping software. As with all the other advice offered here, only go down this route if there are good reasons for doing so. GPS simply offers an additional location-finding tool for that small proportion of projects where satisfactory secondary sources are not available, and where reading locations off a map is problematic. Available texts do not tend to focus on the needs of the human geographer, but do provide helpful background material (see for example, Kennedy, 2002).

In many situations the most appropriate approach will be to georeference (assign specific locations to) research data by use of an existing locational data set. Many data types relevant to human geography can be georeferenced by use of census zones, postal geography, streets or addresses. If the objects of the study can be associated with one of these objects, it should be possible to find an existing digital data set that contains grid references for each object. For example, if you have conducted a postal survey in the UK and want to examine in detail the locational characteristics of the responses, directories of census geography and postcodes created as part of the 1991 and 2001 censuses provide a grid reference and census zone identifiers for each postcode in the country. There is also UK look-up table software that allows you to convert your data between the different georeferencing systems (see Table 16.1). UK postal geography is explained fully in Raper *et al.* (1992) while the most comprehensive collection of reference material on the census data sets up to 2001 will be found in Rees *et al.* (2002). Where small numbers of address locations are involved, various web mapping sites will generate maps, including grid references which could be recorded and used in your study, in response to postal codes, other components of street addresses and place names. These have the advantage (for some research projects) that they generally provide international cover, albeit at varying spatial resolutions. Again, examples are cited in Table 16.1.

Important geographical data such as census boundaries, postcode locations and general topographic information are created by different organisations with some particular interest in each product. In the UK, many of these are available to the academic community because of national purchasing agreements for educational use. Students at subscribing institutions should usually be able to gain access to these data by making initial contact with their local computing support staff. Many of these sources are nationally accessible through networked data services once you have registered to use the data. Registration for UK census data is via a central online service. Once registered, users can obtain data from a number of Census Data Support Units including MIMAS and UKBORDERS, both of which include digital boundary data and other products of use in GIS projects. Further, locationally-referenced information, including key statistics from the census, is also available as part of the government's Neighbourhood Statistics Service. Ordnance Survey, Britain's national mapping agency, produces a wide range of digital map products at different notional mapping scales and a range of important Ordnance Survey data sets is available through the Digimap initiative, including postcode centroids and boundaries, medium-scale and large-scale topographic mapping, a place-name gazetteer and digital terrain model (DTM) data. The Digimap portfolio is subject to continual development and its website (Table 16.1) should be checked for the latest specifications. Digimap provides a facility for online production of simple maps that you may wish to use directly in your project, as well as the ability to download digital geographical data for use in GIS. UKBORDERS contains a range of zone boundary data relevant to human geography projects, and the Bartholomew digital map data held at MIMAS provides an alternative to Digimap for medium-scale topographic map data. The GIGateway service represents one attempt to catalogue the digital geographical information available in the UK, although access arrangements for students are not always the same as those for commercial users, particularly in the case of the national data collections noted here.

In the USA, various tabulations of census results, including TIGER (Topologically Integrated Geographic Encoding and Referencing) digital boundary data are freely available via the web. Much US topographic information has been digitised by the United States Geological Survey (USGS), and a useful listing of other US data providers is given as Appendix B in Longley and Clarke (1995b). There is an increasing range of geographical data (and even GIS software!) available free over the Internet, and some of the resource sites are given in Table 16.1. In addition to external data sources, most geography departments will have their own data collections which are available for project use, and which would be particularly expensive to obtain directly. Such data holdings frequently include locally digitised data and remotely sensed images purchased for other research. For most projects in human geography, such images will not provide the most appropriate data sources, but they can yield valuable information on land-use patterns and may be worth considering in this context. Those wishing

BOX 16.1

Potential sources of error in geographical data sets

- Age of data (out of date, incompatibilities between different layers)
- Differing definitions of objects (e.g. coast defined as high or low water marks)
- Quality of source document (poor-quality source causing geometric distortion)
- Coverage of data (not all data layers cover the whole study area)
- Locational recording errors (e.g. poor digitising, duplicated or missing lines, poor matching at edges of source map sheets)
- Attribute recording errors (e.g. wrongly labelled zones, census under-enumeration, missing values)

to use remotely sensed data should refer to some of the introductory texts in that field such as Mather (1999) or Campbell (2002), as this kind of data usually requires considerable preparation work. It is worth mentioning that some commercial and local government organisations may be prepared to make available small quantities of their own digital data for specific projects, although you should not make any prior assumptions that data will be available from this source. Whatever your data sources, it is particularly important that you make appropriate acknowledgement in your final report: many data sets are supplied with specific licence conditions and the use of a specific acknowledgement or copyright statement is a frequent requirement.

A very important set of issues relating to digital data are those concerning data quality. This involves more than just the accuracy of grid references. Particular problems arise when different data sources are brought into the same system and used together in analysis, as it is extremely difficult to trace the likely propagation of errors and inaccuracies from the source data through subsequent operations. Box 16.1 provides a checklist of potential sources of error, and the likely effects of these should be considered carefully in relation to your own project. In most cases, it is not possible to alter the quality of input data, but appropriate account should be taken when conducting analysis and interpreting results. There is a helpful chapter dealing with different aspects of data quality in Burrough and McDonnell (1998).

Thinking geographically: GIS research projects

In this section, we shall explore how GIS can be used in a research project context by taking a look at three undergraduate geography projects submitted at the University of Southampton during the last few years. Brief synopses of these projects will be found in Boxes 16.2 to 16.4, and it will be useful to refer to these alongside our consideration of the ways in which they used GIS. Stuart's project was concerned with the impact of noise pollution that had resulted from a change to the flight path of aircraft using Southampton Airport (Gibson, 1999); Neil examined the environmental conditions that would allow successful grapevine cultivation in the UK (Kaye, 1997), and Jenny explored the effects of the

BOX 16.2

Noise contour mapping around an international airport

Stuart discovered that aircraft flight paths around Southampton Airport had recently been re-routed in an attempt to moderate the impact of aircraft noise on local residents. He found planning documents which included a professional study including noise contour maps for the locality, and his basic research question was to evaluate the extent to which noise exposure was actually affected by the flight path pattern.

The literature review revealed that most official figures for noise around airports are based on computer simulation of the noise characteristics of different types of aircraft, and that there are many different ways of measuring and expressing exposure to environmental noise. Stuart also discovered that there is an extensive range of statistical tools for the interpolation of data values between geographical sample points. Noise recording equipment was borrowed from another department within the university and 39 sample locations each surveyed for at least 45 minutes and 5 aircraft to establish background and average maximum noise levels. A control site was measured at different days and times in order to understand the amount of variability that might be present in the whole data set. Meteorological reports were obtained and associated with each of these measurements, allowing the impact of weather conditions to be considered. The study data were entered into Arc/Info as point locations with associated attributes, and background mapping extracted from the Bartholomew digital data set. The pattern of maximum noise was then interpolated using Kriging within Arc/Info, which creates a noise surface by distance weighting, the weights being chosen by analysis of the data itself. Lattices of interpolated values of varying resolution were created and these converted into contour lines specified in terms of decibel levels and presented against a simplified background map.

In discussion, Stuart considered carefully the limitations of his study in terms of the number of sample points used, the impact of lattice resolution and parameters to the Kriging algorithm, and the difficulties in comparing the work with other studies. He noted that working with smaller cell sizes produced visually more appealing contour maps, but with increases in residual error. Nevertheless, his maps clearly reflect the change in flight paths. His study produced very similar noise levels to those appearing in the published noise contour maps, and confirmed the value of the change.

socio-economic characteristics of school catchment areas on pupil performance (Turnbull, 2001). Clearly, no discussion based on only three projects can hope to be comprehensive in terms of the types of manipulation and analysis operations that might be required, but they do provide us with a framework for identification and consideration of the main stages that are likely to be encountered in any research project involving GIS, and they may also prove helpful in turning your initial research question into a sequence of GIS operations.

The first thing to recognise about these projects is that none of them were explicitly *about* GIS and, further, each of them concern topics which could have been addressed in some respects without using GIS. A common characteristic is

BOX 16.3

Climatic warming and UK grapevine cultivation

Neil was interested in the fact that at present commercial grapevine cultivation is only really possible in a few locations in the south of Britain, but that with future climate change, many more locations might become suitable with the northward shifting of growing regions. The study involved both regional (England and Wales) and local aspects, aiming to understand the spatial characteristics of vineyard location in England and Wales, to consider climatic factors in explaining this distribution and then to use this information to produce a suitability map for grapevine cultivation at the local scale. Background literature included a review of approaches to climatic modelling and a consideration of the climatic conditions required by different types of vine.

A regional study of climate was undertaken using a general circulation model (GCM) using software provided by a member of physical geography staff in the department, and the results were transferred to Excel in order to calculate a measure known as latitude temperature index (LTI), devised specifically for assessing grapevine suitability. The relationship between LTI and density of grapevines was modelled and mapped using a combination of software packages, including Excel and Minitab for multivariate regression analysis. LTI was computed for two different temperature-increase scenarios and the possible northward migration of two categories of vine predicted.

A separate study of South Hampshire and the Isle of Wight was undertaken, as this was one of the most suitable areas revealed by the regional analysis. The most suitable sites were considered to be those on south facing slopes below 100 metres with light sandy soils, outside major urban and forest areas. A significant amount of digitising was undertaken using SPANS GIS software in order to capture all the necessary information, and the data were then rasterised and manipulated in IDRISI to find the sites best meeting the criteria suggested. Overlay with the existing vineyard locations suggested a very strong relationship, confirming the utility of the site suitability model. Neil's discussion was able to comment both on the interaction between regional climatic and local topographic factors affecting vine cultivation, and on the likely effects of climate change.

that they each deal with research questions, the careful investigation of which involves the handling of geographically referenced data. Each dissertation thus began with the students becoming interested in a fundamentally spatial question – i.e. they started *thinking geographically*. How has the geographical distribution of noise changed over time and how localised is it? Why are vineyards found where they are, and if the climate is getting warmer, how might the pattern change? What are the effective catchment areas for schools, and how does this impact on their pupils' performance? At an early stage, each project involved inspection of a map: visualisation is a very important aid to exploratory data analysis and you should be sure to allow yourself time to simply study maps of your data and to question what you see. Are there overall trends? Are there outliers? Does the pattern generally conform to your expectations? For this reason, it is worth considering the use of exploratory mapping tools

BOX 16.4

Primary school performance and socio-economic characteristics

Jenny became interested in the extent to which the socio-economic characteristics of a primary school's catchment area would affect its overall pupil performance. She discovered a substantial literature on the role of individual and neighbourhood factors in pupil performance, and a number of methodological papers on the ways in which deprivation could be assessed using census data, and the most appropriate ways to define catchment areas.

Having chosen a study area with which she was familiar, Jenny sought the help of the local education authority and was able to view a series of dot maps that showed the residential locations of pupils associated with each school. For reasons of confidentiality, these data could not be used, but she was able to aggregate the numbers of dots associated with each school within each census enumeration district (ED) in the study area. This procedure embodied some known difficulties, such as the inability to identify situations where more than one pupil resided at the same location, but the use of proportions rather than absolute numbers seemed an appropriate way of overcoming this inevitable information loss. Further literature review identified several multivariate indicators based on census data that have been used to indicate levels of deprivation, and one of these, known as the Townsend score, has been used in several studies of school performance. However, Jennifer's review led her to consider that an alternative measure, known as the Breadline Britain Index, would be more appropriate for her purposes. Census data from 1991 and associated ED boundaries were downloaded from the MIMAS service and the Breadline Britain Index computed for the study area. A series of relatively simple analyses were undertaken to explore the relationships between the index and its component variables and overall school performance. In order to gain additional insights for her interpretation of the data, Jennifer discussed her maps and results with teachers in the study area and was able to learn about local population changes and school admissions practices that were not evident in the data used for her mapping.

such as Descartes (see Table 16.1) that provide some powerful alternative ways of visualising pattern in human geography data, and can help carry your thinking much further than conventional shaded-area maps. Each of these projects also involved the students getting to grips with technical and analytical issues that they had not previously encountered, but in none of the examples was this the main focus of the project.

Inevitably, when beginning a project that uses GIS to investigate a substantive research question, it will be necessary to undertake background research that falls in more than one domain. Typically, it will be necessary to find the academic literature that deals with the research question, but also that part of the GIS and spatial analysis literature that addresses the particular techniques to be used. In some cases this may be readily obtained from the standard GIS textbooks, but it will often also involve tracking down papers that deal with the more specialised techniques relevant to the research. Stuart, for example, found that the measurement and modelling of noise contours is itself a sub-area of the literature in which there are many alternative approaches. The airport study involved the use of local planning documents and academic journals on noise measurement, and also GIS literature concerned with the interpolation of values between sample locations to create continuous surface models. Neil dealt both with a general literature on climate modelling and a more specialised one in relation to the conditions for grapevine cultivation. Jenny found an extensive literature dealing with the factors affecting school performance in league tables, debating the role of socio-economic conditions, the scales at which they operated and the best ways to measure them. Her analytical work was more concerned with the construction of deprivation indicators than with complex GIS processing.

Strictly, only Stuart's project involved primary collection of geographical data. In the case of the noise data, this was a major undertaking, involving the borrowing of noise recording equipment from another department in the university and a considerable time spent in the field taking measurements. Neil collated vineyard locations and supporting information on local conditions for digitising. Jenny followed up her secondary data by some investigative work of her own, involving informal interviews with education staff in order to verify the suitability and accuracy of the data she was using. All three projects used secondary data, and it is actually very unusual to encounter a GIS research project that involves only self-generated data. For Stuart and Neil's projects, the secondary data sources related primarily to the physical environment and were therefore drawn from existing sources of topographic (Bartholomew) data, while Jenny's project was more concerned with existing data drawn from administrative records and the 1991 census (obtained from MIMAS). Neil and Jenny's projects both involved the identification and georeferencing of data about vineyards and school catchment areas that already existed but were not available in GIS form. The vineyard data were obtained from published directories and the Internet, while the distribution of pupils was mapped with the help of the local education authority.

Although the school project did not involve the use of postcoded data, that would have been a common scenario in a project of this type. Data sets that relate to survey respondents, customers, patients etc. are very frequently referenced, if not by a full address, by a postcode. Look-up tables are available for the association of census and administrative area codes or grid references with postcoded records, as discussed in the preceding section. The result of adding grid references to a postcoded list would have been very similar in that it would have created a set of point-referenced records containing information on the characteristics of the individuals in the study.

These students chose to use different GIS and analysis software. The noise contour study required interpolation tools that were most readily available in ArcInfo. The vineyard study required climate modelling which was undertaken in specialist software, combined with map overlay and presentation that was conducted using IDRISI, although SPANS GIS was also used because it was the locally available software connected to a digitising table. Jenny's use of GIS primarily involved data linkage and thematic mapping, and was again conducted using IDRISI. IDRISI was chosen because it had been used in the introductory

GIS course that they had taken, and it provided all the standard functions that they required without the need to learn an entirely new system.

Whatever manipulation and analysis are performed, it is very important that they should relate back to the original research question, avoiding the temptation to use techniques simply because they are there or because they result in attractive maps. Stuart's data manipulation consisted primarily in interpolation of noise contours from his survey data, combined with ancillary topographic information. Kriging was chosen as the most appropriate interpolation approach in this context. Interpolation is required in most situations where we have sample locations representing values of phenomena which are actually present at all locations, and it is necessary to construct an appropriate representation – most usually as a grid model, contour map or triangulated irregular network (TIN). His final analysis was essentially a visual comparison between his new maps and those found in local planning documents.

Neil's regional study was primarily concerned with multivariate modelling of climatic variables, with mapping simply for visualisation of the results. His local study, however, was based on a fairly extensive series of raster map overlay and manipulation and is typical of many site suitability analyses based on cartographic modelling with multiple input layers. DeMers (2002) is particularly helpful as a guide to thinking through and setting out your grid-based cartographic modelling scenario, and the importance of planning before doing cannot be overemphasised.

Jenny's main tasks involved calculation of deprivation indicators using the census data she had retrieved via the web. The original point data relating to pupils were aggregated to give proportions of each school's pupils in each census enumeration district (ED) before being entered to the GIS and all analysis was then at the ED level. Working with data aggregated to areas in this way requires an appreciation of something that geographers call the modifiable areal unit problem (Openshaw, 1984), which is basically a recognition that the observed data, and relationships between them, are highly dependent on the zone boundaries chosen. They could be expected to change if larger or smaller zones were used, or with a different set of zones at the same scale.

If point data had been available with individual grid references, alternative approaches might have included point pattern analysis of the pupils themselves, automated aggregation to areas (either by overlay with the EDs in the GIS or use of look-up table software) or generation of school catchment area boundaries by the creation of Thiessen polygons around known point locations. Further, analysis of data for two incompatible sets of areal units will often involve some form of areal interpolation (Flowerdew and Green, 1991). Each of these approaches might have been used in consideration of the same general research question, but the detailed methods used must be adapted to the data available in the specific case. If point data are to be evaluated in relation to a background population, it is a straightforward task to compute the rates of point events relative to zonal populations. In their crude form, however, such rates tell us little about the underlying pattern. Mortality and morbidity rates, for example, are conventionally standardised by calculating the rate in the entire population, dividing the observed number of cases in each zone by the number that we would expect to find in a population of that size. Unfortunately, there is still the problem that large variations in observed rates for zones with small populations may not be significant, particularly where the number of cases is also small, and there is a significant literature concerned with these issues (Bailey and Gatrell, 1995).

Both Neil and Jenny used other software in addition to their main GIS package. Although most GIS software will provide some basic descriptive statistical information and may offer simple regression analysis, it remains the case that the statistical capability of GIS software is generally limited. Both used a spreadsheet as the most convenient way of undertaking recalculation of multiple variables and the creation of specialist indicators derived from the literature surrounding their particular studies. Neil also used statistical software in order to undertake a multivariate regression analysis. The results of all these operations were then reimported to the GIS for mapping and consideration of spatial pattern in the results.

In considering the role which GIS can play in more complex data analysis, it is helpful to distinguish between what Fotheringham (Chapter 11, this volume) terms 'weakly' and 'strongly' geographical research using spatial data. Weakly geographical analysis uses spatially referenced data, but fails to make much use of the spatial component. In theory, at least, GIS should provide a framework within which it is possible to conduct strongly geographical analysis, utilising explicitly geographical concepts such as clustering, dispersion, distance, contiguity, adjacency, etc. For those wishing to undertake more advanced spatial analysis making use of these concepts, Bailey and Gatrell (1995) and Fotheringham *et al.* (2000) are particularly recommended. In order to assess the most appropriate analytical techniques for your specific project after reading this chapter, there is no real alternative to careful reference to more specialised texts such as those identified here. The data manipulation and query tools available within GIS are able to provide many of the input measures required by more specialised spatial statistics such as proximity matrices, quadrat counts or nearest neighbour distances, but you are most likely to have to transfer this information into more specialised software.

Conclusion

In this chapter we have described the major components of GIS and stressed that these systems should be thought of as a research toolkit and not a methodology *per se*. The first task is to discern whether the proposed research project is amenable to processing with a GIS. Indications in favour of GIS use include large volumes of spatially referenced information, and a need for explicitly geographical data processing and analysis. It is necessary to make a realistic assessment of the effort involved in conducting the work, and particularly to discover whether appropriate data are available, and whether the GIS software can provide all the functions that are required.

All three students whose works have been discussed here were able to draw conclusions from their analysis that both addressed the initial research questions and reflected on the methodologies they had used: it is to be hoped that these elements will be present in any well-rounded discussion of a GIS-based project. In closing, it is necessary to consider one further set of issues. By necessity, the focus of this chapter has been primarily on the practical aspects of GIS use. A broader set of questions concerns the limitations of GIS technology in addressing substantive research problems, and an intelligent consideration of these issues will also be of interest to your examiners.

A frequently encountered criticism of quantitative approaches in geography is that we are only able to address those aspects of the problem that are represented in the data, and these may be only partial. A book edited by Pickles (1995) was particularly significant in turning the focus of this discussion on the socio-economic impacts of GIS. It is incumbent on the user of GIS to think broadly around their research questions in order to understand whether their results will be fundamentally limited because the data or analysis do not capture some of the most important dimensions (often the unmeasurable individual, human ones) of the situation being investigated. The human impacts of disturbance by noise, the value placed by different families on education and the interpretation of educational achievement scores are themselves examples of these ambiguities. A further responsibility in many 'real world' GIS applications is to address the ethics of the GIS use, such as the treatment of data relating to individuals (for example with regard to confidentiality), and the purposes to which the analysis may be put. The best research projects, as with the best real-world practice, will not only involve technical competence, but will demonstrate insight in the consideration of these broader issues, as applied to the specific focus of the research.

Acknowledgements

I am particularly grateful to Stuart Gibson, Neil Kaye and Jenny Turnbull for permission to refer to their undergraduate dissertations in producing this chapter.

BOX 16.5

Further reading

Bailey, T C and A C Gatrell 1995 *Interactive spatial data analysis,* Harlow: Longman – an excellent introduction to explicitly geographical data analysis, making reference to the use of GIS

Longley, P A, M F Goodchild, D J Maguire and D W Rhind (eds) 1999 *Geographical information systems: principles, techniques, applications and management,* 2nd edn, Chichester: Wiley – a major collection of papers on GIS covering general principles and technical issues (Volume 1), example applications and organisational considerations (Volume 2)

Martin, D 1996 *Geographic information systems: socioeconomic applications,* 2nd edn, London: Routledge – an introductory GIS text which focuses particularly on applications relevant to human geography

Section D Producing the report

The final section of the book is devoted to producing the actual written document that represents the final output from the research in which you have been engaged. It is easy to underestimate the time that will be absorbed by the process of writing up, and its importance to the overall enterprise. Our advice would be to think about the structure and content of the finished report from the very beginning of the project, and certainly not to dismiss it as a boring and unimportant task after all the interesting parts of the research have been done. This is for at least three reasons.

First and most obviously, presentation is important (it will often be explicitly taken into account in the marking process: you should be able to obtain details of the assessment scheme from your institution). You may have put a lot of effort and imagination into a highly successful project, but if you do not make a good job of presenting the material, you are unlikely to receive the full credit that you deserve. The visual appearance of the report, its neatness, layout and design may greatly influence the impression it gives, and to some extent the impact of its message. Good-quality printing, an attractive font, appearance of maps and diagrams, are all important. Possibly more important are the standards of written style, such as correct spelling, punctuation, referencing style, and so on, including observance of the standard graphic and cartographic conventions. More important still is the success of your report in conveying what you have to say. It is worth taking a great deal of care to make sure your text is easy to read and your illustrations legible. You must think of the reader, and what he or she will need to have explained. By the time you finish you will be an expert in your field; you will probably know more about it than your professor, and you have to provide sufficient context for the work to be intelligible. You should bear in mind that these presentation skills are highly transferable and will be much in demand by potential employers, so getting this right is an important aspect of your personal development.

A second reason why producing the report is important is because the writing process itself is very creative. You may think before you start to write that your conclusions are firm and you know exactly what you are going to say. However, it is a common experience to find that, as you are writing, new ideas or new linkages between them occur to you, sometimes more interesting than the ones you had to start off with. You may also find that things that seemed perfectly clear to you when you started are not so clear when you have to set them down in detail. Producing the report forces you to think through what you want to say, and frequently leads to re-evaluation of your ideas.

Thirdly, producing the report gives you a much clearer idea of what needs to be done, and what is not really necessary. You should try to get an idea of the structure of the report at an early stage, together with ideas about how much you will write on each topic. You may well find that you have collected far too much detail on certain topics; tables or maps that initially seemed a good idea may appear excessive when you have to think about what *has* to go in. If you start working on the report at an early stage, it will help you save time by showing you what things are essential and what things can easily be left out. In addition, it is a big boost to your confidence when you have a substantial proportion of the report already complete.

Paul Boyle and Robin Flowerdew discuss the structure of a dissertation or research project report in Chapter 17. Their aim is to help you clarify the purpose and objectives of the different sections that are usually found in a dissertation or research report. These include some components that you may not previously have been asked to write, such as an abstract, literature review or a detailed account of methodology. In human geography, there is probably no single best way to organise the contents of a dissertation, but there are certainly bad ways to do it. Boyle and Flowerdew do not want to make everybody structure their dissertations in the same way, but for those unsure how to do it, they recommend the so-called 'wine glass' structure.

In Chapter 18, Paul Boyle discusses the details of writing up your dissertation. He warns about some of the mistakes that students often make in writing up and some of the issues of style that are often inadequately explained, and makes some suggestions about the best way to tackle what can potentially be a frightening task. Writing well will make a great deal of difference to the effectiveness of your argument, and it is worth taking pains to do justice to the quality of your ideas. Perhaps the overall message, however, is to be realistic and pragmatic. It is more important to get it done than it is to be a perfectionist.

Geography, probably to a greater extent than other subjects, has a long tradition of graphicacy. 'Geography is about maps' according to the rhyme, and many geographical projects require maps to show the location of the study area, to illustrate the places being written about, and as a powerful way of representing data. In addition, geographers make extensive use of diagrams, charts and other visual aids. Illustrations can not only enliven an otherwise turgid mass of text, they can also be very effective ways of getting your information across. A good diagram or map may help you and your readers to see the world in a new way. In Chapter 19 Christine Dunn discusses a number of issues concerned with the use of maps and illustrations in a research report. The near-universal availability of computer mapping and graphics software makes certain kinds of illustration much easier to produce than before, but principles of good graphical and cartographic design still apply.

17 Designing the report

Paul Boyle and Robin Flowerdew

Synopsis

This chapter deals with how to organise a dissertation. It compares better and worse examples of how to structure a report and describes in detail what should be covered in each of the sections. While the structure should be well planned in advance, you must be flexible enough to include new findings as they arise. It is also stressed that there is no single structure that will suit all dissertations and you should not feel obliged to use the tried and tested approach described here. However, even should you decide to use a more innovative structure, there are certain core issues that must be included in any dissertation.

Introduction

Usually there is a good deal of freedom in how a dissertation is organised, although you should pay attention to any recommendations given to you in your own institution. If you have worked out an effective way to present your work, that is fine, but the comments in this chapter are intended to help those who are not sure how the material should best be presented. Suggestions about organisation here are not intended to deter you from experimentation, just to present some well-established and generally appropriate ideas.

Structure

The structure and presentation of the written product are vitally important to a successful project. The report needs to follow a logical sequence and this requires a great deal of thought from an early stage. Deciding upon this structure can be quite daunting initially and it is tempting to postpone it in order to focus on practical issues, such as data collection, which may appear to be more important in the early stages. However, the early production of a detailed and

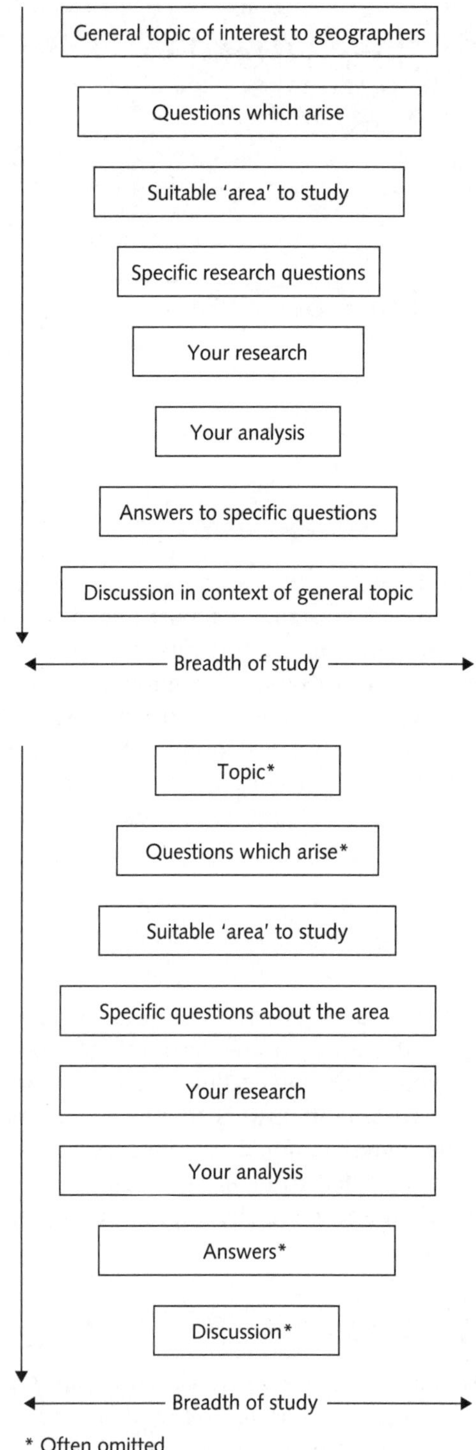

Figure 17.1 The 'wine glass' and 'splodge' approaches to writing up
Source: Kennedy, 1992

comprehensive dissertation structure is extremely useful as it highlights the errors of omission that may exist in the research strategy. By formally recording your anticipated structure you will also have identified a series of targets which need completion during the research enterprise.

The structure that you decide upon will not be set in stone, of course, and there are likely to be substantial changes made to it as the research progresses. Flexibility is an inevitable requirement of any researcher as new and interesting findings alter the weight given to different aspects of the project or, just as likely, promising ideas turn out to lead down blind alleys. As long as the structure is continually updated, this flexibility will improve the dissertation. Also, the constant updating of the structure will help you form a clear picture of what you are trying to achieve in your own mind.

Various formal aspects of the structure can be identified. Perhaps most important is the advice that the project should be a clear and logically constructed piece of work – a story that is both informative and enjoyable to read. Kennedy (1992) depicts research plans as either 'wine glasses' or 'splodges' (Figure 17.1) where the former is a well focused study situated within a broad background framework and the latter is an over-detailed analysis providing little or no general context. This useful distinction can also be applied to the production of different types of write-up. It emphasises the importance of developing and acknowledging a clear and comprehensive understanding of the previous work on the chosen topic, to which your results can be related. You should not be tempted to overemphasise the findings that you have produced at the expense of discussing other people's work. Clearly, projects situated firmly within the general context of the broad topic are preferable, but in reality many are likely to fall somewhere in between these two scenarios.

Dissertation chapters

The aim of your write-up should be to communicate your research ideas and findings to the reader in a clear and logical manner. A sound structure is therefore essential (as suggested earlier and displayed in Figure 17.1) but, once again, there are no strict rules to follow, except those provided in your institution's guidelines. Usually, the structure will be similar to that presented below, but some dissertations may not follow such a rigid arrangement. Some topics or approaches may require material that does not fit into any of the categories suggested and, for others, some of these headings will not apply. In particular, quantitative and qualitative dissertations are often structured differently, with the former following a more traditional 'scientific' approach.

Preliminary material

When I can read my title clear . . .
I bid farewell to every fear
And wipe my weeping eyes.
(Isaac Watts, 1707)

Give the *title* of your dissertation some thought, as it is the first experience of your work that the reader will get. It should be clear and concise and should reflect the work that you have undertaken. A dissertation is not an essay, and you should avoid titles that resemble essay or examination questions. A better model is a journal article, where the title (usually) gives a clear indication of the content. Do not feel you have to cram all pertinent details of the study into the title: 'A comparative study of women's age at marriage based on parish register information from East Middlesham, Buckinghamshire, and Little Thistlefield, East Riding of Yorkshire, from 1732 to 1785' could be shortened to something like 'Getting married in the eighteenth century', provided that the detailed information about the topic is included in the abstract. Original titles that play on words may be fun, but should be supported by a subtitle that explains what the dissertation is about. You should not use them if they merely confuse the reader, and you should not be worried if you cannot think of anything clever. Clarity is probably a safer bet than originality.

It is useful to choose your title at an early stage, preferably while the dissertation is still at the planning stage, and it is likely that your institution will force you to do this. Having a title is useful not only to fend off questions from well-meaning friends and relatives about your dissertation topic and how it is progressing, but also for helping you to keep your aims in focus and to evaluate how relevant particular reading or research activities are going to be. Even so, flexibility is again important. Choose a title, but do not be afraid to change it to reflect changes in your interests, your research activities and your results.

The *title page* itself will usually include the title, your name, the year and the institution. Keep the text to a minimum and emphasise the title, which is the most important information on the page. You can achieve this most effectively by breaking the page into ideal proportions. The *golden section* is a classical architectural device for designing appealing buildings. Basically, this means that at first glance the eye is naturally drawn to a position about one-third of the page down and it makes sense to position the title there. Nowadays, students are usually good at producing professionally produced reports, including the title page, and a general rule is that it should tempt the reader to want to open up the dissertation and find out more. Therefore do not make the title page gaudy, over-elaborate or unclear.

The *acknowledgements* section provides you with the opportunity to thank those who have aided your research endeavour and this is often the first part of the dissertation. In a book or journal article, you have to acknowledge the holders of copyright for any illustrations or lengthy quotations you include. This is not expected in an unpublished dissertation, but it should be borne in mind that it will be necessary if your work is eventually published. Those who do need to be acknowledged may range from a relative who typed up the manuscript for you to the supermarket managers who allowed you to carry out interviews with their staff. Any 'gatekeepers' who have allowed you access to people, places or documents for your research should be credited. You may want to thank people you have interviewed. If your supervisor has been helpful, it may do no harm to acknowledge his or her help. You should also identify those who provided useful ideas that you followed up in the research; verbal ideas need acknowledging as much as published ideas. Again, there are no rules about acknowledgements, but it is sensible to keep them relatively brief and not too personal.

The reader will constantly refer to the *contents page* and therefore you should make sure that it summarises the dissertation structure adequately. As readers progress through the work they may want to refer back to a point you made earlier and this is easier if the contents page is relatively comprehensive. Divide chapters into sections and identify them in the list along with the relevant page number. The headings that you use here should be consistent with those in the body of the text and there are various systems for identifying different levels of the structure. Three tiers of headings should suffice, although you should not expect to use this range in every chapter; do not fall into the trap of providing a minor heading for every paragraph in the text. The chapter titles should be most visually dominant followed by the sub-headings and then the minor headings. You can distinguish between headings by their font size, the use of underlining or bold text, or by a numbering system.

Page numbers conventionally use lower-case roman numerals for the material that comes at the start of the dissertation – acknowledgements, table of contents, and the like. Ordinary (arabic) numerals are used once you get on to the main body of the dissertation. The pagination of the dissertation requires care. Minor changes and revisions can cause the page numbering to change. Either use your word processor's ability to produce tables of contents automatically, or postpone doing the table of contents until you are sure that you will be making no more changes.

Each of the figures and tables that you use in the dissertation requires a title that neatly summarises what it shows. In addition, it is conventional for each of them to be numbered sequentially through each chapter; Figure 5.3 is the third diagram in the fifth chapter, for instance. After the contents page you need to include separate pages listing the figures and tables, showing the number, title and page number for each. These will act as useful guides to any reader who may remember some results that you mentioned earlier in the dissertation, but is not sure precisely where they were. Once again, remember that this is one of the last jobs to do as adding any additional material to the dissertation at a late stage may change the page numbers of tables and figures. Of course, some dissertations will not include figures or tables and these pages can be omitted.

The abstract

A good beginning makes a good ending.
(proverb)

Many people who pick up your dissertation will not read any more than the acknowledgements and the abstract. The abstract should provide a clear idea in reader-friendly language of what the dissertation is about (what is the research 'gap' that you are trying to fill?), a brief summary of the methods you used, and a description of the major findings. Readers can then make an informed decision about whether the topic interests them. Turk and Kirkman (1989) refer to this summary as a *map*, as readers can absorb information better if they are aware of what the main conclusions will be. Without a carefully structured abstract, any slight digressions from the main topic in the body of the text may lead the reader down the wrong road.

Consequently, producing the abstract is not a simple task and the fact that it must be short exacerbates the problem. You should give it careful attention as anyone examining the work critically will continually refer back to the abstract to make sure that you have a clear idea of what your work contributes to the research topic.

A practical approach to writing the abstract is to leave it until the end. As described above, the writing experience is a thinking experience and, as you will reinterpret your findings as you go along, you should expect the contents of the abstract to change during the process. Having completed a first draft of the dissertation you can work through it underlining the major points for inclusion in the abstract. However, if you are determined enough, there is no better way of forcing you to think seriously about what you have achieved than to write the abstract first, even if you revise it at a later stage.

The introduction

Parsons and Knight (1995) stress that the introduction is an essential part of the logical argument you are putting forward. Clearly state the aims of the dissertation by briefly describing what the main findings on the subject have been previously. You can then highlight why you think that more work is needed and how your research is going to achieve this, perhaps through the use of a new data source, an original methodology or a different study area. The identification of the research gap is a critical part of this process of producing a dissertation, so make sure your introduction helps the reader see how you identified it. You may, of course, have a series of aims, rather than a single one. These should link together to produce an overall aim for the project and this will protect you from being criticised for lacking a clear direction. Providing such a bold statement about what you are aiming to do may be a little unnerving initially, but it will undoubtedly impress the reader as long as you have not been hopelessly ambitious. Although an obvious point, it is worth emphasising that you should not identify aims that are not examined subsequently – too many students provide a list of grandiose aims that are never actually realised in the body of the work.

You must also remember that the introduction precedes the detailed literature review that follows and therefore the reader may not yet be familiar with the previous work on the topic. The level at which you pitch the introduction must reflect the reader's potential lack of knowledge on the subject. At the same time you must not over-simplify the introduction so that the more expert reader finds it naive. Imagine a reader who is intelligent and knows something about geography, but has no specific knowledge of your topic, and will not know what you are trying to do unless you explain it carefully.

Above all, the introduction must be interesting, so that it encourages the reader to continue with the remainder of the dissertation. Try to convince the reader that your topic is important and interesting. Relate your topic to issues of interest in geography, or to issues of public interest. You may want to justify the methodology that you used, perhaps by contrasting it with the methods used by others. You may also want to explain briefly how the work was carried out and justify the logical structure of the remainder of the dissertation by referring to the format of the following chapters.

Literature review

If you steal from one author, it's plagiarism; if you steal from many, it's research.
(Wilson Mizner, 1953)

Undoubtedly, there will be a wealth of published material that is relevant to your topic. Your original choice of topic will probably have been influenced by work that you have read or been taught and your analysis is an addition to these previous studies. If your research is especially original you might find that few studies have dealt with the precise topic, but there will be work in the broad field of your research that you should cite. The literature review cannot include every relevant publication. You should aim to cite those which are the most pertinent and those which have been most influential in the field (you can identify these by the regularity that they are cited by others, or by using citation indexes – see Chapter 4). It should also show that you are aware of the most up-to-date literature on the subject. Your intention should be to prove that you have become something of an expert in your chosen field, and one way is by being knowledgeably selective in your decision about which literature to cite. If there was a single study that influenced your work substantially, discuss it and draw attention to those aspects that were different to your research, rather than running the risk of being accused of plagiarism.

Like the rest of your write-up, it is important that your literature review section has a coherent and sensible structure. Do not provide brief and unconnected paragraphs devoted to each book or article you wish to comment on (a literature review is not a string of book or article reviews). Instead, you should aim to relate the various sources to each other. Group together sources which take similar approaches or which discuss similar material. If possible, organise the discussion so that you can link each source you discuss to the previous one, for example by commenting on how it develops or counteracts arguments you have already discussed. Sometimes a chronological approach is useful in showing how research themes have developed over time, but in many cases the best organisation of the material may not be chronological. Organisation by theme may be appropriate, or a progression from the more general to the more specific. Alternatively it may be sensible to separate out reviews of material of theoretical importance, of empirical interest, and of methodological relevance. With such an approach, different aspects of the same book or article may be discussed at different points within the chapter.

The ideal literature review might be one that begins with a discussion of the topic in general terms, and progresses through the development of ideas on the topic, showing how individual contributions have led to partial answers and to new questions being asked. Sometimes the findings from different research will be contradictory and this should be acknowledged, rather than ignored. Such a review should culminate by persuading the reader that the logical next step should be exactly the study that the writer has done. Even if you cannot quite carry this off, do try to show how your study is linked to earlier work and, if possible, that it will tackle some of the questions unresolved by this work, or that it will add a new perspective to the debate.

How long should a literature review be? Perhaps longer than you think. Students may underestimate the importance of setting their own work in context, and if you are reasonably successful in finding relevant material, you may find it takes quite a lot of space to describe other studies in sufficient detail. The literature review may be one of the sections that you can write first, at least in draft, and it is a good idea to plan out how long it should be. If you know the target length for your whole dissertation, it should be possible to estimate what proportion of the total length the literature review should take up. Then, if you find you are not writing enough (or, more likely, that you are writing far too much) you can adjust the level of detail at an early stage.

Literature reviews are not always easy tasks for an inexperienced researcher. Help is available, however: Hart (1998) has produced a book-length study of literature reviews that may be of use in structuring your arguments.

Methodology

Though this be madness, yet there is method in't.
(William Shakespeare, 1601)

Describe the procedures that you used in your research in the methodology chapter. This may include a number of sections, depending on the methodological complexity of the study. You may even need more than one chapter, for example if it is important to describe the study area in some detail, or if you have used a mixed methods approach. In any case it is vital to include a section that discusses the philosophical approach and the methods you decided upon (see Chapter 2). You may want to explain why a quantitative approach was preferable to a qualitative one (or vice versa) and then whether the approach was deductive or inductive, for example. Practically, the details of how the study was conducted need to be described and justified. For example, the description of a questionnaire study may need sections describing: the study area; the sampling strategy used; details of the pilot-study questionnaire; the layout and content of the main questionnaire; where, when and how the information was gathered; the response rate; and the techniques used to analyse the data.

You should also use this opportunity to identify problems that may have influenced your study and any errors that you identified in your data. Resource constraints may have prevented you from collecting enough questionnaires, or adopting a particular sampling strategy. There may be important ethical issues raised by your study that require discussion. You may also have realised that other more complicated statistical techniques have been used elsewhere, but that rather simpler techniques were implemented in your study. Acknowledging the improvements that may have resulted from these advanced techniques indicates that you are at least familiar with them, even if you did not implement them. It is this section that will convince the reader that your choice of methodology was informed by a good understanding of the potential approaches that are available.

It is important to be honest and specific in describing your methodology. Readers will want to know how you selected respondents for questionnaires or interviews, and how you chose your study area. If you are vague about your response rate, your sampling procedure or your interview schedule, people marking it may assume the worst. If you admit the problems, they will probably be sympathetic. Many students choose to work in their own home area for good practical reasons. By all means say why it is well suited to the research question being addressed, but do not be afraid to give the real reason for choosing it too. Likewise, everybody knows that a rigorous random sampling design may not be implemented in practice for very good (or not so good) reasons including weather, fear, health or lack of time.

Results

If we do not find anything pleasant, at least we shall find something new.
(Voltaire, 1759)

The results chapter is often one that causes difficulty as it is tempting to include too much material. It should not include methodological discussions, which should precede the results chapter, nor should it include the discussion and interpretation of the results if you have a dedicated 'discussion' chapter as suggested below. Even in quantitative dissertations the interpretation of results may vary between researchers and consequently it is helpful to separate these two aspects of the research. This is not to say that it is unacceptable to interweave the results and discussion: just make sure that you have not sacrificed interpretation at the expense of an over-elaborate presentation of the results.

There are various ways of presenting the results depending on the style of the dissertation. Results may vary from long quotations collected from respondents in in-depth interviews to detailed cross-tabulations of questionnaire data or the results from statistical tests, but in every case they should be logically structured. If you are using statistical testing, do not feel obliged to explain the details of the chi-squared statistic or to include step-by-step calculations in the results; generally it should be adequate to assume that readers understand the material taught in undergraduate practical classes. More advanced analysis, whether qualitative or quantitative, should have been discussed in the methodology chapter, so that the results chapter can concentrate just on results. If you have a lot of results, concentrate on the most interesting and relevant. Do not try to squeeze everything in. If you want to demonstrate how much work you have done, you can summarise the results from the less important analyses in a table or perhaps relegate them to an appendix.

Organising the presentation of the results can be difficult. In general it is more interesting to group the results according to themes or to the research questions being posed, rather than according to how questions were ordered in a questionnaire or interview schedule. However, it may also be useful to have a section describing the sample before getting into more detailed issues. If you have a mixture of qualitative and quantitative results, it may be interesting to integrate the two rather than to discuss them separately. Often some pertinent quotes can liven up a dry presentation of the results of hypothesis testing.

Positive results are usually more interesting than negative results, but this does not mean that you should ignore the latter. Negative findings are often interesting in themselves, particularly if they fail to conform to conventional ideas, and researchers often fail to recognise their importance because they are unexpected. Even if the results of speculative hypotheses turn out to be incorrect, it is worth presenting the results so that the reader is aware that you did examine the question.

Discussion

If a separate discussion chapter is used it should bridge the gap between the results section and the conclusion and provide a personal interpretation of the results. Even if the research question was interesting and original, the fieldwork was undertaken sensibly, and the analyses of the results were correct, this is where you have the opportunity to shine. Interpreting the results is not a simple task and a clear and logical discussion will confirm to the examiner that your work is of a high standard.

You may want to include references to other work as you describe how your results confirm, or conflict with, previous findings and you are well advised to think about the structure of your discussion chapter long before you begin writing it. Inevitably, you will alter this structure as the write-up progresses, but having a clear picture of the final product at an early stage is a useful guide. In particular, the points you raise should be consistent with the presented results.

As discussed above, this section is a useful place to identify the limitations of the study and, given the resource constraints upon dissertation work, it is inevitable that the research will not be perfect. Acknowledging these problems confirms to the reader that you have considered them, rather than being unaware of them.

What goes into the conclusion

O most lame and impotent conclusion!
(William Shakespeare)

There are two meanings of the word conclusion. It can mean *concluding remarks* which are simply the final statements, or it can mean a *logical outcome* implied from the previous work (Turk and Kirkman, 1989). In order to avoid Shakespeare's harsh verdict, you should aim to produce the latter of these, remembering that the reader will refer to the conclusions more than once. The conclusion is definitely not the place to report findings not mentioned previously, but it is appropriate to summarise the most important points made earlier in the text. It is also a good place to broaden out the discussion (as suggested by the 'wine glass' structure); the implications of the results outside the specific context of the research problem can be discussed. This would be the place to discuss how the results may be of interest to policy-makers, for example. Even though it is the final part of the dissertation, it is not necessary for it to be long. Concise conclusions get the major points across more effectively, displaying the author's ability to synthesise material.

The style of the conclusion also needs some thought and the type of research that you have done will influence this. In some cases you may feel that your findings are original and interesting enough to build the dissertation into a climactic ending but, as Wolcott (1990) points out, this is not necessary. He warns against advancing beyond *what is* to *what ought to be* because the leap from the descriptive to the prescriptive imposes someone's judgement, whether originating from the researcher or one of the respondents. Personal opinions are not frowned upon necessarily, but it is important to label them as such and to back them up by your findings where possible.

Additionally, the conclusion can also include avenues for future work that would extend logically out of the work that you have presented. Interesting results you have identified might usefully be explored in another context. Perhaps you think it would be important to replicate your study in a different area or under different conditions. You may recognise that the methods you adopted were not as technically advanced as they may have been, for example. However, the conclusion should primarily be a confident discussion of the work you have completed and it is important not to allow the sections on limitation and possible future work to dominate.

References

You will find it good practice always to verify your references, sir!

(Martin Joseph Routh, 1888)

Bibliographies include lists of publications that are relevant to the subject, while reference lists only contain those references that you have cited. Some institutions may expect you to provide both, although most only require a reference list. Clearly, markers will be looking at this list, to see how much of the relevant literature you have identified and to check whether there are any important sources missing. The list therefore should be extensive, including general background material and specific studies that are particularly relevant to the work you did. However, you do not have to read everything from cover to cover in order to include it. If a particular study is relevant to your argument, include it if you have only read the relevant part of it, or even if you only know of the work from a discussion in another source.

If your institutional guidelines do not specify a format for referencing, you may wish to take the reference list that appears at the end of this book as a guide. Papers from journals should appear as for Bondi and Domosh (1992);

chapters in edited books as for Dorling (1994), and authored books as for Agar (1980). Order these references alphabetically and include all of the relevant information so that readers will be able to find the reference easily if they wish. If you use more than one reference from the same author these should be ordered with the earliest publication first. If two or more were published in the same year, you need to attach letters to the date, both in the text and in the reference list, as for example O'Brien (1989a, 1989b). It is particularly important to ensure that your reference list exactly matches the references made in your text, and that all the bibliographic details are complete and correct – otherwise future readers will curse you for sending them on wasted trips to the library! References to Internet material should include, where available, the name of the author and the title of the piece referred to. If the website belongs to an organisation, this should be included. Most important, the exact URL needs to be stated, together with the date at which you accessed the website. Kneale (1999, Chapter 14) supplies more detail on these issues, including a 'nightmare reference list' containing examples of many common mistakes in referencing.

Appendices

Reserve appendices for information that is peripheral to the argument, or that would interrupt the flow of the text unnecessarily. You may decide that you can effectively describe the mathematical models you were using in English in the text, but that the detailed equations are needed in an appendix, or you may want to provide one example of the questionnaire that you used. Copies of important correspondence which aided your research should also be included here. However, do not use the appendix as a place to jettison all of the additional material that you accumulated during the research, but could not justify including elsewhere – reams of statistical output that are not referred to in the text will be ignored by the reader, for example.

Conclusion

The layout and design of your dissertation can be done in a variety of ways. You are at liberty, with your supervisor's approval, to write it up in whatever way you think best, bearing in mind that you will probably receive credit for a successful innovative approach, but lose marks for one which the markers feel is incoherent or inappropriate. Our recommendations in this chapter are intended to provide an approach that most people would regard as sensible and effective. You must not feel constrained by our advice; ignoring some or all of our advice may be appropriate for some topics and some approaches.

The most important advice on the structure of the dissertation is probably to make it logical and coherent, and to make the structure obvious to the reader. The 'wine glass' structure is tried and trusted, and it is almost always important to set your own work within the context of other people's ideas. As with anything you write, an introduction and a conclusion are vital. The Methodology – Results – Discussion sequence does not have to be approached in precisely this way. Sometimes it may be better to divide the dissertation into sections according to theme, and to write about methods and results together within each section. This might be especially sensible if you have used a mixture of qualitative and quantitative methods. Other valid and perhaps better ways to organise your material may occur to you.

It is worth repeating that organising how you will present your work should be done at an early stage. Having a working title and a set of research questions is actually helpful in carrying out your research, provided you retain flexibility to change things if the project develops in an unexpected direction. In a similar way, having a provisional table of contents in mind, and a rough idea of how long each chapter should be, is likely to be helpful in writing up your material, allowing you to decide as you write whether something is relevant or not, and if so where it should go. Organising your dissertation is not trivial, but it should not be too big a headache if you decide on your structure early and use it as a framework for your subsequent work.

BOX 17.1

Further reading

Bradford, M 2003 Writing essays, reports and dissertations. In N J Clifford and G Valentine (eds) *Key methods in geography*, London: Sage, pp. 515–532

Kennedy, B A 1992 First catch your hare . . . Research designs for individual projects. In A Rogers, H Viles and A Goudie (eds) *The student's companion to geography*, Oxford: Blackwell, pp. 128–134

Parsons, T and P Knight 1995 *How to do your dissertation in geography and related disciplines*, London: Chapman and Hall

Turk, C and J Kirkman 1989 *Effective writing: improving scientific, technical and business communication*, 2nd edn, London: E & F N Spon

Wolcott, H F 1990 *Writing up qualitative research*, London: Sage

18 Writing the report

Paul Boyle

Synopsis

This chapter deals with writing up a dissertation. Although it is tempting to assume that the write-up is a relatively simple part of the research process, it is stressed that careful thought should be given to the style and presentation of the final product. Consequently, the write-up should be well planned in advance, but you must be flexible enough to include new findings as they arise. The chapter also discusses when you should begin writing, which sections should be dealt with first and the balance that needs to be drawn between presenting your own material and summarising the work of others. It also highlights the importance of thorough editing and recommends alternative ways of achieving this. More practically, the chapter advises on issues of style and layout and stresses the importance of writing clearly, concisely and interestingly.

Introduction

Many of the suggestions in this chapter should be obvious to most students because they are as relevant to essays and other less substantial pieces of written work as they are to dissertations. However, dissertations will involve substantially more time than ordinary essays and they will contribute far more to the assessment of the final degree. Consequently, you should give special care to the writing-up procedure; the examiner has little else to go on except the final product and it would be unfortunate if a careless write-up masked the good work put into the project as a whole.

It is important to acknowledge from the start that there is no one right way of writing dissertations, for which many theoretical and methodological approaches, described in the preceding chapters, may be adopted. Indeed, originality throughout the various aspects of the research endeavour, including the write-up, will be celebrated. If taken seriously, writing is a creative and fulfilling experience and you will be rewarded if you give as much careful thought to this part of the project as you do to the theory and methods that underlie the research.

It is also important to stress that the final product will never be perfect in the author's eyes. We can always improve the work that we have done (or at least we believe we can) and by being too perfectionist there is the risk that the project will never be finished at all. The remainder of this chapter aims to highlight some of the more important aspects of the writing-up process that need consideration.

What needs explaining and what does not

It is important that you pitch the project at the correct level when writing up and, perhaps surprisingly, most problems concern work that is pitched too high, rather than too low. Aim to write for your peers and do not assume that the reader knows a great deal about the topic already. It is common to find students missing out important descriptive material because they assume that the reader is familiar with the background information. Even if the reader (or examiner in many cases) knows a great deal about the subject you must *show* that you have done the preliminary work. Broadly speaking, you must identify: the questions that you wish to answer; the manner in which you went about answering them; the findings that you produced; and the importance, or implications, of these findings. More specifically, there are many aspects of the work which students often ignore, perhaps because they fail to recognise their importance. If you analysed your data statistically, you must explain why the particular technique was appropriate for the data you were analysing, for example. It is also important to acknowledge the limitations of the study, which, because of the time and cost constraints imposed upon you, may be considerable. Thus, it may not be as easy to generalise your results as you had hoped because your study area or your sample of respondents was too small. It is good practice to have a single section that explains the various limitations, preferably in the discussion section, as this removes the need to continually repeat these limitations at the beginning or end of each section of the report.

Of course, you should exclude many details of your study from the write-up. Ignore the physical process of walking around your catchment, even if you want to demonstrate how hard-working you have been, and do not include the 400 questionnaires that you collected, even if it does fatten up the report. Leave the reams of output that you generated using a statistical package to one side as you can summarise the most important aspects of this information in diagrams and tables.

Nothing to fear!

The hardest step is that step over the threshold.
(James Howell, 1659)

How can I know what I think till I see what I say?
(Graham Wallas, 1926)

One of the hardest aspects of producing a dissertation is putting the first words on paper. Many set aside a period of time near the deadline for the work to be handed in as writing-up time, but delaying the inevitable can often be seriously detrimental to the written product. Even when you finally sit down to the task it is amazing how many jobs become more important than putting pen to paper. That cup of coffee seems particularly tempting and the pencil you are using appears to be in desperate need of sharpening! You are not alone if you find it difficult to begin what appears to be a daunting piece of work and it is well worth putting some time into defining your optimum writing environment. For some this will simply be a quiet place, but for others the exact conditions may be far more elaborate. Usually, it helps to equate a particular location, such as a particular spot in the library, with writing.

There is universal agreement among those who advise on this subject that it is essential to begin writing early and to continue writing as the research develops. Many argue that writing reflects thinking but, as Wolcott (1990) suggests, writing *is* a form of thinking. If you wait until your ideas are completely clear, you may never start writing at all, and anyway the act of writing will change many of your ideas as inspiration comes to you during the process. Often, it is only when you actually get your ideas onto the paper in front of you that potential improvements become apparent and leaving the writing-up stage until the last minute forces you to ignore them.

Once you have started it is important to continue. Do not allow yourself to relinquish the task because of some minor gap in the data that you are describing, or because you are not entirely clear what one author was trying to say in their work. You can clarify these points later; once you are actually writing you should take advantage of the moment. If you find yourself suffering from *writer's block* it is probably because you are attempting to perfect the product too early. All writers suffer long periods of time agonising over the wording of single sentences, but in the early stages this is especially inefficient as the modification of material can occur at a later date; nothing needs to come out right the first time. Although some of your wording may seem clumsy initially, it often appears better at a later stage when you return to it. You may even want to set yourself daily word targets. As you accomplish them, the whole process becomes readily achievable. As final encouragement, the fact that you are reading this chapter means that you are probably in a position to begin writing now.

It is simplistic to suggest that the best place to begin writing should be at the beginning. Many do find that writing a brief abstract is a useful way to focus their ideas, even if the abstract ends up being revised subsequently. For others, who are struggling with the writing-up process, there are less daunting parts of the project that you can attempt relatively easily at an early stage. The methods are fairly easy to describe, as you will have a clear idea of what this involves. Few projects deal with methods as their focus (although this is not discouraged); rather it is generally a practical issue that is not too difficult to summarise. Another aspect is the literature review that you should have completed before you undertook the research and that requires the careful synthesis of the ideas of others, rather than the development of your own original ideas. If you are really struggling to begin the writing process, you might even begin with the reference list, as there will be many publications that are relevant enough to mention or quote from in the body of the work. Whichever aspect you choose, it is worth reiterating that the most important point is that you actually *begin*, rather than wasting time deciding where to begin.

Practical considerations

Institutional guidelines

> *We started off setting up a small anarchist community, but people won't obey the rules.*
>
> (Alan Bennett, 1972)

There are relatively few rules in producing a dissertation, but it is important to follow the guidelines laid down by your institution. These will usually specify a maximum length (generally somewhere between 6,000 and 12,000 words) that you should not exceed. Some may also include a minimum length, although this is rarely necessary as students soon realise that they have more than enough to write about. Other regulations relating to the production of the final product are also common. These may include general guidelines about binding conventions and more specific guidelines about lettering size, line spacing, the width of margins, the format of the title page, rules concerning the use of footnotes and a declaration explaining that the dissertation is the student's own work. Guidelines may also give more specific advice regarding the balance between literature review and investigative work. Whatever your guidelines say, they should be followed strictly, and given precedence over the more general recommendations made here.

Style

There is no single correct style that you must adopt, although some elements of style will be more conventional than others. You should not be afraid of using an original approach, but you should not adopt an eccentric writing style for the sake of it. Just as there are debates about how to do research, there are alternative views about writing geographical research. Rogers (1992: 251) draws attention to Edward Soja's (1989) use of 'irony and juxtaposition to bring out the simultaneity of what he sees' in his study of Los Angeles, as an example of an original approach. A useful first step is to read similar works written by others, focusing on the writing style they used rather than the content.

Although the distinction between quantitative and qualitative dissertations need not be clear-cut, it is often the case that quantitative research is written as a logically structured story, similar to that described below. The structure is formal and the chapter headings are common to many dissertations. The write-up is usually in the third person and part of the reason for this stems from the neutrality of the analysis implied by much quantitative research. On the other hand, qualitative dissertations may vary considerably. There is no need for the qualitative researcher to feel obliged to justify the use of qualitative methods as a research technique but, as in quantitative projects, the specific choice of methodology needs justification. In qualitative writing the reader is often invited to see the research endeavour through the eyes of the researcher. The product may simply provide a description of the research events in the order they took place followed by detailed interpretation, or the description and interpretation may be woven together.

Van Maanen (1988) describes three narrative conventions in his 'survey of ethnographic voices' – the *realist, confessional* and *impressionist* and these relate to the qualitative researcher's role in the text. Often it is easiest to act as a third-person narrator, dispassionately relating a realist account of the group that is being studied and this is closest in style to that used in most quantitative studies. Precise attention to detail will abound as descriptions of apparently mundane details of everyday life inform the researcher of the rites, practices and beliefs of the cultural group under study. In contrast, confessional studies are highly personalised accounts that elaborate on the process of fieldwork in detail. This style stems from unease with the scientific status of the work, and accounts of unusual events in the fieldwork and the impacts that the fieldwork had on the researcher are often reported in these confessions. The use of the first person is not frowned upon as the views presented are accepted to be those of the researcher. Finally, an impressionist style involves dramatically relating unique, rather than familiar, events that occurred during the fieldwork, allowing the reader to experience events much as the researcher did. Interpretation of these events is unimportant; the aim is to present the situation to readers who can evaluate the story for themselves. We can add to these ethnographic styles the most extreme, and rarely used, style where the entire work is delivered in the words of the informants. Lewis's (1962) dramatic tale of a poor family in Mexico City is a good example.

It is important, however, to recognise that both qualitative and quantitative research involve a conscious and unconscious process of selectively focusing on the most important aspects of the work (Wolcott, 1990). Decisions about which data to collect, which interviews to conduct, or which landscapes to examine are crucial to the final product, as are decisions about which aspects of this research to include in the write-up.

Language

As stated above, the content is far more important than the language used to describe the content. However, a poorly produced write-up will antagonise the reader and, in circumstances where the grammar and spelling are poor, may result in reduced marks. Most importantly, it is necessary that the reader can understand the points that you are trying to make; readability is essential. Some of the simple suggestions below may improve the write-up considerably without necessitating a great deal of extra work, and problems that you are unaware of initially should become clear during the editing process.

Research on writing procedures shows that long sentences are harder for the reader to absorb. When reading, it is natural to remember a sentence in your short-term memory and then to decode the meaning when you reach the end. Inevitably it is harder to retain the information from long sentences, especially if more than one point is being made (Turk and Kirkman, 1989). Of course, a write-up that consists merely of a string of short sentences becomes monotonous and the best advice is to use a variety of sentence lengths. There are some simple measures of readability that can be calculated manually, such as Gunning's (1962) 'fog index', or by using tools provided in word-processing packages, but a critical rereading of your own work will usually suffice.

The grammatical construction of the written product is also worth careful consideration. Nominalisation, which is the substitution of verbs with nouns, should generally be avoided. In the example below, using the verb *calculated,* rather than the noun *calculation* makes the word *performed* redundant:

Calculation of the regression model was performed using a statistical package.

The regression model was calculated using a statistical package.

The former sentence is unnecessarily long while the latter is more succint.

Certain types of words should be used with caution. Generally, you can remove the words *very* and *much,* which often add needless emphasis, and it is also judicious to use the words *thus, therefore* and *however* infrequently. You should avoid contractions, such as *don't, can't* and *haven't,* and colloquialisms. You should also be consistent in those cases where alternatives such as *judgement* and *judgment* are acceptable and make the choice between *s* and *z* in words like *realise* and *realize.*

Punctuate documents correctly and be aware that punctuation is a means of adding variety to the prose. Question marks should be rare and exclamation marks are even less appropriate in research writing, despite their occasional use here! Colons are useful for preceding lists and semicolons are intermediate between a comma and full stop. Dashes can be used in pairs to replace brackets, which can be unappealing, or singly to add emphasis to the end of a sentence.

Your choice of words is clearly crucial to the salience of the written product. Despite the fact that many academic writers appear to use long words to impress their readers, you will not be penalised for stating your argument simply and clearly. Often long words are necessary to the argument that is being made and knowledgeable writing will be commended, but do not incorporate long words for the sake of it. Readers who continually need to consult a dictionary will become irritated and readers who should consult a dictionary, but fail to do so, will not understand your argument.

Acknowledging your sources

Almost all academic research is an incremental process, in which each new study adds something to the work of previous researchers, drawing on the ideas, data or other material that they have created. Additionally, respondents during the course of your study may provide important information. It is most important that in your own work you fully acknowledge the sources that you have used. This is a matter of etiquette, ascribing to others the credit for their efforts; of convention, allowing your reader to trace an argument back through the literature and sources which you have used; and in some cases, of copyright law. To incorporate the work of others without acknowledgement is termed plagiarism and, if detected, will always result in severe penalties. This section deals with three related situations concerned with the acknowledgement of sources: identification of survey or interview respondents, reference to sources from the literature, and use of data or previously published material such as maps and diagrams.

In the first of these situations, it may actually be necessary to introduce some measures in order to protect the identity of those who have participated in your study. A requirement of any social scientist is the maintenance of their respondents' privacy and this is essential as the final report will be publicly accessible. Qualitative researchers may be especially tempted to name individual respondents, but they should be given pseudonyms except in special circumstances. It may also be necessary to change details of respondents' location, history or circumstances if these might make it possible to guess who a respondent is. The same principles apply if your respondents are companies or any other organisations. Occasionally, you might request the views of a person in a particular occupation whose knowledge of the subject matter is valuable, for example a study of housing might include the views of a local politician – and in such a case it might be reasonable to name the individual. Under no circumstances should you identify respondents without their explicit permission, and you should make clear whether they are responding on behalf of their organisation or as individuals. You should also bear in mind that your findings might be affected by whether or not respondents expect confidentiality to be maintained.

The second situation concerns the use of previously published written material. Even if you have chosen a highly original research area, there will be a considerable amount of published work that is relevant to your topic. Many of your ideas will have been influenced by the work of others and it is important to acknowledge this by referencing these authors. (The precise format that should be used is discussed in Chapter 17.) Most commonly you might attribute a particular idea, methodology or theoretical approach to a particular author. Do not worry that by referencing thoroughly you are giving the impression that you have simply borrowed other people's ideas and that the work lacks originality: you are actually demonstrating to the examiner that you have read widely and understand the arguments put forward by others.

In those cases where an author summarises a point well, you may want to include a direct quotation. It is good practice to try to keep these as short as possible, even when you are quoting interview material from respondents, and these may usually be incorporated into the body of your text. When the quotation is more than a few words in length (your local guidelines may specify how many), it is usual to separate it from the main body of the text by indentation. The motivation for including quotations should be analytical; you provide the quotation in order that you can analyse its meaning, and the longer the passage, the greater the need to make sure that your reader can grasp the point you are making.

A simple rule to follow is that plagiarism is only plagiarism if it is unacknowledged (Watson, 1987), and a common problem is that plagiarism can often occur innocently. During the lengthy period of reviewing the relevant literature it is inevitable that you will make notes of what others have said and in some cases these notes will be direct quotations. It is important that you make it clear in your notes where the point came from and indicate when it is a direct quotation.

The third situation concerns the direct reproduction of previously published data, maps or diagrams and may apply to both printed and computer-readable media. Copyright policies are established to protect the interests and investments of organisations supplying information but it is also important to consider the needs of potential users of such information, and a balance needs to be maintained between the requirements of both parties (AGI Copyright Working Party, 1993). There is considerable variety between different countries in the likely sources of material and the precise requirements of copyright, but these comments based on the UK situation are illustrative of the issues which should be considered. Of particular relevance to many research projects in geography will be the use of maps or map data from existing sources, as discussed in Chapters 16 and 19. The most usual source of map products in the UK is the Ordnance Survey (OS). OS mapping products are subject to Crown copyright and educational establishments not funded by a local education authority who wish to copy OS maps must hold a copyright licence or obtain prior permission from OS. Under such an agreement, OS material can be reproduced for teaching purposes, projects or examinations. The licence includes digital map data as long as these data are used for teaching, lecturing or university research projects. However, in order to digitise information from an OS map, written permission must first be obtained, and additional royalties may then be payable. Each copy of OS material or presentation of digital data must acknowledge OS as follows:

> *Reproduced from the (1) . . . Ordnance Survey (2) . . . map with the permission of The Controller of Her Majesty's Stationery Office © Crown Copyright*

where (1) is the date of the map and (2) is the title and scale. More generally, it is important to acknowledge clearly the source of all illustrative material that was not originally your own work. Many computer-readable data sets such as those produced from the census of population are available to the UK academic community under special purchasing arrangements which also require specific forms of acknowledgement. You should always ensure that you register personally for the use of such data, and that you take careful note of the acknowledgement statements which must appear on any resulting maps, tables or diagrams.

Editing and presentation

I aim to be content with what I produce. It's an aim I never achieve.

(Graham Greene, 1983)

One reason why it is important to begin writing early is that it is useful to have a lengthy period before the deadline when you can put the project to one side and forget all about it. When you return to it you will have forgotten the precise line of thought you were following when you wrote it and this may help you to identify glaring mistakes. You will also spot habitual uses of certain phrases or words that you missed previously. As Wolcott (1990: 52) admits: 'I've always written with too many *buts*, but I have a hard time eliminating them.'

Being able to step back from the research is the most useful result of carefully planning your timetable. Most people are surprised by the improvements that come from this strategy. Distinguishing between *revising* and *editing* the manuscript is helpful, as the former relates to content and the latter to issues of style. Bear in mind that 'writing right' is not possible the first time and a careful editing process will be necessary however hard you worked on the first draft. Tightening up the manuscript is an important process where you reduce unnecessarily complex sentences and remove repetition. This is also the time when you can reduce the length of the dissertation as in most cases the first draft will be too long.

Editing strategies will vary between individuals. One method is to read it out quickly or loudly, rather than in your head, which is how you have dealt with the manuscript until this stage. Wolcott (1990) uses a mechanical strategy where he aims to remove an unnecessary word from each sentence and an unnecessary sentence from each page. Three helpful techniques are: to identify and remove irrelevant *diatribes* where you go off on a tangent from the main point; omit *duplication* paying particular attention to quotations which repeat the same point; and make sure that the *beginnings* of sections are to the point and do not ramble.

You may find it helpful to produce a posterior plan noting the main point from each paragraph. This allows the logic of the structure to be investigated and identifies repetition. It is often useful to ask someone else to read the final product before you submit it. This may be a fellow student, although someone without a geographical background will still be able to spot problems of style and clarity. You should bear in mind that there will be diminishing returns from editing and you should not fall into the trap of changing words for the sake of it.

Nowadays most students use word processors, which are invaluable aids to the writing process. Their main advantage is the ease with which a document can be edited and altered and they are particularly useful for producing plans because of the ease of moving sections of text around in the document using the *cut* and *paste* facilities. The final product should be stylish and most packages will provide an automatic formatting system for you if you wish to use it. The tools provided in these packages are also extremely helpful. Spelling errors can be spotted and easily corrected, electronic thesauruses enable you to replace commonly used words with alternatives and even the grammar can be checked.

You should also be wary of the wonders of the word processor, however. The ease with which you can move text means that awkward links between points can result if the altered version is not reread carefully. You should not be tempted to copy and paste parts of the text which seem relevant in more than one part of the document, as this repetition will be frowned upon when it is spotted by the reader. Although the package will check spelling for you, it will not identify places where an inappropriate but correctly spelt word was used. More practically, you should also refrain from printing numerous copies of the document after every slight alteration that you make.

The layout and physical style of the dissertation need some thought as first impressions unfortunately influence readers. Generally, a clearly laid out dissertation which is not over-elaborate is best. Use diagrams and tables to replace words. They are useful aids for breaking up the text and they should be correctly numbered and referred to in the list of figures and tables. The diagrams may include photographs. This is fine, but bear in mind that reproductions of your work will probably not be in colour. You may produce the tables using a statistical package, but often it is sensible to reproduce them in your word-processing package as the style will match the text better.

Conclusion

> *But when thoughts and words are collected and adjusted, and the whole composition at last concluded, it seldom gratifies the author . . . he is still to remember that he looks upon it with partial eyes.*
>
> (Samuel Johnson, 1754)

Finally, it is worth emphasising that your work will never be perfect in your own eyes. You will be aware of certain pitfalls in the design and analysis, but this does not mean that the work is not commendable. Some of the points that seem obvious to you, because you have been so close to the study, will not be obvious to those reading the work. The whole project, including the writing-up stage where the final product finally takes shape, should be a rewarding experience. This is probably the largest, most important and most individual piece of work that you have done and you should be proud of your achievement. The final, and perhaps most important, piece of advice that can be offered is that once you have done all of this work, make sure you hand it in on time!

BOX 18.1

Further reading

A number of texts are available that provide more detail about how to write up research. A good general introduction is:

Watson, G 1987 *Writing a thesis: a guide to long essays and dissertations*, London: Longman

A book concerned with scientific writing is:

Turk, C and J Kirkman 1989 *Effective writing: improving scientific, technical and business communication*, 2nd edn, London: E & F N Spon

An excellent discussion of writing up qualitative work is provided by:

Wolcott, H F 1990 *Writing up qualitative research*, London: Sage

19 Illustrating the report

Christine E Dunn

Synopsis

Illustrative material is likely to form an important part of many, if not all, dissertations and research reports in human geography. A fundamental aim of this chapter is to communicate the key role that maps and illustrations play, not only in student projects but also in the discipline of geography more widely. Within this context, the chapter gives practical guidance on how to plan, design and produce high quality illustrations that convey the key information clearly and simply. The chapter provides examples of good practice as well as describing some of the pitfalls to be avoided in creating illustrations. A number of types of illustration are discussed including graphs, photographs and tables although the focus is on producing maps. The chapter outlines the major sources of maps and discusses how to make the most effective use of mapping software packages.

Introduction: why maps and illustrations matter

Many good dissertations and project reports can be badly let down by the poor quality of their illustrations. Even the best dissertations will lose some of their impact, and marks, if they are poorly illustrated. If you check your institutional guidelines on producing a dissertation you should find that part of the criteria for assessment is given over to 'presentation'. This can include assessment of literacy, editing and proofreading, documentation, bibliography and overall organisation; but a critical component is evaluation of your ability to produce high quality illustrative material and to integrate this into the rest of the report. As research projects and dissertations are almost always word-processed, it is often the quality of the illustrations, rather than the layout of the text, which helps to convey much of a 'first impression' and which may therefore ultimately influence your final mark or grade. Whilst you should always adhere to the specific instructions that are stipulated by your own institution, important general guidelines for producing and incorporating illustrations also apply. This chapter aims to provide you with some advice and guidance to ensure that your finished report contains a set of high quality illustrations which will impress those that read, and grade it.

Illustrative material in a geographical dissertation could include any or all of: maps, diagrams, graphs, charts, photographs, satellite images or aerial photographs. In addition, tables may be seen as a way of illustrating geographical information, either in the form of raw data, or as summaries of key information or data analyses. Mention will be made of most of these here although much of the focus of this chapter is on producing maps, since this tends to be the main type of illustration for which specific guidance is required. Examples of most of the other forms of illustration are also provided throughout this book. See, for example, Chapter 12 on using graphs to explore statistical relationships between variables, Chapter 11 on using linked window displays in data exploration and visualisation, and Chapter 9 on using participatory diagramming as a way of eliciting qualitative information from small groups of people.

For some human geographers, conventional maps and map-making no longer hold the central place within the academic discipline of geography that they once did. Indeed, the idea of using a map as a representation of geographical knowledge or human activity in a particular place has come under a good deal of criticism from some writers in recent years (Dorling and Fairbairn, 1997). Maps are certainly inherently selective representations of 'reality' and this lies behind much of this disapproval. Maps are, usually, two-dimensional representations on a flat plane, of larger, complex three-dimensional space and, as such, immediately serve to alter and simplify the 'real' geography of a specific location. A map is selective in terms of what is and what is not included, and the map-maker holds considerable power in deciding how to represent specific features. Maps contain special symbols that need to be interpreted by the reader and, to those who are unfamiliar with handling maps, this may create barriers to interpretation and use. That said, 'mental' maps (Figure 19.1) or 'participatory' maps can be created by almost anyone, both literate and non-literate (Chambers, 1994a) while large-scale aerial photographs form a commonly used secondary source in participatory mapping (Abbot *et al.*, 1998; Chambers, 1994b). Representations of space by distorting real locations and spatial relationships on a map can, in some circumstances, form a valid way of using maps as navigation tools. One of the best examples of this transformation of space is that portrayed in the map of the London underground (subway) system. Here, only the relative positions of stations, but not their absolute locations and proximity, are preserved. Maps can also be used for negative, as well as positive, purposes, however. They can be used, for example, as tools to mislead or to tell lies (Monmonier, 1996; Wood, 1993), and for political gerrymandering where administrative boundaries are manipulated in order to secure unfair influences.

So why do maps and other types of illustration matter in a human geography dissertation? Whatever their specific purpose, illustrations are normally used in a general sense to communicate ideas. Maps in particular are inherently spatial and thus communicate spatial relationships and awareness to the viewer. Sometimes an illustration is used simply to provide backdrop or to set the context, as in the case of a simple study-location map, a study-site photograph or a diagram from a secondary source. In other cases illustrations take on a much more

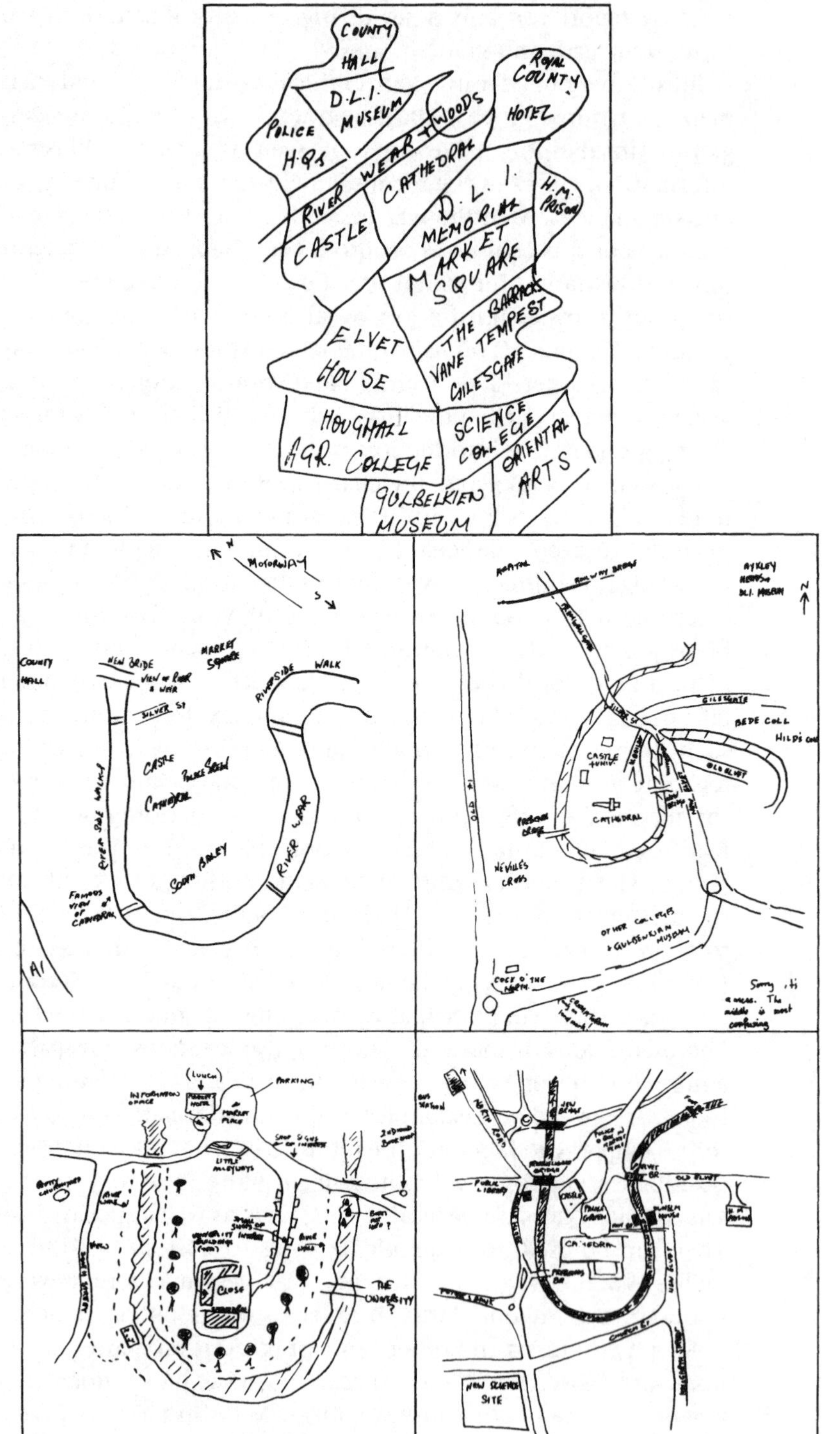

Figure 19.1 Mental maps of Durham, England
Source: After work by DCD Pocock, 1975

significant role and are used actively to demonstrate a particular analysis or to serve as a foundation for subsequent analyses or interpretations. This may be the case, for example, with graphs resulting from a statistical operation, or mental maps in a study of sense of place or in participatory appraisal work.

The idea of map-making has taken on renewed importance over the last fifteen years or so with the onset of geographical information systems (GIS). This has led to fresh interest in the interpretation of maps, and to new and exciting opportunities for visualising space and place. A GIS serves as a 'toolbox' for integrating and analysing diverse types of spatially-referenced information where a particular map may serve as much as a starting point for data interpretation as an end-product in itself. For a dissertation in which GIS or spatial analysis plays a central role in analysing data then, illustrations will be an important focus of the work both from the outset and throughout the execution of the project. In these cases a range of illustrations will be required and may include maps, diagrams, flow charts and graphs. For other types of project, for instance a postmodern analysis of film, music or text (see Chapter 14), different types of illustrative material such as photographs, paintings or drawings are likely to be more important. Indeed, in this type of research, historical maps may themselves serve as source materials or cultural 'texts' to be analysed as, for example, in a cultural critique of colonialism. In some projects, for example, a study of the historical geography of land-use change, *interpretation* of map features is also critical. This type of project will probably involve some original map design, perhaps involving ground survey and will entail some elements of creating and drawing maps (see below).

For many geographers then, the map remains a primary multipurpose research tool of considerable power and sophistication. Maps can serve to generate ideas and hypotheses, and they allow us to explore spatial variations and relationships, and to ask 'What if . . . ?' type questions. When used appropriately a map forms an effective and succinct means of communicating spatial thinking to others. Recent efforts have been made to capitalise on the language of the map to represent 'multiple realities'. Thus Weiner *et al.* (1995) integrated into one illustration two different sources of spatial thinking about the environment in Kiepersol, South Africa (Figure 19.2). One is an 'official', 'expert' representation of land types using area shading on a 1:50,000 topographic map and the other is a mental map derived from participatory research methods (village workshops using conventional maps, aerial photographs and interviews) to explore local geographical knowledge and to represent it as simple labelled polygons. This approach bestows equal importance on each type of geographical information and demonstrates some of the contradictions and similarities in spatial thinking about the natural resources of the study area. Maps can therefore be used in exciting and novel ways not only to illustrate geographical knowledge and information but also to aid our understanding.

In most, if not all, cases then, production of a geographical dissertation is likely to involve at least one form of illustration. Illustrations can be used to summarise in a small amount of space on the page what would otherwise be cumbersome or difficult to explain by using text (hence the adage 'a picture is worth a thousand words'). In planning your illustrations try to capitalise on this

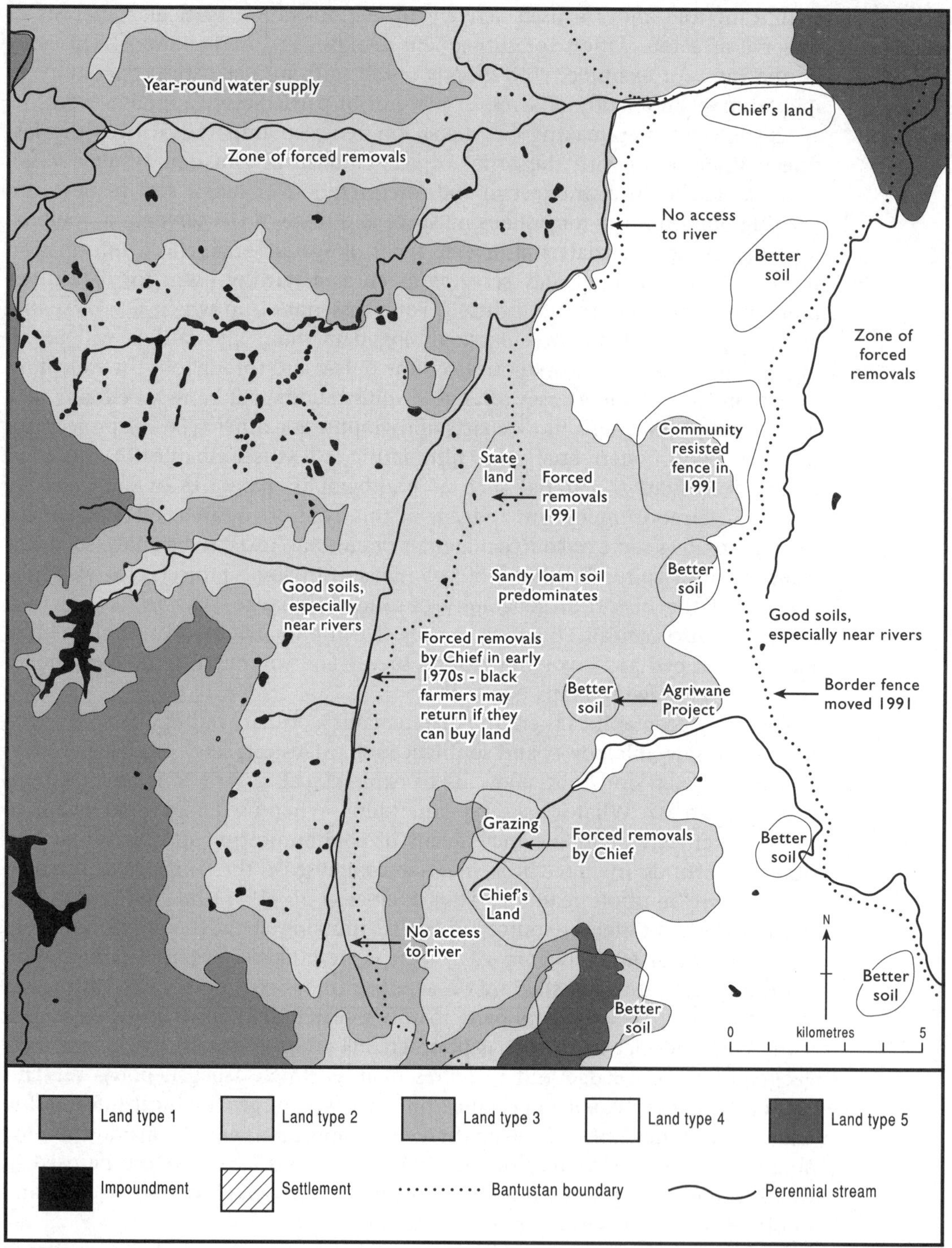

Figure 19.2 Integration of official data and local knowledge to create multiple realities of agro-ecological space in Kiepersol, South Africa

Source: Weiner *et al.*,1995

idea of an illustration as an effective way of conveying a message. Remember also that there is no need to repeat in the text information that is contained in an illustration. Rather use the text to refer to and interpret the information.

A further and final argument which may help to convince you of the value of being able to produce good quality maps and illustrations is that potential employers often look for graphicacy and visualisation skills in prospective employees. When demonstrated in a substantial piece of work such as a dissertation these skills can be used subsequently to make yourself more 'marketable'. Employers often appreciate the broad, all-round aptitude and expertise which geography students acquire during their degree programmes; specific abilities in graphicacy should help you to make even more of an impression than the competing applicants from other disciplines.

Planning and thinking ahead

Many of the recommendations for time management of the writing-up process which are discussed in Chapters 17 and 18 are also pertinent, perhaps more so, to the production of illustrations. One of the potential pitfalls in producing a major piece of work such as a dissertation or project report is that the production of illustrative material will be regarded as somehow one of the least important aspects of the whole process and that this is something that can be 'left until the end'. This can often prove to be a risky strategy since it is easy to underestimate the length of time that is required to produce a set of good quality illustrations. Properly produced illustrations represent an effective and powerful way of representing data or an idea. They therefore require careful thought in terms of their source, design and presentation. In other words, their production should be regarded as an integral part of the work – not something that can be relegated to the last few days before the submission deadline.

Time invested in planning and starting to produce your illustrations at an early stage will subsequently pay off both in academic terms and in a practical sense. It will enable you to think of your illustrations as a central part of your research materials; something to be carefully considered, interpreted and discussed rather than something that has been hurriedly produced, tagged on at the end of the research process and which subsequently results in an amateurish and untidy finished product. Thinking carefully about which illustrations are important will also help you to avoid wasting time on producing finalised versions of diagrams that you subsequently decide not to use in the report. Planning ahead will also assist in allowing for unforeseen difficulties later on such as lost or corrupt computer files or photographic film, printer failures or general pressure on computing resources around the time of report submission.

You should give some thought to the type and number of illustrations from both primary and secondary sources and try to assemble these as you go along. In some cases illustrative material will form a critical part of the primary data collection. In a dissertation on the fear of crime, for example, it will nearly always prove more convenient to take photographs of important spaces or neighbourhoods at the time of data collection in those specific settings rather than having to return at a later date. Adopting the latter strategy is not only less efficient in terms of time, but in some circumstances it may be less possible to have the support of 'gatekeepers' or research participants and you may be more likely to arouse suspicions or concerns from onlookers to whom the purpose of the photography is not clear.

Production of maps and other illustrations by computer software rather than by 'traditional' means of pen-and-ink cartography has advanced immensely in recent years and we have come a long way since the first attempts at producing maps by computers in the 1960s which used rows of text symbols produced by line printers. Computer-assisted cartography greatly facilitates the process of illustrating a piece of work, helps to provide a professionally finished end-product and can also save time. Be aware though that it can also consume time and can easily serve as a way of procrastinating to avoid other tasks, for example, writing up the text of the project. Try to avoid falling into the trap of constant experimentation in the desire to 'perfect' an illustration by making numerous minor changes to font sizes, shading styles, line widths etc. Whilst it is essential to produce a good quality illustration that is appropriate to the purpose, the law of diminishing returns can soon set in. You can often save time, and paper, by making use of a 'print preview' (or similar) facility before printing off the final hard-copy version. This is particularly important if expensive colour illustrations are involved. You need, therefore, to find a balance between, on the one hand, using the software to serve your requirements and timescale and, on the other, letting the software control your use of time. The next section aims to help you optimise the planning process by providing information on some useful starting points for map production.

Sources of maps

In most cases sources of maps for use in a dissertation fall into two main categories: existing maps (paper or digital) and wholly new maps created entirely by the author. When using paper maps (or other illustrations) from published sources it is rarely acceptable simply to reproduce these, for example, by photocopying them. This not only produces potentially rather scruffy results and gives the impression of laziness, it may also infringe copyright regulations. Remember also that presenting other people's work as your own (i.e. without proper acknowledgement) is plagiarism (see Chapter 18). Digital map data are perhaps more easily integrated into the finished report but again, a wholesale 'download' of a particular map or image is usually less satisfactory than a map that you have adapted from an existing source. A more professional result can be obtained by redrawing, or by scanning or digitising and then editing if the facilities are available to you. This allows you to abstract those features that are most relevant to your application, to generalise and add information and to keep all your illustrations in the same 'house style'. This process can be carried out by using traditional, 'low-tech' means such as tracing paper and drawing pens, while in the case of digital map sources the process may be facilitated by allowing the user to select 'tick boxes' for inclusion of different map features.

Table 19.1 Sources of digital map data

International coverage

http://www.geographynetwork.com/
http://www.mapblast.com
http://www.multimap.com
http://www.lib.utexas.edu/maps/

Data for UK and Europe

http://www.map24.com
http://www.mimas.ac.uk/spatial/maps/barts/

Data for UK

http://www.edina.ac.uk/digimap
http://www.landmark-information.co.uk
http://www.ordnancesurvey.co.uk/oswebsite/getamap

Data for USA

http://www.topozone.com

Remember also to provide the appropriate attribution with the figure caption, e.g. 'adapted from . . .' or 'redrawn from . . .' (see the section on questions of copyright). Since the proper design and use of illustrations is usually an important component of the assessment of a dissertation, you will earn credit for producing a professionally finished product in this way.

A range of digital maps and atlases is available on the web, some providing international coverage (See Table 19.1). For UK users one important source is Ordnance Survey's digital map data for Great Britain that can be accessed through Edina's online 'Digimap' service (Table 19.1). This service, to which all UK higher education institutions are eligible to subscribe, enables users to access a range of map features for Britain including roads, rivers, settlements, administrative boundaries, place names, land use and contours at a variety of source scales (from 1:1250 to 1:250,000). Users can select specific local areas of interest, choose to include or exclude particular features and then either print the digital map, or download it for subsequent use with GIS or mapping software (Figure 19.3). An alternative source is provided through Bartholomew's digital map data that are available through MIMAS (Manchester Information and Associated Services) (Table 19.1). This source includes digital topographic and thematic data for the UK and Europe at a range of scales (e.g. 1:1 million for Europe, 1:200,000 for GB and 1:50,000 for the London area). MIMAS also hosts a range of other useful digital spatial data including population census boundaries and recent satellite image data for the UK. The Digimap initiative is discussed further in Chapters 5 and 16, the latter of which also discusses important issues relating to data quality in digital map data.

Ordnance Survey's 'Get-a-map' service is another useful source of digital map data and is free of charge for personal use. This provides map data for scales of up to 1:50,000 for the UK. Digital map data can also be accessed through commercial companies such as Landmark who supply both current and historical

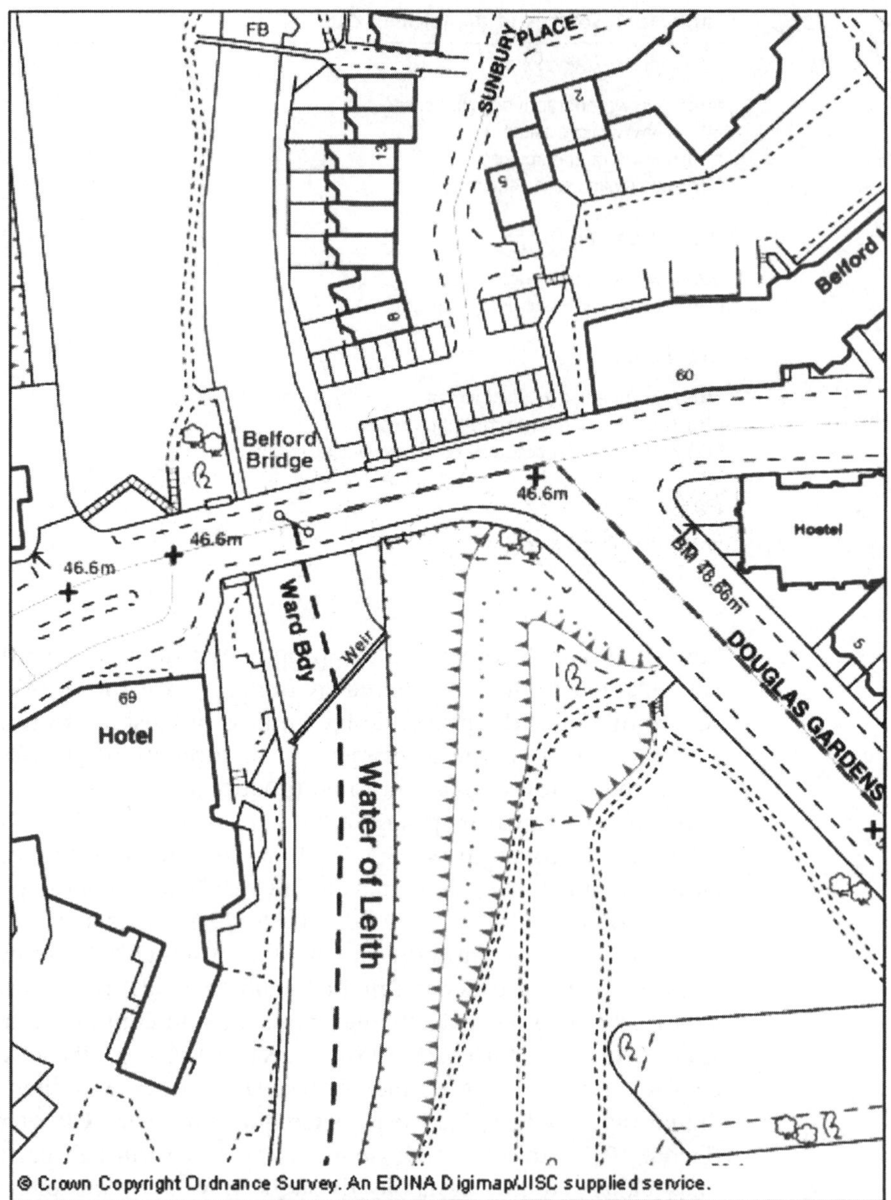

Figure 19.3: An example of Ordnance Survey's digital map data for Great Britain ('Digimap') showing part of Leith at a scale of 1:1250

Source: Ordnance Survey © Crown Copyright. Crown copyright material is reproduced with the permission of the Controller of HMSO and the Queen's Printer for Scotland

digital maps for the UK although clearly there are cost implications with these sources (Table 19.1).

Where it is necessary to create your own original maps this process may involve field survey perhaps using GPS (Global Positioning Systems). Hand-held GPS receivers are available at relatively low cost and can be extremely valuable for collecting spatial information in situations where large-scale secondary mapped data are not yet available, as in remote parts of some lower-income countries, for example. Locational (e.g. latitude/longitude) and associated attribute data then need to be downloaded to a PC for subsequent mapping. For further information on using GPS in the field see Chapter 16. In some studies, production of original maps may form an integral part of the research process itself, for example where mental maps are created for you by your research participants and which form a key source of primary data. This serves as a reminder of the importance of planning ahead since if your project involves groups of people generating participatory maps on large pieces of paper or even with sticks in the ground (Chambers, 1994b), you will need to consider how these are to be transferred to the appropriate page size for presentation in your report. Mental or participatory maps are perhaps best transferred from original paper format to your dissertation by scanning since this retains the impression of the original image (which a process such as digitising does not) or, if this is not possible, by photocopying (and if necessary reducing) to page size. Note though that the larger the original, the more likely it is that detail will become illegible on reduction.

Questions of copyright

Copyright is an important but sometimes complex issue and with the advent of digital sources of geographical information, it is becoming more important for users of map-based data to have an understanding of what is and what is not allowed under copyright law. It is essential to acknowledge clearly the sources of all illustrative material where it is not your own, and if such material is based upon a published source but has been used in a modified form, the acknowledgement should include the phrase 'based upon'. Copyright issues are discussed in more detail in Chapter 18. In general terms all maps and diagrams require some form of acknowledgement unless they are your own original designs, and the appropriate wording is readily found alongside most web-based sources of map data.

Creating and presenting illustrations

In those projects where a map, or other illustration, takes on more significance than simply acting as a backcloth, and even in these cases, it is unlikely that an existing published map will provide exactly the right features at the right scale for your purposes. It is therefore expected that at least one of your illustrations will involve some original creative input from you whether this is designing your own original map from surveyed data, combining digital map data with geographical information of your own or modifying an existing illustration to better meet your own needs. This, then, entails thinking more seriously about the cartographic principles of map design and production. The demise of 'traditional' cartographic techniques – producing illustrations by using drawing pens and ink – means that many geography students are now less familiar with the rudiments of cartographic conventions. This is not to say that computer-assisted cartography requires no prior knowledge of such conventions (indeed, the design principles are the same however an illustration is produced) but rather that because computer software facilitates the process of producing maps and diagrams, there is now less need for courses in traditional cartography to form a distinctive component of a degree programme.

The intention here is not to describe the numerous detailed functions of individual cartographic or mapping software packages. Software undergoes frequent improvements and development so any description of individual features or capabilities rapidly becomes out-of-date. In addition, to some extent familiarity with one software package can facilitate use of others. It is also becoming more straightforward to exchange data sets between packages since software will often accept data in more than one data file format, and files are interchangeable between commonly used software packages such as MapInfo, ArcView and IDRISI. Rather, the aim here is to give general guiding principles on how to produce a good quality illustration which conveys the key information clearly and simply. For more in-depth detail you should refer to the relevant user guide for the specific software packages that are available in your own department or IT service and for which some technical support should also be available.

It is important to consider at the outset which type of software might best meet your needs, although this decision may have already been made for you on pragmatic grounds in terms of what is actually available in your own institution. A range of software exists for producing illustrations, from purpose-written graphic design packages such as Adobe Illustrator to other types of software that can produce illustrations though this is not their primary purpose. Microsoft Excel, for example, is primarily a spreadsheet package but can also be used to create maps and charts, while simple line diagrams, flow charts and illustrations can be constructed within word-processing packages such as Microsoft Word. Although the boundaries between different types of software are becoming increasingly blurred, drawing or graphic design packages generally contain more sophisticated and flexible functions for presenting illustrative material than do, say, GIS which are designed primarily for manipulation, analysis and management of spatial data. Nevertheless, for many projects GIS provides sufficient functionality to produce high quality maps. If you have undertaken some training in GIS and have perhaps used it for analysing your data it makes sense to continue to use this type of software to produce your finished maps. If, on the other hand, you are not familiar with GIS and your data exist as paper maps and/or digital sources in other formats you may want to create your finished illustrations by using a specialised drawing package such as Macromedia Freehand or Adobe Illustrator. Very high quality finished products can be achieved by using these packages based, for example, on hand-sketched maps which are subsequently scanned into the software and then enhanced by using appropriate line styles and thicknesses, and by adding symbols, shading, text labels and a legend (Figure 19.4). Alternatively, more widely available drawing packages such as Microsoft Paintbrush, designed primarily as software for designing graphics and diagrams, can also be used successfully to produce simple maps. Much then, depends on what is available in your own institution, which software you are already familiar with and which package most closely meets your needs. If

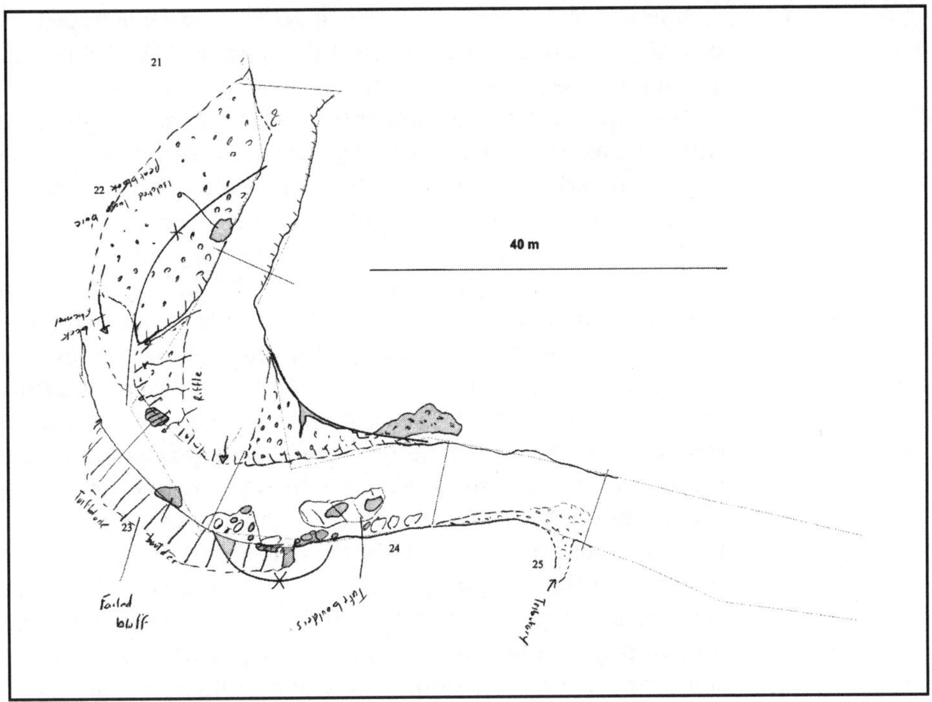

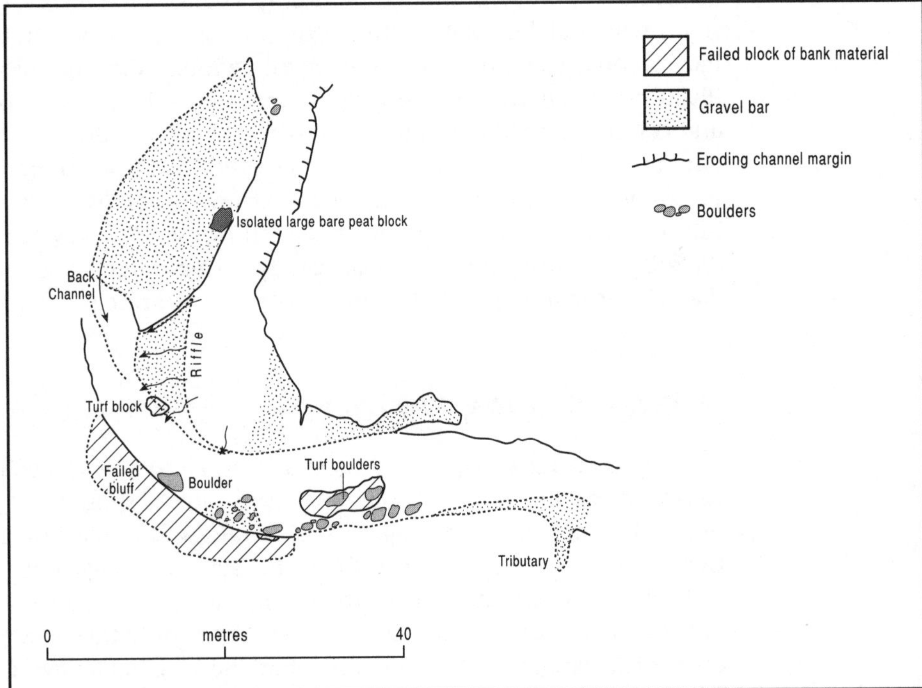

Figure 19.4 Converting sketches to finished illustrations: (a) hand-sketched draft of a river channel; (b) final illustration after digital scanning and enhancement

Source: A J Mills and J Warburton, unpublished research, Department of Geography, University of Durham

computer-assisted cartographic tools are not available to you, traditional methods of map drawing using pens and ink, or transfer lettering, can be used instead, although these too require training and practice.

Although it is relatively straightforward to produce high quality output by using computer-assisted means, an attractive finished product does not automatically imply scholarly content or high quality data, and we remind the reader of the potential misuse of maps in communicating ideas, referred to in the first section of this chapter. One of the key benefits of any form of automation is the ability to save time and labour. The reduction in time spent in producing a map or illustration by automated cartography should mean that more time is available both at the map design stage and for interpreting the results.

In addition, even though cartographic or mapping software packages take away much of the labour-intensiveness of producing illustrations, it is still important to be aware of the established cartographic conventions of map design and production. Indeed, the broad principles of cartography are equally, if not more, applicable when designing maps with the aid of computer packages than when using manual methods. With computer technology there is a danger of misusing the facilities provided by the package. With a plethora of symbols, colours, line types, pattern fills and text characteristics available in most, the opportunity to rotate maps, produce prism maps or present them in three dimensions, and with invitations to combine different map types and graphics on the same display area, the temptation is to select inappropriate and flamboyant styles which at best detract from the purpose of the illustration, and at worst may produce something which is plainly wrong. Although production of colour illustrations is becoming more commonplace, with access to colour laser printers and colour photocopiers, colour illustrations are usually more expensive to produce and colour, in any case, should be used with discretion. The best maps and illustrations are often the simplest: a 'clean' appearance with clear and simple linework and legible annotations. By following a few key principles when producing your illustrations you can readily turn a rough sketch into a professional diagram or map (Figure 19.4) and it is to these principles which we now turn.

Principles of map design

When producing a map, the subject should always be placed on the page so as to obtain the maximum size possible; the subject should never appear isolated or small within the frame (Figure 19.5). Questions of scale quickly arise: usually, the study area will be larger than the standard page size required by your institution for the finished report. Here, standard UK paper sizes are discussed, but students in other countries will be able to apply the same principles to their own local requirements. Providing that the original drawing does not exceed A3 it can be reduced to A4 using a photocopier without losing essential detail and definition. Alternatively, an A3 sheet can be inserted into the standard A4 dissertation format with imaginative folding into three panels of decreasing size and folding out to the right. Larger maps, necessitating more complex folds, are generally not recommended, and maps that are presented as separate entities

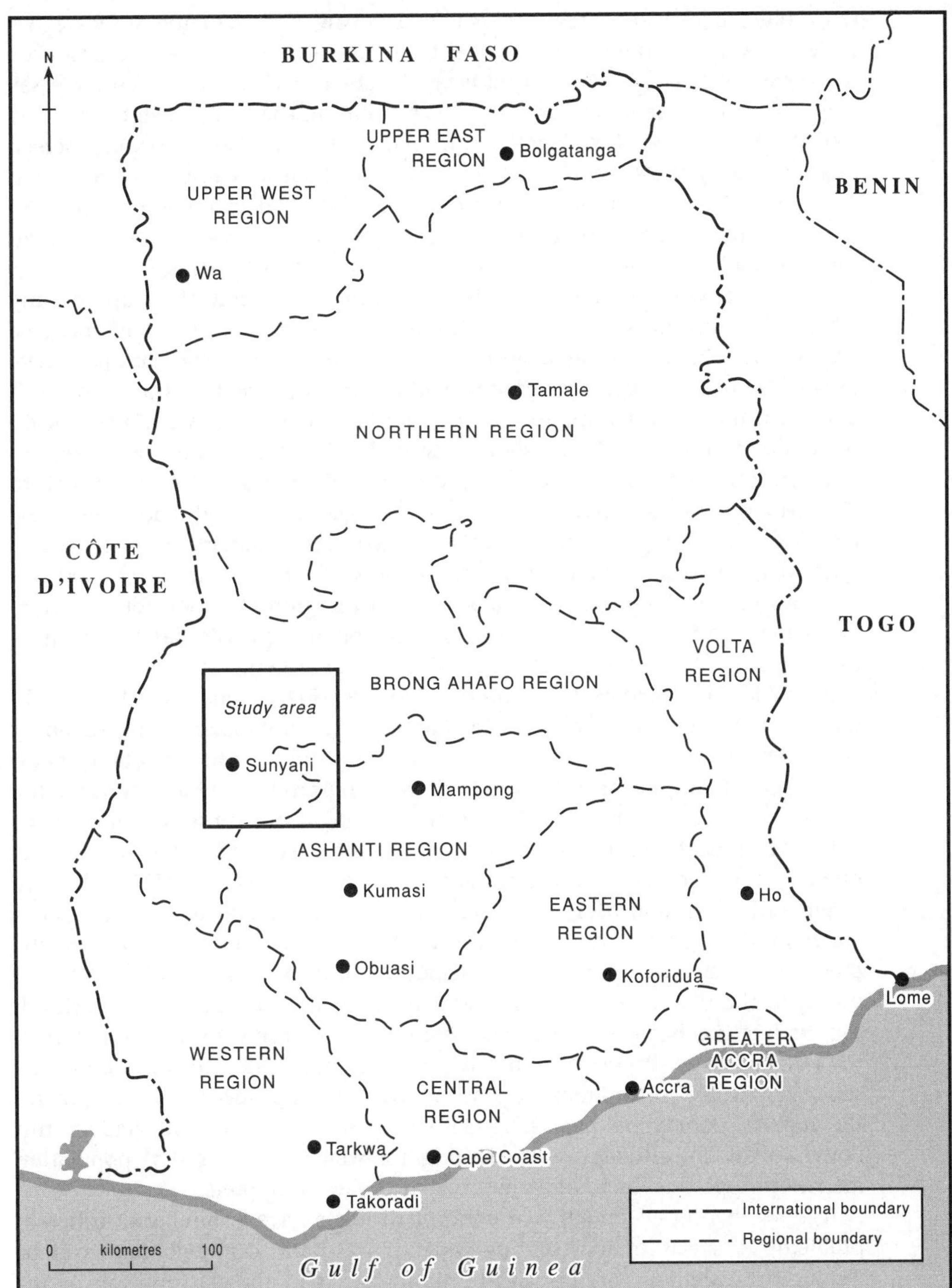

Figure 19.5 An example of a well-proportioned map with effective use of simple line styles, shading and annotation and indicating study area location in a wider country context (Ghana)

Source: Lyon, 2000

(e.g. rolled maps) are rarely acceptable. Drawing and mapping software has a clear advantage here where apparently maps of any scale can be produced from the same data set by 'zooming' in or out to the desired extent. We say 'apparently' because it is important to bear in mind the scale at which the data were originally created in digital form. Producing a map on a computer screen or as hard copy at a scale of 1:50,000 from data which were originally derived at 1:10,000 will result in a lot of unnecessary detail if the data are not 'generalised' in some way, while reversing the process (displaying a map at a larger scale than that originally intended) will create a false impression of accuracy.

The frame, which should comprise a bold line delimiting the map and any surrounding information, helps to improve the overall design and must be proportioned in such a way as to fit within the format of a standard page size, usually A4, in either portrait or landscape format. The title, key, scale and north arrow can then be fitted in, around the subject, where space is available within the frame (Figure 19.5). Scales should always be simple and linear, i.e. comprise a drawn line with units of distance indicated (kilometres, miles or both) since if maps are used elsewhere, they may be enlarged or reduced, and a line scale remains true throughout these processes, whereas a statement such as 'one centimetre to one kilometre' or a representative fraction such as 1:25,000 (i.e. one unit on this map equals 25,000 units on the ground), does not. For more detailed technical guidance on the concepts of cartographic design and map drawing see Robinson *et al.* (1995) or Monmonier (1993).

Above all, it is perhaps the choropleth, or area-shaded, map which is a hallmark of much geographical work in human geography dissertations, mapping spatial variations in a specific variable (e.g. population density, mortality rates) by simply shading the areas (Figure 19.6). Shading can be used to emphasise the idea of contrasts, as in maps of land ownership, with the use of solid colour, hatching or dots. Solid black and white are effective when plotting a single phenomenon for a large area, say the woodland in Europe, the USA or Britain, leaving all other land uses white. Shading can also be used to show breaks in a statistical range, choosing four distinctive types of shading to indicate the presence of the four parts of a distribution broken into quartiles. 'Natural' breaks in the distribution can also be used to create the mapped categories. If the shading on choropleth maps is carefully created or sensitively selected, then the range of intensity used, from the darkest to the lightest, can bring out the subtleties in the distribution: choosing the darkest category of shading for the highest population per square kilometre, for example, and grading this down to the lowest density of shading for the lowest levels of population marries cartographic technology with what the eye interprets.

It is worth noting though that choropleth maps have a number of inherent problems as a way of representing geographical data. Different maps can be produced by slightly varying the threshold breaks in the classification or the number of classes. In addition, the size and shape of areal units used in some choropleth mapping may vary over space and time e.g. electoral wards in Britain. Since the definition of unit boundaries is often essentially arbitrary, different results can be obtained by varying the boundary locations (the 'modifiable areal unit problem', see Chapter 11 and Openshaw, 1984). Some authors reject the use of choropleth maps, especially in human geography, in favour of cartograms

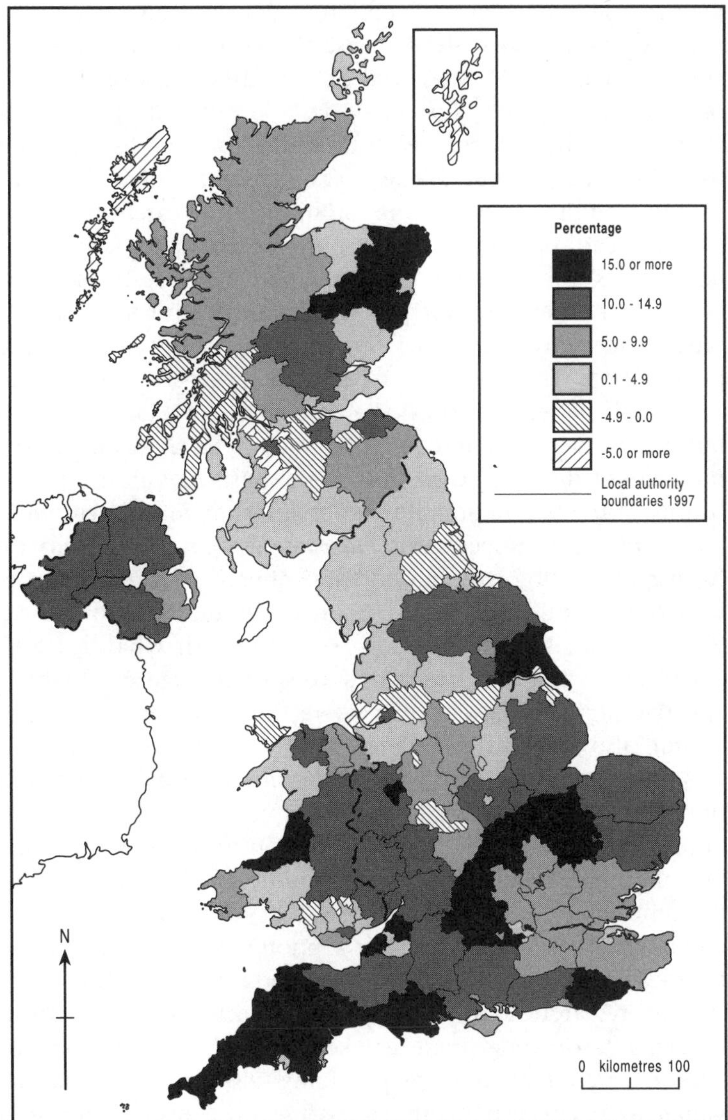

Figure 19.6 Choropleth map of percentage change in population in the UK, mid 1981–1999

From National Statistics (2001). Crown copyright material is reproduced with the permission of the Controller of HMSO and the Queen's Printer for Scotland

(e.g. Dorling, 1994; Tobler, 1973) in which the spatial units used are altered or transformed such that their area is proportional to the underlying variable of interest, e.g. population size.

Other types of illustrative material

The principles that underlie the creation of a good map are equally valid when the subject is a *graph* or a *diagram*. These can be generated at the same time as your maps by using graphic design or drawing software, although you may find it more convenient instead to use the functions of the software that contains your raw data, e.g. a statistical or spreadsheet package. Produced on a printer of reasonable quality (usually a laser printer) illustrations from this type of software will usually suffice for an undergraduate dissertation without the need for re-drawing (although see earlier comments on house style). When presenting data in graphical form, it is important to think carefully about the type of graph that is most appropriate to the purpose and data. Using straight lines to join data points, for example, is acceptable where the lines themselves take on an appropriate 'meaning' such as with data relating to time trends but they are often incorrectly used when a histogram or bar chart is the correct form of presentation. Thought also needs to be exercised over shading those portions of a graph between two associated curves; for example, when plotting annual births and deaths it is common practice to shade those portions where the numbers of deaths exceed births, emphasising catastrophic events within the history of the population. As with maps, though, the temptation to experiment, and to produce unnecessarily elaborate graphs, is a potential pitfall and excessive use of shading, colour and three-dimensional effects on graphs serves only to detract from, rather than enhance, the message contained within the data. Figure 19.7(a) shows an example of this with what Tufte (1983) refers to as 'chartjunk' while an alternative version in Figure 19.7(b) illustrates a more effective and appropriate way of representing the same data. This improved version also reveals the inherent simplicity of the original data set.

Appropriate use of black and white or colour *photographs* enhances a dissertation where the image is used to convey an immediate message, and which otherwise would entail a long description in the text, e.g. stages of urban decay, neighbourhood characteristics, or examples of public and private spaces. Bear in mind that in certain settings (e.g. lower-income countries or 'deprived' communities) the act of taking a photograph may present its own set of difficulties where there are sensitivities about your position as 'researcher' in relation to 'researched'. In this type of project, check first with relevant research participants or appropriate gatekeepers and take their advice. In other work, your research may involve providing disposable cameras to groups or individual research participants themselves as a useful way of eliciting views on key social, cultural or environmental aspects of your study area. Examples of appropriate use of photographs in participant observation work are covered in Chapter 10.

In terms of presentation the best results for including photographs in a report are probably obtained by using a digital camera. This not only allows photographic images to be downloaded and incorporated directly into the text but also allows you to experiment with the appropriate software (e.g. Adobe Photoshop) to enhance colour or contrast, to edit or crop unwanted features and so on. As an alternative, reasonable results can be obtained by scanning hard-copy print photographs into your work or, failing that, by using good quality colour photocopying. Use of adhesive tape, photo corners or similar materials to attach photographic prints to the page will result in a much poorer effect. Situate your photographic images squarely in the appropriate place in

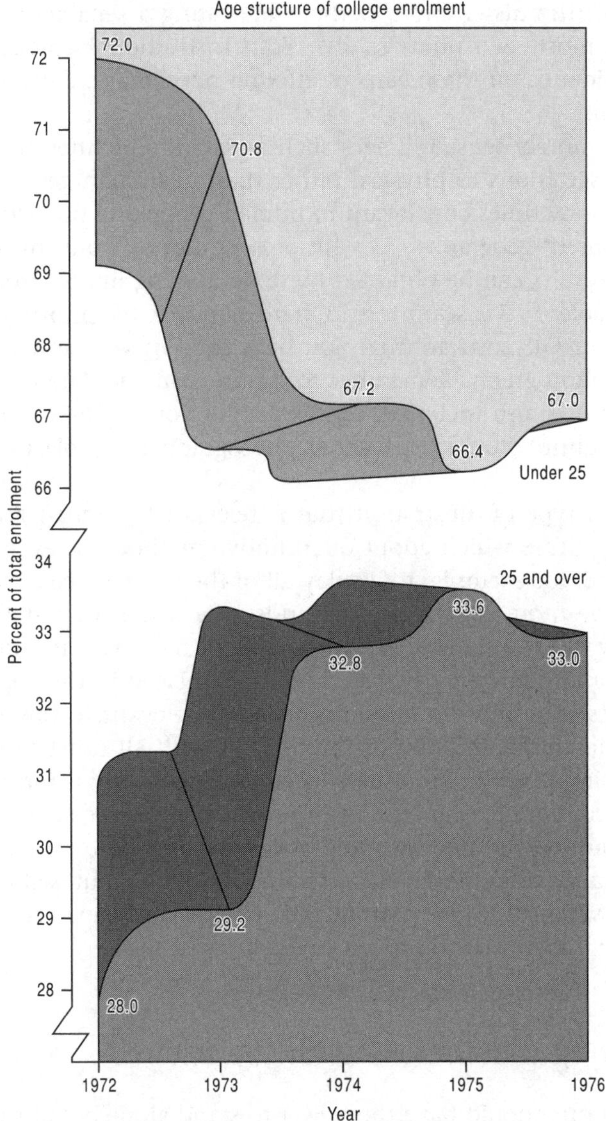

Figure 19.7(a) Excessively elaborate graphical representation of a simple data set

1972 . . . . 28.0

1973 . . . . 29.2

1974 . . . . 32.8

1975 . . . . 33.6

1976 . . . . 33.0

Age structure of college enrolment: *percent of total enrolment 25 and over*

Figure 19.7(b) Alternative version of the same data as in Figure 19.7(a) presented as a dot chart

From Michael Friendly, Psychology Department, York University, Toronto, personal communication the body of the dissertation, perhaps allocating a separate page for displaying one or, at most, two photographs. Your institutional guidelines may provide explicit guidance on the means of affixing photographic material into the final dissertation.

While remotely sensed images such as satellite pictures are more commonly seen in dissertations in physical rather than in human geography, *aerial photographs* can sometimes be relevant to human geography projects, e.g. in historical or development geography. As with presentation of conventional photographs, good end results can be obtained by downloading images from existing digital sources (Table 19.1), scanning in hard copy or by printing from the image processing or GIS software that you have used in your initial interpretation of the aerial photograph. Remember to ensure that the date on which the image was captured is also included, e.g. as part of your figure caption, and that any additional annotation to the aerial photograph is legible in the final printed version.

One final type of illustration that is relevant to human geography projects, particularly those which adopt quantitative methodologies, is the *table*. Tables may be used, for example, to display all of the raw data collected in the field, or they may be more concise and comprise the results of a statistical analysis or a summary of information, as with a table of response rates for a questionnaire survey. The former type of table is normally placed in the appendix of a report while tables of results are more immediately relevant to the reader and should be placed in the main body of the text. As with all other types of illustration, clear and simple tables are usually preferable to overly complex ones. Judicious use of spacing and/or horizontal lines will help to create an effective finished product. Numbers in tables should always be aligned by the decimal point and each row and column should contain a succinct but self-explanatory label. Where possible try to use portrait rather than landscape orientation and never break a table across two or more pages.

Integrating illustrations with the text

All illustrations should have the power to stand alone but at the same time they will need to be properly situated and integrated with your text. Thus, it is generally not acceptable for an illustration to appear in the body of the dissertation without any accompanying text that discusses or refers to it. These references in the text should *always* include the figure number: for example, 'Figure 1.1 shows the location of the study area . . .' rather than 'the following figure shows . . .'. You will also need to include separate 'lists' of figures and tables at the start of the dissertation, usually immediately after the list of contents. These lists should include the figure (or table) number, the same caption as attached to the figure in the main text and the number of the page on which it appears. Your own institution may provide specific guidelines on labelling and listing illustrations that you should follow in the first instance, but if these are not available, browse the chapters in this book for an idea of an appropriate format to follow.

Illustrations should be positioned as close as possible to, and after, their *first* mention in the text. It can be irritating for the reader if you refer, for example, to a 'Figure 6' which the reader is expecting to find on the following page but which in fact appears several pages later. Grouping a series of illustrations together at the end of a chapter or at the end of the report should be avoided unless guidelines specifically indicate otherwise. Illustrations should be labelled sequentially throughout the report perhaps by using numbering related to each chapter (e.g. Figure 2.1, Figure 2.2 as in this book, or simply Figure 1, Figure 2 etc.). Whilst photographs can be labelled either as 'figures' or 'plates' it is generally advisable otherwise to confine yourself to the terms 'figure' and 'table' and to avoid labels such as 'map' or 'graph'. Maps are a kind of figure and should be labelled as such. Tables should be allocated a separate series of labels (Table 1 or Table 1.1 etc).

Having produced an illustration with which you are happy it is then easy to overlook the need to ensure that it has a succinct and appropriate caption. Try to ensure that the caption captures the essence and purpose of the illustration without being too long-winded; there is no need, for example, to use phrases such as 'a map of . . .' or 'a diagram which shows . . .'. Remember also to include appropriate acknowledgement where relevant; for example, if you have reproduced a map from a secondary source you need to give proper credit to the original source. For further discussion of acknowledging sources and issues of copyright see Chapter 18.

Conclusion

A dissertation or project report constitutes a substantial piece of work that often accounts for a significant proportion of the final degree mark and, as such, it deserves a professional approach from you. The production of illustrations should not be seen as somehow less important in this process. Indeed, a set of high quality, polished illustrations can help to persuade the reader, and examiner, of the serious way in which you have engaged with the work. It is therefore important from the outset to regard your illustrations as a fundamental part of the research process. Illustrations are essentially used to communicate ideas succinctly while maps convey ideas about spatial relationships. While some illustrations form an end result in themselves, others serve as a starting point for further analyses or interpretations.

This chapter has provided a number of ideas for potential sources of maps and illustrations although the need for creative input from you in adapting existing illustrative material has also been emphasised. Some of the guiding principles of designing and presenting maps and other types of illustration have been provided in order to help you achieve this. It is worth reiterating that the best maps and illustrations are often the simplest. Mapping software can help you to provide a professionally finished design in this way although this does not necessarily imply that the underlying content is of equally high quality.

Some human geographers have criticised the idea of using conventional maps as inappropriate and highly selective ways of representing 'place'. In contrast, mental and participatory maps provide highly effective ways of eliciting geographical information from research participants who may not be familiar with the rudiments of 'official' maps such as those produced by national mapping agencies. Parallel to this, the advent of computer-assisted cartography and GIS has opened up fresh opportunities for visualising human and physical environments. This chapter has shown how, for example, different types of spatial representation can be integrated in innovative and exciting ways to create 'multiple realities' of space. This is only one example of the way in which maps can be used convincingly to demonstrate that you have given serious thought to the best way of 'illustrating your report'.

BOX 19.1

Further reading

Dorling, D and D Fairbairn 1997 *Mapping: ways of representing the world*, Harlow: Longman

Monmonier, M 1993 *Mapping it out*, Chicago: University of Chicago Press

Robinson A H, J L Morrison, P C Muehrcke, A J Kimerling and S C Guptill 1995 *Elements of cartography*, 6th edn, New York: Wiley

20 Conclusion

Robin Flowerdew and David Martin

This book is intended to help you design, conduct and write up a research project in human geography. Human geography is vast, diverse and exciting, and the results of your endeavours may look very different depending on the type of project you have undertaken. You will almost certainly not have achieved all that you set out to do, but we hope that you have achieved enough to satisfy both yourself and your examiners!

Letting people know what you have found

Your institution will require you to submit a number of copies in a specified form, but your responsibilities for reporting on what you have done may go beyond that. People who have helped you along the way may be interested in what you have found out. In some cases they may have specifically requested to see your findings; if so, it is very important to keep any promises you have made. Otherwise, use your judgement about who to show it to. The most obvious question to ask is 'Would they be interested?'; another might be 'Could it do any good?'

If your results have policy implications, and if you feel your work is good enough to withstand critical scrutiny, it may be worth passing on to a community group, local planners, or the press. It may also be worth letting the academic community (and the students of the future) know what you have done. Colossal amounts of energy are poured every year into research projects and undergraduate dissertations; archivists, officials, survey respondents and interviewees give of their time; many interesting conclusions are reached; but very little of this work is ever read or consulted after it has been marked, simply because nobody knows about it. Findings from student research projects sometimes reach the academic journals, but on far more occasions they are lost because nobody takes the initiative and the time to take them any further.

You may be keen to pass on the results of your research, but unfortunately it is seldom appropriate just to photocopy your dissertation and send it off. Most people will prefer a much shorter and non-technical account than the one demanded by the examiners. Theoretical debates, methodological technicalities and references to previous literature will usually be of little interest to the people you have been studying or who have helped you. Just as it is important to learn how to set out what you have done and how you have done it in the form of a project report or dissertation, so it is important (probably more so in the long run) to be able to express the key aspects of what you have done in a clear, readable and concise manner. The abstract or 'executive summary' for a research project is probably the most important part (except for the title!), if only because more people are likely to read it, only progressing to the main body of text if the abstract has convinced them it is going to be worthwhile.

The appropriate form for publication in an academic journal is different again. Here the most important difference is likely to be the need to focus closely on one specific point – 'have something to say, and say it'. As in the dissertation itself, however, it will be necessary to introduce the topic and to link it to related academic literature. The design of the study and the methods used in analysing the data must be included. If you think that it might be worth developing your work in this way, you should begin by suggesting the idea to your supervisor and seeking his or her guidance on the next steps to take.

Learning from doing a research project

Apart from the findings that your research project may have come up with, what else have you accomplished? Geography departments ask you to do a research project not (usually!) because it is a cheap way of conducting the research but because they hope you will learn a lot from doing it. Think about what you have learned.

First, you will have learned a lot about whatever you have been studying – your initial impressions may have turned out to be right, wrong or irrelevant, but there will be details, constraints and new aspects of the situation that you were not aware of before the study; if your study involved working with people, you will have learned much about how they see their world and what is important to them.

Secondly, you will have learned much about methods for research. A keen student can learn a lot from a course in research methods, but almost everybody will learn a lot more when applying those methods for themselves. Obviously you will have learned most about the methods you have used, but you may well have also learned a good deal about methods you considered and rejected.

Thirdly, you will have learned about conducting a research project. Especially if the project is large and involves substantial independent research, it is likely to be the most demanding piece of work you will have undertaken. The number of separate activities undertaken (project design, literature review, data collection, data analysis, writing up) distinguishes it from other tasks students are asked to do; so does the need for planning your activities, perhaps months or even a year in advance; so does the requirement that it is independent and original research; so does the mere length of the finished project.

Finally, you will have learned something about yourself. Doing a dissertation or major research project is a demanding task. Completing it successfully is an achievement, even if (like most people) you have not done everything you intended. Many people go through dramatic changes of mood – perhaps trepidation when starting out, enthusiastic anticipation as the project takes shape, alternating excitement and depression when things go well or go badly, frustration and boredom when it seems to have been dragging on for ever, and finally a mechanical determination to get it out of the way before the final deadline. It may take a while after you have finished before you can put the experience in perspective. Many people (not all) look back on the research project as the most satisfying part of their course. For some, it creates or feeds an appetite to do more research, an ambition that may be fulfilled in many walks of life, not just the academic.

Clearly the best research project is one which comes up with interesting findings and fulfils your original goals. If it does not, however, it is not a complete failure. You may learn as much if not more from failure as from success. A project that has failed to do what was originally intended may still have taught you a great deal, especially if you can work out what went wrong and how, if there were to be a next time, you could avoid the problems.

Using your dissertation

Research projects in one form or another are common in many walks of life. A high proportion of geography graduates are likely to end up in jobs which from time to time require them to find out about some topic and to report what they have found out, usually in written form, sometimes in a verbal presentation. This may take the form of summarising the activities they have been involved in, it may present the case for exploring a new opportunity, it may be intended to help in taking a decision or evaluating organisational procedures, or it may be producing copy for public relations. Such activities are common in industry, commerce, public services, central and local government. In comparison with the academic project report, most work-related reports are likely to require much tighter timescales, and to be much shorter and to the point. Library resources will be far less adequate, and questions will have to be answered and opinions formed even when adequate information is not available on which to base these answers and opinions.

Successful completion of research projects is one of the more saleable achievements that a geography graduate can offer a potential employer. Obviously it is more relevant to job performance than the ability to answer examination questions in an hour without referring to outside sources. A human geography project is particularly relevant as it usually involves some degree of encounter with the world beyond academia, and in most cases offers the student plenty of scope for originality and independent work. The knowledge gained of research methods may be of direct relevance to many jobs, as will the ability to use appropriate word-processing, database, spreadsheet or graphics software. Possibly more important is the experience gained of planning and carrying through a major project on your own. Perhaps the main drawback of most research projects from a prospective employer's point of view is that they are usually carried out by individuals working alone and do not involve the vital skills of teamwork required in almost all job environments.

The most obvious way in which a research project can be of relevance to the job market is where the topic is of direct relevance to the job. Students may choose a topic with no thought for future employment, but studying it may lead to interest in the area, contact with relevant organisations and perhaps knowledge of employment opportunities. Some students end up working with organisations they first contacted as part of a research project, and others with organisations working in a related field. Clearly knowledge of a topic is helpful to getting a job with such an organisation, though applicants should recognise that members of the organisation may know (or think they know!) more than a student can find out in one research project. Our experience suggests that your research project may make a good topic for discussion at job interviews as it provides an opportunity to talk about the aspect of your university course about which you now have the greatest depth of understanding.

For some, the process of conducting independent research as part of a taught course in geography will have raised the possibility of undertaking a research degree. Many good Masters' and doctoral theses have had their roots in undergraduate dissertations which have raised interesting research questions or prompted an interest in a particular methodology. Project supervisors should be able to guide you towards possible research supervisors and funding sources, and may even be keen to get involved themselves. Your experience with this project will have provided a glimpse of the frustrations and considerable rewards to be gained from a major piece of independent study lasting two or three years. We certainly hope that some of you will be encouraged to take the plunge!

The corpus of research methods

The central sections of this book have presented a variety of different methods, all of which were in reasonably common use by human geographers in their research at the time of writing. Not all methods have been covered, and some have not been discussed in much detail; we hope that you have not been disappointed that the methods you particularly wished to use are not covered in sufficient depth. The choice of methods, and the way in which they are covered, is based on our perception, informed by consultation with the chapter authors and many other colleagues in different institutions, of what was most relevant to writers of dissertations and research projects in the many research traditions of contemporary human geography. The coverage of the book is based on our reconstruction of how geographers of the recent past have interpreted, modified, supplemented and used the sets of methods handed on to them by their predecessors.

A book of this nature published in 1955 would have looked very different, perhaps containing chapters on map interpretation or regional synthesis, and very little on either qualitative or quantitative methods as they are now understood. One published in the 1970s might have been entirely quantitative, like Daly (1972) or Haggett, Cliff and Frey (1977). Even in the 1980s, there would have been less emphasis on qualitative methods, with the 'Analysis of texts' chapters and the material on GIS and participatory methods being the most unlikely to have been present. As you read this in the twenty-first century, the selection of material may already seem odd. Think about which chapters no longer seem appropriate, and what other chapters might be added. If we are still around, let us know what you think!

References

Abbot, J, R Chambers, C Dunn, T Harris, E de Merode, G Porter, J Townsend and D Weiner 1998 Participatory GIS: opportunity or oxymoron?, *PLA Notes* 33, 27–34

Adams, P 1992 Television as gathering place, *Annals of the Association of American Geographers* 82(1), 117–135

Agar, M 1980 *The professional stranger: an informal introduction to ethnography*, New York: Academic Press

Agar, M 1986 *Speaking of ethnography*, Beverly Hills: Sage

Aitken, S C 1991 A transactional geography of the image-event: the films of Scottish director, Bill Forsyth, *Transactions, Institute of British Geographers* New Series 16(1), 105–118

Aitken, S C 2001 Tuning the self: city space and SF horror movies. In R Kitchin and J Kneale (eds) *Lost in space: geographies of science fiction*, London and New York: Continuum Press, pp. 104–122

Aitken, S C 2002 Public participation, technological discourses and the scale of GIS. In W Craig, T Harris and D Weiner (eds) *Community participation and geographic information systems*, London and New York: Taylor and Francis, pp. 357–366

Aitken, S C and J Craine 2002 The pornography of despair: lust, desire and the music of Matt Johnson, *ACME: An International E-Journal for Critical Geographers* 1(1), 91–116

Aitken, S C and L E Zonn 1993 Weir(d) sex: representation of gender–environment relations in Peter Weir's *Picnic at Hanging Rock* and *Gallipoli*, *Environment and Planning D: Society and Space* 11, 191–212

Aitken, S C and L E Zonn (eds) 1994 *Place, power, situation and spectacle: a geography of film*, Lanham, Maryland: Rowman and Littlefield

Aitkin, M, D Anderson, B Francis and J Hinde 1989 *Statistical modelling in GLIM*, Oxford: Clarendon Press

Allport, G W 1954 Attitudes in the history of social psychology. In N Warren and M Jahoda (eds) 1979 *Attitudes – Selected Readings*, Harmondsworth: Penguin

Althusser, L 1972 *Politics and history: Montesquieu, Rousseau, Hegel and Marx*, London: New Left Books

Amit, V (ed.) 2000 *Constructing the field: ethnographic fieldwork in the contemporary world*, London: Routledge

Anselin, L 1988 *Spatial econometrics, methods and models*, Dordrecht: Kluwer Academic

Appleton, K and A Lovett 2003 GIS-based visualisation of rural landscapes: defining 'sufficient' realism for environmental decision-making, *Landscape and Urban Planning* 65(3), 117–131

Area 28(2) 1996 Special issue on focus groups

Association for Geographic Information 1993 *Report of the copyright working party*, London: Association for Geographic Information

Bailey, C, C White and R Pain 1999 Evaluating qualitative research, dealing with the tensions between 'science' and 'creativity', *Area* 31, 169–183

Bailey, T C and A C Gatrell 1995 *Interactive spatial data analysis*, Harlow: Longman

Barbour, R and J Kitzinger (eds) 1999 *Developing focus group research*, London: Sage

Barnes, T and J Duncan (eds) 1992 *Writing worlds: discourse, text and metaphor*, London: Routledge

Barthes, R 1972 *Mythologies*, translated by A Lavers, London: Hill and Wang

Bateman, I J, K G Willis, G D Garrod, P Doktor, I Langford, and R K Turner 1992 Recreation and environmental preservation value of the Norfolk Broads: a contingent valuation study. Report to the National Rivers Authority, Norwich: School of Environmental Sciences, University of East Anglia

Baxter, J and J Eyles 1997 Evaluating qualitative research in social geography: establishing 'rigour' in interview analysis, *Transactions. Institute of British Geographers* 22, 505–525

Bedford, T and J Burgess 2001 The focus-group experience. In M Limb and C Dwyer (eds) *Qualitative methodologies for geographers: issues and debates*, London: Arnold, pp. 121–135

Bell, C and H Roberts (eds) 1984 *Social researching: politics, problems and practice*, London: Routledge

Bell, D, P Caplan and W J Karim 1993 *Gendered fields: women, men and ethnography*, London: Routledge

Bell, J 1999 *Doing your research project: a guide for first-time researchers in education and social science*, 3rd edn, Buckingham: Open University Press

Benko, G and U Strohmayer (eds) 1997 *Space and social theory*, London: Blackwell

Benton, L 1995 Will the reel/real Los Angeles please stand up?, *Urban Geography* 16(2), 144–164

Berry, W D 1993 *Understanding regression assumptions*, Quantitative Applications in the Social Sciences 92, London: Sage

Berry, W D and S Feldman 1985 *Multiple regression in practice*, Quantitative Applications in the Social Sciences 50, London: Sage

Bertin, J 1983 *Semiology of graphics*, translated by W Berg, Madison, WI: University of Wisconsin Press

Black, I S 2003 Analysing historical and archive sources. In N J Clifford and G Valentine (eds) *Key methods in geography*, London: Sage, pp. 477–500

Black, T R 1993 *Evaluating social science research: an introduction*, London: Sage

Bloor, M, M Frankland, M Thomas and K Robson 2001 *Focus groups in social research*, London: Sage

Bondi, L 1990 Progress in geography and gender: feminism and difference, *Progress in Human Geography* 14(3), 438–445

Bondi, L 1993 Locating identity politics. In M Keith and S Pile (eds) *Place and the politics of identity*, London: Routledge

Bondi, L 1999 Stages on journeys: some remarks about human geography and psychotherapeutic practice, *The Professional Geographer* 51(1), 11–24

Bondi, L and J Davidson 2003 Troubling the place of gender. In K Anderson, M Domosh, S Pile and N Thrift (eds) *Handbook of cultural geography*, London: Sage, pp. 325–343

Bondi, L and M Domosh 1992 Other figures in other places: on feminism, postmodernism and geography, *Environment and Planning D: Society and Space* 10, 199–213

Booth, B and A Mitchell 2001 *Getting started with ArcGIS*, Redlands, CA: ESRI

Boots, B N and A Getis 1988 *Point pattern analysis*, London: Sage

Bourdieu, P 1984 *Distinction: a social critique of the judgement of taste*, London: Routledge

Bourdieu, P 1990 *The logic of practice*, Cambridge: Polity Press

Bowlby, S and J Silk 1982 Analysis of qualitative data using GLIM: two examples based on shopping survey data, *Professional Geographer* 34, 80–90

Bowlby, S, J Lewis, L McDowell and J Foord 1989 The geography of gender. In R Peet and N Thrift (eds) *New models in geography: the political economy perspective*, London: Unwin Hyman, Vol II, 157–175

Bradford, M 2003 Writing essays, reports and dissertations. In N J Clifford and G Valentine (eds) *Key methods in geography*, London: Sage, pp. 515–532

Bryman, A 1988 *Quantity and quality in social research*, London: Unwin Hyman

Bryman, A and D Cramer 1994 *Quantitative data analysis for social scientists*, London: Routledge

Bryman, A and D Cramer 2001 *Quantitative data analysis with SPSS Release 10 for Windows: a guide for social scientists*, Hove: Routledge

Buckley, A, M Gahegan and K Clarke 2000 Geographic visualization. In *Emerging themes in GIScience research*, University Consortium for Geographic Information Science White Papers

Budniakiewicz, T 1992 *Fundamentals of story logic: introduction to Greimassian semiotics*, Amsterdam: J Benjamins

Burgess, J 1990 The production and consumption of environmental meaning in the mass media: a research agenda for the 1990s, *Transactions of the Institute of British Geographers* 15, 139–161

Burgess, J 1996 Focusing on fear: the use of focus groups in a project for the Community Forestry Unit, Countryside Commission, *Area* 28, 130–135

Burgess, J and J Gold (eds) 1985 *Geography, the media, and popular culture*, New York: St Martin's Press

Burgess, J, M Limb and C Harrison 1988a Exploring environmental values through the medium of small groups: 1 Theory and practice, *Environment and Planning A* 20, 309–326

Burgess, J, M Limb and C Harrison 1988b Exploring environmental values through the medium of small groups: 2 Illustrations of a group at work, *Environment and Planning A* 20, 457–476

Burgess, R 1982 *Field research: a sourcebook and field manual*, London: Routledge

Burgess, R 1984 *In the field: an introduction to field research*, London: Routledge

Burrough, P A and R A McDonnell 1998 *Principles of geographical information systems*, 2nd edn, Oxford: Oxford University Press

Butlin, R A 1993 *Historical geography: through the gates of space and time*, London: Edward Arnold

Buttimer, A 1982 Musing on Helicon: root metaphors and geography, *Geografiska Annaler* 64B, 89–96

Campbell, D and M Campbell 1995 *Doing research on the Internet*, Reading, MA: Addison-Wesley

Campbell, J B 2002 *Introduction to remote sensing*, 3rd edn, New York: Guilford

Carney, G (ed.) 1994 *The sounds of people and places: a geography of American folk and popular music*, 3rd edn, Lanham, MD: Rowman and Littlefield

Carr, D 1986 *Time, narrative and history*, Bloomington: Indiana University Press

Cassell, J 1988 The relationship of observer to observed when studying up, *Studies in Qualitative Methodology* 1, 89–108

Chalfen, R 1987 *Snapshot versions of life*, Bowling Green OH: Bowling Green State University Popular Press

Chambers, R 1994a Participatory Rural Appraisal (PRA): analysis of experience, *World Development* 22, 1253–1268

Chambers, R 1994b The origins and practice of Participatory Rural Appraisal, *World Development* 22, 953–969

Chambers, R 1997 *Whose reality counts; putting the last first*, London: Intermediate Technology Publications

Chouinard, V 1997 Structure and agency: contested concepts in human geography, *Canadian Geographer* 41, 363–377

Chouinard, V 2000 Getting ethical: for inclusive and engaged geographies of disability, *Ethics, Place and Environment* 3, 70–79

Clark, G 1992 Data sources for studying agriculture. In I R Bowler (ed.) *The geography of agriculture in developed market economies*, Harlow: Longman, pp. 32–55

Clark, G 1995 Enterprise dissertations revisited, *Journal of Geography in Higher Education* 19, 207–211

Clark, H A J 1989 Conservation advice and investment on farms: a study in three English counties. Unpublished PhD thesis, Norwich: University of East Anglia

Clayton, D 1998 Captain Cook and the spaces of contact at Nootka Sound. In J Brown and E Vibert (eds) *Documenting native histories: texts and contexts*, Peterborough, Ontario and Calgary, Alberta: Broadview Press

Cliff, A D, P R Gould, A G Hoare and N J Thrift (eds) 1995 *Diffusing geography*, Oxford: Blackwell

Clifford, J and G Marcus (eds) 1986 *Writing culture: the poetics and politics of ethnography*, Los Angeles and Berkeley: University of California Press

Clifford, N J and G Valentine (eds) 2003 *Key methods in geography*, London: Sage

Cloke P, C Philo and D Sadler 1991 *Approaching human geography*, London: Paul Chapman

Cloke, P, I Cook, P Crang, M Goodwin, J Painter and C Philo 2004 *Practising human geography*, London: Sage

Connolly, W E 2002 *Neuropolitics: thinking, culture, speed*, Minneapolis, MN: University of Minnesota Press

Conover, W J 1999 *Practical nonparametric statistics*, 3rd edn, New York: John Wiley

Cook, I 1995 Constructing the exotic: the case of tropical fruit. In J Allen and D Massey (eds) *Geographical worlds*, Oxford: Oxford University Press, pp. 137–142

Cook, I and M Crang 1995 *Doing ethnographies*, Concepts and Techniques in Modern Geography 58, Norwich: Environmental Publications

Cooke, B and U Kothari (eds) 2001 *Participation: the new tyranny?*, London: Zed Books

Cornwall, A 1992 Body mapping in health, RRA/PRA, *RRA Notes* No. 16, 69–76

Cornwall, A and R Jewkes 1995 What is participatory research?, *Social Science and Medicine* 41, 1667–1676

Cosgrove, D 1989 Geography is everywhere: culture and symbolism in human landscapes. In D Gregory and R Walford (eds) *Horizons in human geography*, Basingstoke: Macmillan, pp. 118–135

Cosgrove, D 2003 Landscape and the European sense of sight – eyeing nature. In K Anderson, M Domosh, S Pile and N Thrift *Handbook of Cultural Geography*, London: Sage, pp. 249–268

LIVERPOOL
JOHN MOORES UNIVERSITY
AVRIL ROBARTS LRC
TITHEBARN STREET
LIVERPOOL L2 2ER
TEL. 0151 231 4022

Cosgrove, D and S Daniels 1988 *The iconography of landscape: essays on the symbolic representation, design and use of past environments*, Cambridge: Cambridge University Press

Cosgrove, D and M Domosh 1993 Author and authority: writing and the new cultural geography. In J Duncan and D Ley (eds) *Place/culture/representation*, London: Routledge, pp. 25–38

Coshall, J 1989 *The application of nonparametric statistical tests in geography*, Concepts and Techniques in Modern Geography 50, Norwich: Environmental Publications

Coyne, M 1997 *The crowded prairie: American national identity in the Hollywood western*, London: IB Tauris & Co. Ltd

Crang, M 1994 Spacing time, timing spaces and narrating the past, *Time and Society* 3 (Feb 1), 29–45

Crang, M 2002 Qualitative methods: the new orthodoxy?, *Progress in Human Geography* 26, 647–655

Crang, M, A C Hudson, S M Reiner and S J Hinchliffe 1997 Software for qualitative research: 1 prospectus and overview *Environment and Planning A* 29, 771–787

Crang, M, P Crang and J May (eds) 1999 *Virtual geographies: bodies, space and relations*, New York and London: Routledge

Crang, P 1994 It's showtime: on the workplace geographies of display in a restaurant in Southeast England, *Environment and Planning D: Society and Space* 12, 675–704

Cresswell, T 1993 Mobility as resistance: a geographic reading of Kerouac's 'On the road', *Transactions of the Institute of British Geographers* 18(2), 249–262

Cresswell, T and D Dixon (eds) 2002 *Engaging film: geographies of mobility and identity*, New York: Rowman and Littlefield

Crick, M 1992 Ali and me: an essay in street corner anthropology. In J Okely and H Callaway (eds) *Anthropology and autobiography*, London: Routledge, pp. 175–192

Cronon, W 1992 A place for stories: nature, history and narrative, *Journal of American History* 78, 1347–1376

Curry, M 1991 Postmodernism, language and the strains of modernism, *Annals of the Association of American Geographers* 81, 210–228

Curry, M 1995 Rethinking the rights and responsibilities of geographic information systems: beyond the power of the image, *Cartography and Geographic Information Systems* 22(1), 58–69

Curry, M 2000 Wittgenstein and the fabric of everyday life. In M Crang and N Thrift (eds) *Thinking space*, London: Routledge, pp. 89–113

Curtis, J 1976 Woody Guthrie and the Dust Bowl, *Places* 3, 12–18

Dale, A and C Marsh 1993 *The 1991 census users' guide*, London: HMSO

Dale, A, S Arber and M Proctor 1988 *Doing secondary analysis*, London: Unwin Hyman

Daly, M T 1972 *Techniques and concepts in geography*, Melbourne: Nelson

Daniel, L, P Loree and A Whitener 2001 *Inside Mapinfo Professional*, Florence, KY: Delmar Thomson Learning

Daniels, S 1993 *Fields of vision: landscape imagery and national identity in England and the United States*, Cambridge: Polity Press

Daniels, S and S Rycroft 1993 Mapping the modern city: Alan Sillitoe's Nottingham novels, *Transactions of the Institute of British Geographers* 18(4), 460–480

Darby, H C 1948 The regional geography of Hardy's Wessex, *Geographical Review* 38, 426–443

Davis, L 1987 *Resisting novels: ideology and fiction*, New York and London: Methuen

Davis, M 2001 Bunker Hill: Hollywood's dark shadow. In M Shiel and T Fitzmaurice (eds) *Cinema and the city*, Oxford: Blackwell

Dear, M 1988 The postmodern challenge: reconstructing human geography, *Transactions of the Institute of British Geographers* 13, 262–274

Dear, M and S Flusty (eds) 2002 *The spaces of postmodernity*, Oxford: Blackwell

de Lauretis, T 1984 *Alice doesn't live here: feminism, semiotics, cinema*, London: Macmillan

DeLyser, D and P F Starrs (eds) 2001 Doing fieldwork, *Geographical Review* 91(1 and 2)

DeMers, M N 2000 *Fundamentals of geographic information systems*, 2nd edn, New York: Wiley

DeMers, M N 2002 *GIS modelling in raster*, New York: Wiley

Denzin, K and Y Lincoln 1998 *Strategies of qualitative enquiry*, London: Sage

Denzin, N 1970 *The research act: a theoretical introduction to social research*, Chicago: Aldine

Desforges, L 1998 'Checking out the planet': global representations/local identities and youth travel. In T Skelton and G Valentine (eds) *Cool places: geographies of youth culture*, New York and London: Routledge, pp. 175–194

Deutsche, R 1991 Boy's town, *Environment and Planning D: Society and Space* 9, 5–30

de Vaus, D A 1991 *Surveys in social research*, London: UCL Press

de Vaus, D A 2001 *Surveys in social research*, 5th edn, London: UCL Press

Devereaux, D and J Hoddinott (eds) 1992 *Fieldwork in developing countries*, Hemel Hempstead: Harvester Wheatsheaf

DiBiase, D, A MacEachren, J Krygier and C Reeves 1992 Animation and the role of map design in scientific visualization, *Cartography and Geographic Information Systems* 19(4), 201–214

Dillman, D A 1978 *Mail and telephone surveys: the total design method*, New York: Wiley

Dixon, C and B Leach 1984 *Survey research in underdeveloped countries*, Concepts and Techniques in Modern Geography 39, Norwich: Geo Books

Dodgshon, R A and R Butlin 1990 *An historical geography of England and Wales*, 2nd edn, London: Academic Press

Domosh, M 1991 Towards a feminist historiography of geography, *Transactions of the Institute of British Geographers* 16(1), 95–118

Donovan, J 1988 'When you're ill, you've gotta carry it': health and illness in the lives of black people. In J Eyles and D Smith (eds) *Qualitative methods in human geography*, Oxford: Polity Press, pp. 180–196

Dorling, D 1994 Cartograms for visualizing human geography. In H M Hearnshaw and D J Unwin (eds) *Visualization in geographical information systems*, Chichester: Wiley, pp. 85–103

Dorling, D and D Fairbairn 1997 *Mapping: ways of representing the world*, Harlow: Longman

Dorling, D and S Simpson (eds) 1999 *Statistics in society: the arithmetic of politics*, London: Arnold

Duncan, J 1993 Commentary on the 'The reinvention of cultural geography', *Annals of the Association of American Geographers* 83(3), 516–517

Duncan, J 1995 Landscape geography, 1993–94 *Progress in Human Geography* 19, 414–422

Duncan, J and N Duncan 1988 (Re)reading the landscape, *Environment and Planning D: Society and Space* 6, 117–26

Duncan, J and N Duncan 2001 Theory in the field, *Geographical Review* 91(1 and 2), 399–406

Duncan, J and D Gregory 1999 *Writes of passage: reading travel writing*, New York and London: Routledge

Duncan, J and D Ley 1982 Structural marxism and human geography: a critical assessment, *Annals of the Association of American Geographers* 72, 30–59

Duncan, J and D Ley (eds) 1993 *Place/culture/representation*, London: Routledge

Dyck, I 1993 Ethnography: a feminist method? *Canadian Geographer* 37, 52–57

Earickson, R J and J M Harlin 1994 *Geographic measurement and quantitative analysis*, New York: Macmillan

Eco, U 1976 *A theory of semiotics*, Bloomington: Indiana University Press

Eliot Hurst, M E 1985 Geography has neither existence nor future. In R J Johnston (ed.) *The future of geography*, London: Methuen, pp. 59–91

Elwood, S and D Martin 2000 Placing interviews: location and scales of power in qualitative research, *Professional Geographer* 52(4), 649–657

England, K 1994 Getting personal: reflexivity, positionality and feminist research, *Professional Geographer* 46(1), 80–89

England, K 2002 Interviewing elites: cautionary tales about researching women managers in Canada's banking industry. In P Moss (ed.) *Feminist geography in practice: research and methods*, Oxford: Blackwell, pp. 200–213

Erickson, B H and T A Nosanchuk 1992 *Understanding data*, 2nd edn, Buckingham: Open University Press

Ethics, Place and Environment (2000) Special issue on Disability, Geography and Ethics 3(1), 61–102

Eyles, J 1988 Interpreting the geographical world: qualitative approaches in geographical research. In J Eyles and D Smith (eds) *Qualitative methods in human geography*, Cambridge: Polity Press, pp. 1–16

Eyles, J and D Smith (eds) 1988 *Qualitative methods in human geography*, Cambridge: Polity Press

Fairbairn, D, G Andrienko, N Andrienko, G Buziek and J Dykes 2001 Representation and its relationship with cartographic visualization, *Cartography and Geographic Information Science* 28(1), 13–28

Feldman, M 1995 *Strategies for interpreting qualitative data*, Beverly Hills: Sage

Ferguson, R 1977 *Linear regression in geography*, Concepts and Techniques in Modern Geography 15, Norwich: Geo Books

Fielding, J L 2000 *Understanding social statistics*, London: Sage

Fielding, N and R Lee 1991 *Using computers in qualitative research*, Beverly Hills: Sage

Fielding, N G and R M Lee 1998 *Computer analysis and qualitative research*, London: Sage

Fine, E 1984 *The folklore text: from performance to print*, Bloomington: Indiana University Press

Flowerdew, R 1993 Poisson regression modelling of migration. In J C Stillwell and P Congdon (eds) *Migration models: macro and micro approaches*, London: Belhaven

Flowerdew, R 1998 Reacting to *Ground Truth, Environment and Planning A* 30, 289–301

Flowerdew, R and M Green 1991 Data integration: methods for transferring data between zonal systems. In I Masser and M Blakemore (eds) *Handling geographical information: methodology and potential applications*, Harlow: Longman, pp. 38–54

Ford, L 1971 Geographic factors in the origin, evolution and diffusion of rock and roll music, *Journal of Geography* 70, 455–464

Fotheringham, A S and D W Wong 1991 The modifiable areal unit problem and multivariate analysis, *Environment and Planning A* 23, 1025–1044

Fotheringham, A S, C Brunsdon and M Charlton 2000 *Quantitative geography: perspectives on spatial data analysis*, London: Sage

Fotheringham, A S, C Brunsdon and M Charlton 2002 *Geographically weighted regression: the analysis of spatially varying relationships*, Chichester: Wiley

Freud, S 1955 Medusa's head. In J Strachey (ed.) *Standard edition of the complete psychological works*, Volume 18, London: Hogarth Press

Fyfe, N (ed.) 1998 *Images of the street: planning, identity and control of public space*, New York and London: Routledge

Fyfe, N 1992 Observations on observations, *Journal of Geography in Higher Education* 16(2), 127–133

Gaile, G L and J E Burt 1980 *Directional statistics*, Concepts and Techniques in Modern Geography 25, Norwich: Geo Books

Gatrell, A C 1991 Teaching students to select topics for undergraduate dissertations in geography, *Journal of Geography in Higher Education* 15, 15–20

Geertz, C 1973 *The interpretation of cultures*, New York: Basic Books

Geiger, S 1986 Women's life histories: method and content, *Signs* 11(2), 334–351

Getis, A and J K Ord 1991 The analysis of spatial association by use of distance statistics, *Geographical Analysis* 23, 189–206

Gibbons, J D 1993 *Nonparametric statistics: an introduction*, Quantitative Applications in the Social Sciences 90, London: Sage

Gibson, S 1999 Noise contour mapping: a case study of Southampton International Airport. Unpublished dissertation, Southampton: Department of Geography, University of Southampton

Gibson-Graham, J K 1996 *The end of capitalism (as we knew it): a feminist critique of political economy*, Oxford: Blackwell

Giddens, A 1984 *The constitution of society* Cambridge: Polity Press

Gilbert, N 1993 *Analyzing tabular data: loglinear and logistic models for social researchers*, London: UCL Press

Gill, W 1993 Region, agency and popular music: the northwest sound, 1958–66, *Canadian Geographer* 37(2), 120–131

Glaser, B and A Strauss 1967 *The discovery of grounded theory*, Chicago: Aldine

Gleeson, B 2000 Enabling geography: exploring a new political-ethical ideal, *Ethics, Place and Environment* 3, 65–69

Glesne, C and A Peshkin 1992 *Becoming qualitative researchers: an introduction*, New York: Longman

Gold, J R 2002 Understanding narratives of nationhood: filmmakers and Culloden, *Journal of Geography* 101(6), 261–270

Goodchild, M F 1987 *Spatial autocorrelation*, Concepts and Techniques in Modern Geography 47, Norwich: Geo Books

Gordon, D and R Behar (eds) 1995 *Women writing culture*, Los Angeles and Berkeley: University of California Press

Goss, J and T Leinbach 1996 Focus groups as alternative research practice: experience with transmigrants in Indonesia, *Area* 28(2), 115–123

Gould, P and G Olsson (eds) 1982 *A search for common ground*, London: Pion

Graham, E 1995 Postmodernism and the possibility of a new human geography, *Scottish Geographical Magazine* 111, 175–178

Graham, E 1999 Breaking out: the opportunities and challenges of multi-method research in population geography, *Professional Geographer* 51, 76–89

Graham, E, J Doherty and M Malek 1992 The context and language of postmodernism. In J Doherty, E Graham and M Malek (eds) *Postmodernism and the social sciences*, London: Macmillan, pp. 1–23

Graham, H 1984 Surveying through stories. In C Bell and H Roberts (eds) *Social researching: politics, problems and practice*, London: Routledge, pp. 104–124

Greenbaum, T 1998 *The handbook for focus group research*, 2nd edn, London: Sage

Gregory, C and J Altman 1989 *Observing the economy*, London: Routledge

Gregory, D 1978 *Ideology, science and human geography*, London: Hutchinson

Gregory, D 1989 The crisis of modernity? Human geography and critical social theory. In R Peet and N Thrift (eds) *New models in geography: the political economy perspective*, Vol II, London: Unwin Hyman, pp. 348–385

Gregory, D 1994a Social theory and human geography. In D Gregory, R Martin and G Smith (eds) *Human geography: society, space and social science*, London: Macmillan, pp. 78–109

Gregory, D 1994b *Geographical imaginations*, Oxford: Basil Blackwell

Gregory, D 1995 Between the book and the lamp: imaginative geographies of Egypt, 1849–50, *Transactions of the Institute of British Geographers* 20, 29–57

Gregory, D and R Walford (eds) 1989 *Horizons in human geography*, London: Macmillan

Gregory, D, R Martin and G Smith (eds) 1994 *Human geography: society, space and social science*, London: Macmillan

Gregson, N 1987 Structuration theory: some thoughts on the possibilities for empirical research, *Environment and Planning D: Society and Space*, 5, 73–91

Greimas, A-J 1985 The love-life of a hippopotamus: a seminar with A-J Greimas. In M Blonsky (ed.) *On signs*, Oxford: Blackwell, pp. 340–362

Greimas, A-J 1987 *On meaning*, Minneapolis: University of Minnesota Press

Griffiths, V 1991 Doing feminist ethnography on friendship. In J Aldridge, V Griffiths and A Williams (eds) *Rethinking: feminist research processes reconsidered*, Manchester: Feminist Praxis Monograph

Grosz, E 1990 *Jacques Lacan: a feminist introduction*, London: Routledge

Guijt, I and M Shah (eds) 1998 *The myth of community: gender issues in participatory development*, London: Intermediate Technology Publications

Gunning, R 1962 *More effective writing in business and industry*, Boston, MA: Industrial Education International

Habermas, J 1974 *Theory and practice*, London: Heinemann

Habermas, J 1978 *Knowledge and human interests*, 2nd edn, London: Heinemann

Hägerstrand, T 1975 Time, space and human conditions. In A Karlqvist, L Lundqvist and F Snickers (eds) *Dynamic allocation of urban space*, Farnborough: Saxon House, pp. 3–14

Hagey, R 1997 The use and abuse of participatory action research, *Chronic Diseases in Canada* 18, 1–4

Haggett, P, A D Cliff and A Frey 1977 *Locational analysis in human geography II: locational methods*, London: Arnold

Hammersley, M 1992 *What's wrong with ethnography: methodological explorations*, London: Routledge

Hammersley, M and P Atkinson 1989 *Ethnography: principles in practice*, London: Routledge

Hamnett, C 1999 The emperor's new theoretical clothes, or geography without origami. In C Philo and D Miller (eds) *Market killing: what the free market does and what social scientists should do about it*, Harlow: Longman, pp. 159–170

Hampson, L 1994 *How's your dissertation going?* Unit for Innovation in Higher Education, Lancaster University

Harding, S 1986 Is there a feminist method? In S Kemp and J Squires (eds) *Feminisms*, Oxford: Oxford University Press, pp. 160–161

Harley, J B 1989 Deconstructing the map, *Cartographica*, 26, 1–20

Harris T, D Weiner, T Warner and R Levin 1995 Pursuing social goals through participatory geographic information systems: readdressing South Africa's historical political economy. In J Pickles (ed.) *Ground truth: the social impacts of geographic information systems*, New York: Guilford Press, pp. 195–222

Hart, C 1998 *Doing a literature review: releasing the social science research imagination*, London: Sage

Hart, C 2001 *Doing a literature search: a comprehensive guide for the social sciences*, London: Sage

Harvey, D 1969 *Explanation in geography*, London: Edward Arnold

Harvey, D 1973 *Social justice and the city*, London: Edward Arnold

Harvey, D 1989 *The condition of postmodernity: an enquiry into the origins of cultural change*, Oxford: Basil Blackwell

Hay, A M 1985 Scientific method in geography. In R J Johnston (ed.) *The future of geography*, London: Methuen, pp. 129–142

Hay, I (ed.) 2000 *Qualitative research methods in human geography*, Oxford: Oxford University Press

Hay, I 2003 Ethical practice in geographical research. In N J Clifford and G Valentine (eds) *Key methods in geography*, London: Sage, pp. 37–53

Hay, I and M Israel 2001 'Newsmaking geography': communicating geography through the media, *Applied Geography* 21, 107–125

Haynes, K E and A S Fotheringham 1984 *Gravity and spatial interaction models*, London: Sage

Healey, M (ed.) 1991 *Economic activity and land use: the changing information base for local and regional studies*, Harlow: Longman

Healey, M 2003 How to conduct a literature search. In N J Clifford and G Valentine (eds) *Key methods in geography*, London: Sage

Heidegger, M 1962 *Being and time*, New York: Harper and Row

Herz, R and J Luber 1995 *Studying elites: using qualitative methods*, London: Sage

Heywood, I, S Cornelius and S Carver 1998 *An introduction to geographical information systems*, Harlow: Longman

Higgitt, D and J E Bullard 1999 Assessing fieldwork risk for undergraduate projects, *Journal of Geography in Higher Education* 23, 441–449

Hinchliffe, S J, M Crang, S M Reimer and A C Hudson 1997 Software for qualitative research: 2 Some thoughts on 'aiding' analysis, *Environment and Planning A* 29, 1109–1124

Hirsch, J 1981 *Family photographs: content, meaning, effect*, Oxford: Oxford University Press

Hoggart, K, L Lees and A Davies 2002 *Researching human geography*, London: Arnold

Holbrook, B 1996 Shopping around: focus group research in North London, *Area* 28(2), 136–142

Hopkins, J 1994 Mapping cinematic places: icons, ideology, and the power of (mis)representation. In S C Aitken and L E Zonn (eds) *Place, power, situation and spectacle: a geography of film*, Lanham, MD: Rowman and Littlefield, pp. 47–68

Horton, J 2001 'Do you get some funny looks when you tell people what you do?' Muddling through some angst and ethics of (being a male) researching children, *Ethics, Place and Environment* 4(2), 159–166

Howard, S 1994 Methodological issues in overseas fieldwork: experiences from Nicaragua's Northern Atlantic Coast. In E Robson and K Willis (eds) *Postgraduate fieldwork in developing areas: a rough guide*, Monograph No. 8, London: Developing Areas Research Group (DARG) of the RGS/IBG, pp. 19–35

Hubbard, P, R Kitchin, B Bartley and D Fuller 2002 *Thinking geographically: space, theory and contemporary human geography*, London: Continuum

Hunt, J 1989 *Psychoanalytic aspects of fieldwork*, London: Sage

IDS (Institute of Development Studies at the University of Sussex) 2000 *Participation Home Page*, http://www.ids.ac.uk/ids/particip/ (accessed 05.09.02)

Ingham, J, M Purvis and D B Clarke 1999 Hearing places, making spaces: sonorous geographies, ephemeral rhythms and the Blackburn warehouse parties, *Environment and Planning D: Society and Space* 17, 283–305

International Labour Organisation (annual) *Yearbook of labour statistics*, Geneva

Irvine, J, I Miles and J Evans (eds) 1979 *Demystifying social statistics*, London: Pluto Press

Jackson, P 1991 The cultural politics of masculinity: towards a social geography, *Transactions of the Institute of British Geographers* 16(2), 199–213

Jackson, P 1992 Constructions of culture, representations of race: Edward Custis's 'way of seeing'. In K Anderson and F Gale (eds) *Inventing places: studies in cultural geography*, Melbourne: Wiley, pp. 89–106

Jackson, P 2001 Making sense of qualitative data. In M Limb and C Dwyer (eds) *Qualitative methodologies for geographers: issues and debates*, London: Arnold, pp. 199–214

Jackson, P and B Holbrook 1995 Multiple meanings: shopping and the cultural politics of identity, *Environment and Planning A* 27, 1913–1930

Jackson, P and S Smith 1984 *Exploring social geography*, London: Allen and Unwin

Jacobs, D 1981 Domestic snapshots: towards a grammar of motives, *Journal of American Culture* 4(1), 93–105

Jameson, F 1981 *The political unconscious*, Ithaca, NY: Cornell University Press

Jameson, F 1984 Postmodernism, or the cultural logic of late capitalism, *New Left Review* 146, 53–92

Jameson, F 1992 *The geopolitical aesthetic: cinema and space in the world system*, Bloomington: Indiana University Press

Jarrett, R L 1993 Focus group interviewing with low-income, minority populations: a research experience. In D Morgan (ed.) *Successful focus groups: advancing the state of the art*, London: Sage, pp. 184–201

Johnson, M 1992 A silent conspiracy? some ethical issues of participant observation in nursing research, *International Journal of Nursing Studies* 29(2), 213–223

Johnston, R J 1986 *Philosophy and human geography*, 2nd edn, London: Edward Arnold

Johnston, R J, D Gregory, G Pratt and M Watts 2000 *The dictionary of human geography*, 4th edn, Oxford: Basil Blackwell

Jones, J P and E Casetti 1992 *Applications of the expansion method*, London: Routledge

Jones, S 1985 The analysis of depth interviews. In R Walker (ed.) *Applied qualitative research*, Aldershot: Gower Press, pp. 56–70

Katz, C 1992 All the world is staged: intellectuals and the projects of ethnography, *Environment and Planning D: Society and Space* 10, 495–510

Katz, C 1994 Playing the field; questions of fieldwork in geography, *Professional Geographer* 46, 54–102

Katz, C and A Kirby 1991 In the nature of things: the environment and everyday life, *Transactions of the Institute of British Geographers* 16(3), 259–271

Katz, J 2001 From how to why: on luminous description and causal inference in ethnography (Part 1), *Ethnography* 2(4), 443–473

Katz, J 2002 From how to why: on luminous description and causal inference in ethnography (Part 2), *Ethnography* 3(1), 63–90

Kaye, N 1997 Climatic warming and the potential for grapevine cultivation in England and Wales. Unpublished dissertation, Southampton: Department of Geography, University of Southampton

Keesing's *Record of World Events* (annual) Cambridge: Cartermill

Keith, M 1992 Angry writing: (re)presenting the unethical world of the ethnographer, *Environment and Planning D: Society and Space* 10, 551–568

Kellerman, A 1981 *Centrographic measures in geography,* Concepts and Techniques in Modern Geography 32, Norwich: Geo Books

Kennedy, B A 1992 First catch your hare . . . Research designs for individual projects. In A Rogers, H Viles and A Goudie (eds) *The student's companion to geography,* 1st edn, London: Blackwell, pp. 128–134

Kennedy, M 2002 *The global positioning system: an introduction,* London: Taylor and Francis

Kesby, M 2000 Participatory diagramming: deploying qualitative methods through an action research epistemology, *Area* 32(4), 423–435

Kesby, M 2005 Retheorising empowerment-through-participation as a performance in space: beyond tyranny to transformation, *Signs* 30(4)

Kesby, M, K Fenton, P Boyle and R Power 2003 An agenda for future research on HIV and sexual behaviour among African migrant communities in the UK, *Social Science and Medicine* 57, 1573–1592

Kindon, S 2003 Participatory video in geographic research: a feminist practice of looking?, *Area* 35(2), 142–153

Kindon, S and A Latham 2002 From mitigation to negotiation: ethics and the geographic imagination in Aotearoa, New Zealand, *New Zealand Geographer* 58(1), 14–22

Kindon, S L 1995 Dynamics of difference: exploring empowerment methodologies with women and men in Bali, *New Zealand Geographer* 51(11), 10–12

Kindon, S L 1998 Of mothers and men: questioning gender and community myths in Bali. In I Guijt and M Shah (eds) 1998 *The myth of community: gender issues in participatory development,* London: Intermediate Technology Publications, pp. 152–164

Kirk, J and M Miller 1986 *Reliability and validity in qualitative research,* Beverly Hills: Sage

Kirk, W 1963 Problems of geography, *Geography* 48, 357–371

Kitchin, R 2001 Using participatory action research approaches in geographical studies of disability: some reflections, *Disability Studies Quarterly* 21(4), 61–69

Kitchin, R and P Hubbard 1999 Research, action and 'critical' geographies, *Area* 31(3), 195–198

Kitchin, R and N J Tate 2000 *Conducting research into human geography: theory, methodology and practice,* Harlow: Prentice Hall

Kneale, P E 1999 *Study skills for geography students: a practical guide*, London: Arnold

Kobayashi, A 2001 Negotiating the personal and the political in critical qualitative research. In M Limb and C Dwyer (eds) *Qualitative methodologies for geographers*, London: Arnold, pp. 55–72

Kobayashi, A and S Mackenzie eds 1989 *Remaking human geography*, London: Unwin Hyman

Kong, L 1996 Popular music in Singapore: exploring local cultures, global resources and regional identities, *Environment and Planning D: Society and Space* 14, 273–292

Kong, L 1998 Refocusing on qualitative methods: problems and perspectives for research in a specific Asian context, *Area* 30, 79–82

Koskela, H 1995 The semiotics of urban fear. Paper presented to the Geography and Gender ERASMUS Intensive Course, Roskilde University

Krueger, R 1998a *Developing questions for focus groups*, Focus Group Kit 3, London: Sage

Krueger, R 1998b *Moderating focus groups*, Focus Group Kit 4, London: Sage

Krueger, R 1998c *Analysing and reporting focus group results*, Focus Group Kit 6, London: Sage

Krueger, R 1998d *Focus groups: a practical guide for applied research*, 2nd edn, London: Sage

La Pelle, M 2004 Simplifying qualitative data analysis using general purpose software tools, *Field Methods* 16(1), 85–108

Lacan, J 1978 *The four fundamental concepts of psychoanalysis*, translated by Alan Sheridan, New York: W W Norton

Lacan, J 1983 *Feminine sexuality*, edited by J Mitchell and J Rose, translated by J Rose, New York: W W Norton

Ladyman J 2002 *Understanding philosophy of science*, London: Routledge

Latham, A and D Conradson (eds) 2003 Special issue on Making place: performance, practice, space, *Environment and Planning A* 35(11)

Laurier, E and H Parr 2000 Emotions and interviewing in health and disability research, *Ethics, Place and Environment* 3, 98–102

Laurini, R and D Thompson 1992 *Fundamentals of spatial information systems*, APIC Series 37, London: Academic Press

Lawson, V 2000 Arguments within geographies of movement: the theoretical potential of migrants' stories, *Progress in Human Geography* 24(2), 173–189

Lee, R M 1995 *Dangerous fieldwork*, London: Sage

Lee-Treweek, G and S Linkogle (eds) 2000 *Danger in the field: risk and ethics in social research*, London: Routledge

Lefebvre, H 1984 *Everyday life in the modern world*, New Brunswick, NJ: Transaction Books

Lefebvre, H 1991 *The production of space*, translated by D Nicholson-Smith, Oxford: Blackwell

Lewis, O 1962 *The children of Sanchez: autobiography of a Mexican family*, London: Secker and Warburg

Ley, D 1982 Rediscovering man's place, *Transactions of the Institute of British Geographers* 7, 248–253

Ley, D and R Cybriwsky 1974 Urban graffiti as territorial markers, *Annals of the Association of American Geographers* 64, 491–505

Ley, D and M S Samuels (eds) 1978 *Humanistic geography: prospects and problems*, London: Croom Helm

Leyshon, A, D Matless and G Revill 1998 Introduction: music, space and the production of place. In A Leyshon, D Matless and G Revill (eds) *The place of music*, New York: Guilford, pp. 1–30

Lieber, E, T Weisner and M Presley 2003 Ethnonotes: an Internet-based field note management tool, *Field Methods* 15(4), 405–425

Limb, M and C Dwyer (eds) 2001 *Qualitative methodologies for geographers*, London: Arnold

Lloyd Evans, S, E Robson and K Willis 1994 The logistics of undertaking field research in 'developing areas'. In E Robson and K Willis (eds) *Postgraduate fieldwork in developing areas: a rough guide*, Monograph No. 8, London: Developing Areas Research Group (DARG) of the RGS/IBG

Loehr, D and L Harper 2003 Commonplace tools for studying commonplace interactions: practitioners' notes on entry-level video analysis, *Visual Communication* 2(2), 225–233

Longhurst, R 1996 Refocusing groups: pregnant women's geographical experiences of Hamilton, New Zealand/Aotearoa, *Area* 28, 143–149

Longley, P and G Clarke 1995a Applied geographical information systems: developments and prospects. In P Longley and G Clarke (eds) *GIS for business and service planning*, Cambridge: GeoInformation International, pp. 3–9

Longley, P and G Clarke (eds) 1995b *GIS for business and service planning*, Cambridge: GeoInformation International

Longley, P A, M F Goodchild, D J Maguire and D W Rhind (eds) 1999 *Geographical information systems: principles, techniques, applications and management*, 2nd edn, Chichester: Wiley

Longley, P A, M F Goodchild, D J Maguire and D W Rhind 2001 *Geographical information: systems and science*, Chichester: Wiley

Lovett, A A and M Bayman 1993 *Public knowledge and attitudes in relation to HIV and AIDS: a case study of North Lancashire*, Lancaster: Department of Geography, Lancaster University

Lukinbeal, C and S C Aitken 1998 Sex, violence and the weather: male hysteria, scale and the fractal geographies of patriarchy. In S Pile and H Nast (eds) *Places through the body*, London: Routledge, pp. 356–380

Lynch, K 1961 *The image of the city*, Cambridge, MA: MIT Press

Lyon, F 2000 Trust and power in farmer–trader relations: a study of small scale vegetable production and marketing systems in Ghana. Unpublished PhD thesis, Durham: University of Durham

Macdonald, G 1990 Third World films: a strategy for promoting geographic understanding, *Journal of Geography*, Nov–Dec, 253–259

Macdonald, G 1994 Third Cinema and the Third World. In S C Aitken and L E Zonn (eds) *Place, power, situation and spectacle: a geography of film*, Lanham, MD: Rowman and Littlefield, pp. 47–68

MacEachren, A, B Buttenfield, J Campbell, D DiBiase and M Monmonier 1992 Visualization. In R Abler, M Marcus and J Olson (eds) *Geography's inner worlds*, New Brunswick, NJ: Association of American Geographers

Macmillan, B (ed.) 1989a *Remodelling geography*, Oxford: Blackwell

Macmillan, B 1989b Quantitative theory construction in human geography. In B Macmillan (ed.) *Remodelling geography*, Oxford: Blackwell, pp. 89–107

Maguire, P 1987 *Doing participatory research: a feminist approach*, Center for International Education, University of Massachusetts, Amherst, Massachusetts

Mallory, W and P Simpson-Housley (eds) 1987 *Geography and literature: a meeting of disciplines*, Syracuse, NY: Syracuse University Press

Manning, P 1987 *Semiotics and fieldwork*, Beverly Hills: Sage

Maranhao, T 1986 The hermeneutics of participant observation, *Dialectical Anthropology* 10(3–4), 291–309

Marsh, C 1988 *Exploring data*, Cambridge: Polity Press

Martin, D 1996 *Geographic information systems: socioeconomic applications*, 2nd edn, London: Routledge

Martin, D, M L Senior and H C W L Williams 1994 On measures of deprivation and the spatial allocation of resources for primary health care, *Environment and Planning A* 26, 1911–1929

Martin, R 2001 Geography and public policy; the case of the missing agenda, *Progress in Human Geography* 25(2), 189–210

Marx, K 1974 *Capital*, Vol. I, London: Lawrence and Wishart

Mason, J 1996 *Qualitative research*, London: Sage

Mather, P M 1999 *Computer processing of remotely sensed images: an introduction*, 2nd edn, Chichester: Wiley

McCrossan, L 1991 *A handbook for interviewers – a manual of social survey practice and procedures on structured interviewing*, London: HMSO

McDowell, L 1992 Valid games? A response to Erica Schoenberger, *Professional Geographer* 44(2), 212–215

McDowell, L 2002 Understanding diversity: the problem of/for 'Theory'. In R J Johnston, P J Taylor and M J Watts (eds) *Geographies of global change*, 2nd edn, Oxford: Blackwell, pp. 296–309

McKendrick, J H 1999 Multi-method research: an introduction to its application in population geography, *Professional Geographer* 51, 40–50

Melville, H 1984 [1857] *The confidence-man: his masquerade*, CEAA edn, H Hayford, H Parker and G Tanselle (eds) Evanston, IL: Northwestern University Press

Menard, S 2002 *Applied logistic regression analysis*, 2nd edn, Quantitative Applications in the Social Sciences 106, London: Sage

Merton, R and P Kendall 1946 The focused interview, *American Journal of Sociology* 51, 541–557

Merton, R, M Fiske and P Kendall 1990 [1956] *The focused interview*, Glencoe: The Free Press

Metz, C 1974 *Film language: a semiotics of the cinema*, Ann Arbor: University of Michigan Press

Mikkelson, B 1995 *Methods for development work and research*, London: Sage

Miles, M and J Crush 1993 Personal narratives as interactive texts: collecting and interpreting migrant life-histories, *Professional Geographer* 45(1), 84–94

Miller, R L, C Acton, D A Fullerton and J Maltby 2002 *SPSS for social scientists*, Basingstoke: Palgrave Macmillan

Mills, C 1992 Myths and meanings of gentrification. In T Barnes and J Duncan (eds) *Writing worlds: discourse, text and metaphor*, London: Routledge, pp. 149–170

Mitchell, D 1995 There's no such thing as culture: towards a reconceptualization of the idea of culture in geography, *Transactions of the Institute of British Geographers* 20(1), 102–116

Mitchell, D 2000 *Cultural geography: a critical introduction*, Malden, MA: Blackwell Publishers

Mitchell, L 1996 *Westerns: making the man in fiction and film*, Chicago: University of Chicago Press

Mohammad, R 2001 'Insider'/'outsiders': positionality, theory and praxis. In M Limb and C Dwyer (eds) *Qualitative methodologies for geographers*, London: Arnold, pp. 101–120

Monk, J 1994 Place matters: comparative international perspectives on feminist geography, *Professional Geographer* 46, 277–285

Monmonier, M 1993 *Mapping it out*, Chicago: University of Chicago Press

Monmonier, M 1996 *How to lie with maps*, 2nd edn, Chicago: University of Chicago Press

Morehouse, B J 2002 Geographies of power and social relations in Marge Piercy's *He, she and it*. In R Kitchin and J Kneale (eds) *Lost in space: geographies of science fiction*, London and New York: Continuum Press, pp. 74–89

Morgan, D 1988a *The focus group guidebook*, Focus Group Kit 1, London: Sage

Morgan, D 1988b *Planning focus groups*, Focus Group Kit 2, London: Sage

Morgan, D 1988c *Focus groups as qualitative research*, London: Sage

Morris-Roberts, K 2001 Intervening in friendship exclusion? The politics of doing feminist research with teenage girls, *Ethics, Place and Environment* 4(2), 147–152

Moss, P (ed.) 2002 *Feminist geography in practice: research and methods*, Oxford: Blackwell

Mulvey, L 1975 Visual pleasure and narrative cinema, *Screen* 16(3), 6–18

Mulvey, L 1992 Pandora: topographies of the mask and curiosity. In B Colomina (ed.) *Sexuality and space*, New York: Princeton Architectural Press

Muzzio, D 1996 'Decent people shouldn't live here': the American city in cinema, *Journal of Urban Affairs* 18(2), 189–215

Nash, D J 2000a Doing independent overseas fieldwork. 1 Practicalities and pitfalls, *Journal of Geography in Higher Education* 24, 139–149

Nash, D J 2000b Doing independent overseas fieldwork. 2 Getting funded, *Journal of Geography in Higher Education* 24, 425–433

Nast, H 1994 Opening remarks on 'Women in the Field', *Professional Geographer* 46(1), 54–66

Nast, H 2000 Mapping the 'unconscious': racism and the oedipal family, *Annals of the Association of American Geographers* 90(2), 215–255

Natter, W and J P Jones 1993 Pets or meats: class, ideology, and space in *Roger and Me*, *Antipode* 25(2), 140–158

Nencel, L and P Pels (eds) 1991 *Constructing knowledge: authority and critique in social science*, London: Sage

Norwood, V and J Monk (eds) 1987 *The desert is no lady: Southwestern landscapes in women's writing and art*, New Haven: Yale University Press

Oakes, T 1997 Place and the paradox of modernity, *Annals of the Association of American Geographers* 87, 509–531

Oakley, A 1981 Interviewing women: a contradiction in terms. In H Roberts (ed.) *Doing feminist research*, London: Routledge

O'Connell Davidson, J 1991 The employment relation: diversity and degradation in the privatised water industry. PhD thesis, University of Bristol

O'Connell Davidson, J and D Layder 1994 *Methods, sex and madness*, London: Routledge

Odland, J 1988 *Spatial autocorrelation*, London: Sage

Office for National Statistics 2000 *Guide to official statistics*, London: HMSO

Office for National Statistics 2001 *Regional Trends* 36, London: The Stationery Office

Ogborn, M 2003 Finding historical data. In N J Clifford and G Valentine (eds) *Key methods in geography*, London: Sage, pp. 101–115

Openshaw, S 1984 *The modifiable areal unit problem*, Concepts and Techniques in Modern Geography 38, Norwich: Geo Books

Openshaw, S (ed.) 1995 *Census users' handbook*, Harlow: Longman

Oppenheim, A N 1992 *Questionnaire design, interviewing and attitude measurement*, London: Pinter Publishers

Ovnick, M 1994 *Los Angeles: the end of the rainbow*, Los Angeles: Balcony Press

Pain, R and P Francis 2003 Reflections on participatory research, *Area* 35(1), 46–54

Pantazis, C and D Gordon 1999 Are crime and fear of crime more likely to be experienced by the 'poor'? In D Dorling and S Simpson (eds) *Statistics in society: the arithmetic of politics*, London: Arnold, pp. 198–211

Parmentier, R 1994 *Signs in society: studies in semiotic anthropology*, Bloomington: Indiana University Press

Parr, H 1998 Mental health, ethnography and the body, *Area* 30(1), 28–37

Parsons, T and P Knight 1995 *How to do your dissertation in geography and related disciplines*, London: Chapman and Hall

Patai, D 1991 US academics and Third World women: is ethical research possible? In S Berger Gluck and D Patai (eds) *Women's words: the feminist practice of oral history*, London: Routledge

Peace, R 2000 Computers, qualitative data and geographic research. In I Hay (ed.) *Qualitative research methods in human geography*, Oxford/Melbourne: Oxford University Press, pp. 144–160

Peet, R 1998 *Modern geographical thought*, Oxford: Blackwell

Peet, R and N Thrift (eds) 1989 *New models in geography: the political economy perspective*, Vol. II, London: Unwin Hyman

Perks, R 1992 *Oral history: talking about the past*, Helps for Students of History 94, London: Historical Association

Phillips, D C 1987 *Philosophy, science and social inquiry*, Oxford: Pergamon Press

Phillips, E and D Pugh 2000 *How to get a PhD: a handbook for students and their supervisors*, 3rd edn, Buckingham: Open University Press

Phillips, R 1997 *Mapping men and empire: a geography of adventure*, New York and London: Routledge

Phillips, R 1999 Writing travel and mapping sexuality: Richard Burton's Sotadic Zone. In J Duncan and D Gregory (eds) *Writes of passage: reading travel writing*, New York and London: Routledge, pp. 70–91

Pickles, J 1992 Text, hermeneutics and propaganda maps. In T Barnes and J Duncan (eds) *Writing worlds: discourse, text and metaphor*, London: Routledge, pp. 193–230

Pickles, J (ed.) 1995 *Ground truth: the social implications of geographic information systems*, New York: Guilford Press

Pile, S 1996 *The body and the city: psychoanalysis, space and subjectivity*, New York and London: Routledge

Pipkin, J 2001 Hiding places: Thoreau's geographies, *Annals of the Association of American Geographers* 91(3), 527–545

Pocock, D (ed.) 1981 *Humanistic geography in literature*, London: Croom Helm

Pocock, D C D 1975 *Durham: images of a cathedral city*, Occasional Publications New Series 6, Durham: Department of Geography, University of Durham

Popper, K 1945 *The open society and its enemies*, London: Routledge and Kegan Paul

Popper, K 1968 *The logic of scientific discovery*, London: Hutchinson

Popper, K 1969 *Conjectures and refutations*, London: Routledge and Kegan Paul

Porteous, D 1985 Literature and humanistic geography, *Area* 17, 117–122

Powell, J C, A L Craighill, J P Parfitt and R K Turner 1996 A lifecycle assessment and economic valuation of recycling, *Journal of Environmental Planning and Management* 39, 97–112

Powell, R, H Single and K Lloyd 1996 Focus groups in mental health research: enhancing the validity of user and provider questionnaires, *International Journal of Social Psychology* 42, 193–206

Power, R and G Hunter 2001 Developing a strategy for community-based health promotion targeting homeless populations, *Health Education Research* 16(5), 593–602

Pratt, A 1995 Putting critical realism to work: the practical implications for geographical research, *Progress in Human Geography* 19, 61–74

Pratt, G 2000 Participatory action research. In R J Johnston, D Gregory, G Pratt and M Watts (eds) *Dictionary of human geography* 4th edn, Oxford: Blackwell, p. 574

Price, M and M Lewis 1993 The reinvention of cultural geography, *Annals of the Association of American Geographers* 83(1), 1–17

Pringle, D G 1980 *Causal modelling: the Simon–Blalock approach*, Concepts and Techniques in Modern Geography 27, Norwich: Environmental Publications

Pringle, T 1988 The privation of history: Landseer, Victoria and the Highland myth. In D Cosgrove and S Daniels (eds) *The iconography of landscape: essays on the symbolic representation, design and use of past environments*, Cambridge: Cambridge University Press, pp. 142–160

Proctor, J D and D M Smith (eds) 1999 *Geography and ethics: journeys in a moral terrain*, London: Routledge

Pulsipher, L 1993 'He won't let she stretch she foot': gender relations in traditional West Indian houseyards. In C Katz and J Monk (eds) *Full circles: geographies of women over the life course*, London: Routledge, pp. 107–121

Radcliffe, S A 1994 (Representing) post-colonial women: authority, difference and feminisms, *Area* 26, 25–32

Raghuram, P, C Madge and T Skelton 1998 Feminist research methodologies and student projects in geography, *Journal of Geography in Higher Education* 22(1), 35–48

Raper, J F, D W Rhind and J W Shepherd 1992 *Postcodes: the new geography*, Harlow: Longman

Rees, P, D Martin and P Williamson (eds) 2002 *The census data system*, Chichester: Wiley

Regional Trends (annual), London: HMSO

Revill, G 2000 Music and the politics of sound: nationalism, citizenship, and auditory space, *Environment and Planning D: Society and Space* 18, 597–613

Revill, G and S Seymour 2000 Telling stories: story telling as a textual strategy. In A Hughes, C Morris and S Seymour (eds) *Ethnography and rural research*, Cheltenham: Countryside and Community Press, pp. 136–157

Richards, L 1999 *Using NVivo in qualitative research*, London: Sage

Ricoeur, P 1971 The model of the text: meaningful action considered as text, *Social Research* 38, 529–562

Ricoeur, P 1980 Narrative time, *Critical Inquiry* 7(1), 169–190

Ricoeur, P 1984 *Time and narrative*, Vol. 1, Chicago: University of Chicago Press

Riessman C 1993 *Narrative analysis*, Beverly Hills: Sage

Roberts, H (ed.) 1981 *Doing feminist research*, London: Routledge

Robins, K 1996 *Into the image*, London: Routledge

Robinson, A H, J L Morrison, P C Muehrcke, A J Kimerling and S C Guptill 1995 *Elements of cartography*, 6th edn, New York: John Wiley

Robinson, G M 1998 *Methods and techniques in human geography*, Chichester: Wiley

Robson, E and K Willis (eds) 1994 *Postgraduate fieldwork in developing areas: a rough guide*, Monograph No. 8, London: Developing Areas Research Group (DARG) of the RGS/IBG

Robson, E J and K Willis (eds) 1997 *Postgraduate fieldwork in developing areas: a rough guide*, 2nd edn, Monograph No. 8, London: Developing Areas Research Group, RGS-IBG

Rocheleau D, R Slocum and B Thomas-Slayter (eds) 1995 *Power, process and participation: tools for environmental and social change*, London: Intermediate Technology Publications

Rogers, A 1985 *Regional population projection models*, London: Sage

Rogers, A 1992 Key themes and debates. In A Rogers, H Viles and A Goudie (eds) *The student's companion to geography*, Oxford: Blackwell

Rogers, A, H Viles and A Goudie (eds) 2002 *The student's companion to geography*, 2nd edn, London: Blackwell

Rogerson, P A 2001 *Statistical methods for geographers*, London: Sage

Rogoff, I 2000 *Terra infirma: geography's visual culture*, London: Routledge

Rose, G 1993 *Feminism and geography*, Oxford: Polity Press

Rose, G 1994 The cultural politics of place: local representation and oppositional discourse in two films, *Transactions of the Institute of British Geographers* 19(1), 46–60

Rose, G 1997 Situating knowledges: positionality, reflexivities and other tactics, *Progress in Human Geography* 21, 305–320

Rose, G 2001 *Visual methodologies: an introduction to the interpretation of visual materials*, London: Sage

Rose, G and N Thrift (eds) 2000 Special Issue on Performance, *Environment and Planning D: Society and Space* 18, 411–518 and 575–652

Rosenberg, A 1988 *Philosophy of social sciences*, Oxford: Clarendon Press

Rowles, G 1978 Reflections on experiential fieldwork. In D Ley and M Samuels (eds) *Humanistic geography: prospects and problems*, Chicago: Maaroufa, pp. 173–193

Sachs, P 1993 Old ties: women, work and aging in a coal-mining community in West Virginia. In C Katz and J Monk (eds) *Full circles: geographies of women over the life course*, London: Routledge, pp. 156–170

Saco, D 1992 Masculinity as signs: poststructuralist feminist approaches to the study of gender. In S Craig (ed.) *Men, masculinity, and the media*, Newbury Park: Sage, pp. 23–39

Said, E 1993 *Culture and imperialism*, London: Chatto and Windus

Sauer, C 1925 The morphology of landscape, *University of California Publications in Geography* 2(2), 19–54

Sauer, C 1956 The education of a geographer. In J Leighly (ed.) 1974 *Land and life: a selection from the writings of Carl Ortwin Sauer*, Berkeley: University of California Press, pp. 389–406

Saussure, F de 1966 *The responsive chord,* Garden City, NY: Doubleday

Sayer, A 1985 Realism in geography. In R J Johnston (ed.) *The future of geography,* London: Methuen, pp. 159–173

Sayer, A 1992 *Method in social science: a realist approach,* 2nd edn, London: Hutchinson

Sayer, A 1993 Postmodernist thought in geography: a realist view, *Antipode* 25, 320–344

Schein, R 1997 The place of landscape: a conceptual framework for interpreting an American scene, *Annals of the Association of American Geographers* 87(4), 660–680

Schmid, D 1995 Imagining safe urban space: the contribution of detective fiction to radical geography, *Antipode* 27, 242–269

Schoenberger, E 1991 The corporate interview as a research method in economic geography, *Professional Geographer* 43(2), 180–189

Schoenberger, E 1992 Self criticism and self awareness in research: a reply to Linda McDowell, *Professional Geographer* 44(2), 215–218

Schroeder, L D, D L Sjoquist and P E Stephan 1986 *Understanding regression analysis: an introductory guide,* Quantitative Applications in the Social Sciences 57, London: Sage

Schuenemeyer, J 1984 Directional data analysis. In G L Gaile and C J Willmott (eds) *Spatial statistics and models,* Dordrecht: Kluwer, pp. 253–270

Schwartz, J and J Ryan 2003 *Picturing place: photography and the geographical imagination,* London: I B Tauris

Seale, C (ed.) 1998 *Researching society and culture,* London: Sage

Seamon, D 1990 Awareness and reunion: a phenomenology of the person–world relationship as portrayed in the New York photographs of André Kertész. In L Zonn (ed.) *Place images in media: portrayal, experience and meaning,* Savage, MD: Rowman and Littlefield, pp. 31–62

Selener, D 1997 *Participatory action research and social change,* Cornell Participatory Action Research Network, Cornell University, Ithaca, NY

Shah, A 1999 Power plays: reflections on the process of submitting an undergraduate dissertation, *Area* 31, 307–312

Sharp, J 1994 A topology of 'post' nationality: (re)mapping identity in *The Satanic Verses, Ecumene* 1, 65–76

Shiel, M 2001 Cinema and the city in history and theory. In M Shiel and T Fitzmaurice (eds) *Cinema and the city,* Malden, MA: Blackwell, pp. 1–18

Shrestha, N R 1997 On 'What causes poverty? A postmodern view': a postmodern view or denial of historical integrity? The poverty of Yapa's view of poverty, *Annals of the Association of American Geographers* 87, 709–716

Sibley, D 1988 *Spatial applications of exploratory data analysis,* Concepts and Techniques in Modern Geography 49, Norwich: Geo Books

Sidaway, J D 1992 In other worlds: on the politics of research by first world geographers in the third world, *Area* 24(4), 403–408

Siegel, S and N J Castellan 1988 *Nonparametric statistics for the behavioural sciences,* 2nd edn, New York: McGraw-Hill

Silverman, D 2000 *Doing qualitative research,* London: Sage

Silverman, D 2001 *Interpreting qualitative data: methods for analysing talk, text and interaction,* 2nd edn, London: Sage

Simonsen, K 2000 Editorial: the body as battlefield, *Transactions of the Institute of British Geographers* 25, 7–9

Skeggs, B 1994 Situating the production of feminist ethnography. In M Maynard and J Purvis (eds) *Researching women's lives*, Basingstoke: Taylor and Francis, pp. 72–93

Skelton, T 2001 Cross-cultural research: issues of power, positionality and 'race'. In M Limb and C Dwyer (eds) *Qualitative methodologies for geographers*, London: Arnold, pp. 87–100

Skinner, C J, D Holt and T M F Smith (eds) 1989 *Analysis of complex surveys*, Chichester: Wiley

Smith, F M 1996 Problematising language: limitations and possibilities in 'foreign language' research, *Area* 28, 160–166

Smith, F M 2003 Working in different cultures. In N J Clifford and G Valentine (eds) *Key methods in geography*, London: Sage, pp. 179–193

Smith, H W 1975 *Strategies of social research – the methodological imagination*, London: Prentice Hall International

Smith, M M 2002 Foreword. In R Kitchin and J Kneale (eds) *Lost in space: geographies of science fiction*, London and New York: Continuum Press, pp. xi–xii

Smith, N 1993 Homeless/global: scaling places. In J Bird, B Curtis, T Putman, G Robertson and L Tickner (eds) *Mapping the futures: local cultures, global change*, London: Routledge

Smith, S 1988 Constructing local knowledge: the analysis of the self in everyday life. In J Eyles and D Smith (eds) *Qualitative methods in human geography*, London: Polity Press, pp. 17–38

Smith, S 2000 Performing (sound) world, *Environment and Planning D: Society and Space* 18, 615–637

Social Trends (annual), London: HMSO

Soja, E W 1989 *Postmodern geographies: the reassertion of space in the social sciences*, London: Verso

Soja, E W 1995 Postmodern urbanisation: the six restructurings of Los Angeles. In S Watson and K Gibson (eds) *Postmodern cities and spaces*, Oxford: Basil Blackwell

Sparke, M 1998 Mapped bodies and disembodied maps: (dis)placing cartographic struggle in colonial Canada. In H Nast and S Pile (eds) *Places through the body*, New York and London: Routledge

Sporton, D 1999 Mixing methods of fertility research, *Professional Geographer* 51, 68–76

Stanley L and S Wise 1993 *Breaking out again: feminist ontology and epistemology*, London: Routledge

Stewart, D and P Shamdasani 1990 *Focus groups: theory and practice*, London: Sage

Strauss A 1987 *Qualitative analysis for social scientists*, Cambridge: Cambridge University Press

Taussig, T 1987 *Shamanism, colonialism and the wild man: a study in terror and healing*, Chicago: Chicago University Press

Taylor, J S 2002 The subjectivity of the near future: geographical imaginings in the work of J G Ballard. In R Kitchin and J Kneale (eds) *Lost in space: geographies of science fiction*, London and New York: Continuum Press, pp. 90–103

Taylor, T and D Cameron 1987 *Analysing conversation: rules and units in the structure of talk*, Oxford: Pergamon Press

Tedlock, B 1991 From participant observation to the observation of participation. *Journal of Anthropological Research* 47(1), 69–94

Telotte, J P 1992 The fantastic realism of film noir: *Kiss Me Deadly*, *Wide Angle* 14(1), 5–18

Tesch, R 1990 *Qualitative research: types and research tools*, London: Falmer Press

Thomas, R 1985 *Citizenship, gender and work: social organisation of industrial agriculture*, Berkeley and Los Angeles: University of California Press

Thomas, R 1999 The politics and reform of unemployment and employment statistics. In D Dorling and S Simpson (eds) *Statistics in society: the arithmetic of politics*, London: Arnold, pp. 324–334

Thomas, R W and R J Huggett 1980 *Modelling in geography: a mathematical approach*, London: Harper and Row

Thrift, N 1983 On the determination of social action in time and space, *Environment and Planning D: Society and Space* 1, 23–57

Tobler, W R 1973 A continuous transformation useful for districting, *Annals of the New York Academy of Sciences* 219, 215–220

Townend, J 2002 *Practical statistics for environmental and biological scientists*, Chichester: Wiley

Townsend, J 1995 *Women's voices from the rainforest*, London: Routledge

Tronya, B and R Hatcher 1992 *Racism in children's lives*, London: Routledge

Tukey, J W 1977 *Exploratory data analysis*, Reading, MA: Addison-Wesley

Turk, C and J Kirkman 1989 *Effective writing: improving scientific, technical and business communication*, 2nd edn, London: E & F N Spon

Turnbull, J 2001 Do primary schools perform consistently well across different socio-economic environments? Unpublished dissertation, Southampton: Department of Geography, University of Southampton

Twyman, C, J Morrison and D Sporton 1999 The final fifth: autobiography, reflexivity and interpretation in cross-cultural research, *Area* 31, 313–325

United Nations (annual) *Statistical yearbook*, New York: United Nations

Unwin, T 1992 *The place of geography*, London: Longman

Urry, J 1995 *Consuming places*, London: Routledge

Valentine, G 1999a Doing household research: interviewing couples together and apart, *Area* 31(1), 67–74

Valentine, G 1999b Being seen and heard? The ethical complexities of working with children and young people at home and at school, *Ethics, Place and Environment* 2(2), 141–155

Valentine, G 2002 People like us: negotiating sameness and difference in the research process. In P Moss (ed.) *Feminist geography in practice*, Oxford: Blackwell

Valentine, G and T Skelton 2003 Living on the edge: the marginalisation and 'resistance' of D/deaf youth, *Environment and Planning A* 35(2), 301–321

Van Maanen, J 1988 *Tales of the field: on writing ethnography*, Chicago: University of Chicago Press

Vidal, J 1997 *McLibel: burger culture on trial*, London: Pan

Wade, J 1984 Role boundaries and paying back: 'switching hats' in participant observation, *Anthropology and Education Quarterly* 15(3), 211–224

Walford, N 1995 *Geographical data analysis*, Chichester: Wiley

Walford, N 2002 *Geographical data: characteristics and sources*, Chichester: Wiley

Walker, A, M O'Brien, J Traynor, K Fox, E Goddard and K Foster 2001 *Living in Britain: results from the 2001 General Household Survey*, London: The Stationery Office

Walton, J 1995 How real(ist) can you get?, *Professional Geographer* 47, 61–65

Watson, G 1987 *Writing a thesis: a guide to long essays and dissertations*, London: Longman

Wax, R 1983 The ambiguities of fieldwork. In R Emerson (ed.) *Contemporary field research*, Boston and Toronto: Little Brown & Co., pp. 191–202

Weiner, D, T A Warner, T M Harris and R M Levin 1995 Apartheid representations in a digital landscape: GIS, remote sensing and local knowledge in Kiepersol, South Africa, *Cartography and Geographic Information Systems* 22, 30–44

Weitzman, E and M Miles 1995 *Computer programs for qualitative data analysis: a software sourcebook*, Thousand Oaks, CA: Sage

White, P 2003 Making use of secondary data. In N J Clifford and G Valentine (eds) *Key methods in geography*, London: Sage, pp. 67–85

White, P 1985 On the use of creative literature in migration study, *Area* 17, 277–283

Whitehead, T 1986 Breakdown, resolution and coherence: the fieldwork experiences of a big, brown, pretty talking man in a West Indian community. In T Whitehead and M Conaway (eds) *Self, sex and gender in cross-cultural fieldwork*, Urbana and Chicago: University of Illinois Press, pp. 213–239

Whyte, W F 1991 *Participatory action research*, London: Sage

Widdowfield, R 2000 The place of emotions in academic research, *Area* 32, 199–208

Williams, R 1993 Culture is ordinary. In A Gray and J McGuigan (eds) *Studying culture: an introductory reader*, London: Edward Arnold

Wilson, E 1991 *The sphinx in the city: urban life, the control of disorder, and women*, Berkeley: University of California Press

Wilton, R 2000 'Sometimes it's ok to be a spy': ethics and politics in geographies of disability, *Ethics, Place and Environment* 3, 91–97

Wolcott, H F 1990 *Writing up qualitative research*, London: Sage

Women and Geography Study Group 1997 *Feminist geographies: explorations in diversity and difference*, Harlow, Essex: Longman

Wood, D 1993 *The power of maps*, London: Routledge

World Health Organisation (annual) *Statistical annual*, Geneva

Wright, J K 1947 *Terrae incognitae*: the place of imagination in geography, *Annals of the Association of American Geographers* 37, 1–15

Wrigley, N 1985 *Categorical data analysis*, London: Longman

Yapa, L 1996 What causes poverty? A postmodern view, *Annals of the Association of American Geographers* 86, 707–728

Yapa, L 1997 Reply: why discourse matters, materially, *Annals of the Association of American Geographers* 87, 717–722

Yeung, H W-C 1997 Critical realism and realist research in human geography: a method or a philosophy in search of a method? *Progress in Human Geography* 21, 51–74

Zelinsky, W 1974 Selfward bound? Personal preference patterns and the changing map of American society, *Economic Geography* 50, 144–179

Zonn, L E 1990 *Place images in the media: portrayal, experience and meaning*, Savage, MD: Rowman and Littlefield

Index

LIVERPOOL
JOHN MOORES UNIVERSITY
AVRIL ROBARTS LRC
TITHEBARN STREET
LIVERPOOL L2 2ER
TEL. 0151 231 4022